# ENERGY AND AGRICULTURE:
## SCIENCE, ENVIRONMENT, AND SOLUTIONS

# ENERGY AND AGRICULTURE:

## SCIENCE, ENVIRONMENT, AND SOLUTIONS

## STEPHEN D. BUTZ

CENGAGE
Learning·

Australia • Brazil • Mexico • Singapore • United Kingdom • United States

**Energy and Agriculture: Science, Environment, and Solutions**
**Stephen D. Butz**

Vice President, Careers & Computing: Dawn Gerrain

Director of Learning Solutions: Steve Helba

Product Manager: Nicole Sgueglia

Director, Development-Career and Computing: Marah Bellegarde

Product Development Manager: Juliet Steiner

Senior Content Developer: Darcy M. Scelsi

Editorial Assistant: Maria Gargulio

Market Development Manager: Scott Chrysler

Senior Production Director: Wendy Troeger

Content Project Manager: Brooke Greenhouse

Senior Art Director: David Arsenault

Media Editor: Debbie Bordeaux

Cover image(s): MAIN IMAGE:
  © Iakov Kalinin/www.Shutterstock.com

INSETS (LEFT TO RIGHT):

CORN: © Zeljko Radojko/www.Shutterstock.com

SOLAR PANELS: © Kletr/www.Shutterstock.com

POPLAR TREES: © Gareth Kirkland/
  www.Shutterstock.com

For product information and technology assistance, contact us at
**Cengage Learning Customer & Sales Support, 1-800-354-9706**

For permission to use material from this text or product,
submit all requests online at **www.cengage.com/permissions.**
Further permissions questions can be e-mailed to
**permissionrequest@cengage.com**

Library of Congress Control Number: 2013950837

Book Only ISBN-13: 978-1-111-54108-8

**Cengage Learning**
200 First Stamford Place, 4th Floor
Stamford, CT 06902
USA

Cengage Learning is a leading provider of customized learning solutions with office locations around the globe, including Singapore, the United Kingdom, Australia, Mexico, Brazil, and Japan. Locate your local office at: **www.cengage.com/global**

Cengage Learning products are represented in Canada by Nelson Education, Ltd.

To learn more about Cengage Learning, visit **www.cengage.com**

Purchase any of our products at your local college store or at our preferred online store **www.cengagebrain.com**

**Notice to the Reader**
Publisher does not warrant or guarantee any of the products described herein or perform any independent analysis in connection with any of the product information contained herein. Publisher does not assume, and expressly disclaims, any obligation to obtain and include information other than that provided to it by the manufacturer. The reader is expressly warned to consider and adopt all safety precautions that might be indicated by the activities described herein and to avoid all potential hazards. By following the instructions contained herein, the reader willingly assumes all risks in connection with such instructions. The publisher makes no representations or warranties of any kind, including but not limited to, the warranties of fitness for particular purpose or merchantability, nor are any such representations implied with respect to the material set forth herein, and the publisher takes no responsibility with respect to such material. The publisher shall not be liable for any special, consequential, or exemplary damages resulting, in whole or part, from the readers' use of, or reliance upon, this material.

Printed in the United States of America
1 2 3 4 5 6 7 18 17 16 15 14

# TABLE OF CONTENTS

# PREFACE

As worldwide demand for energy continues to rise and conventional non-renewable resources dwindle in supply, the need for new, environmentally conscious ways to meet society's energy requirements are becoming increasingly more important. Energy and Agriculture is a science textbook designed to introduce students to the ways energy is generated and used today, and the role agriculture can play in helping to satisfy the world's energy demands. The use of agriculturally based fuel systems, also known as bioenergy, as a means to supply energy to our technological society, provides environmentally safe, renewable energy options for all aspects of life, including industry, transportation, and electrical power generation.

# ABOUT THE AUTHOR

Stephen D. Butz received his Bachelor's and Master's degrees from Cornell University and has taught courses in agriculture and environmental science for more than 15 years. He is also the author of numerous publications dealing with the subjects of science, technology and history.

# ACKNOWLEDGMENTS

## REVIEWERS

Jesse Faber
Pontiac Township High School
Pontiac, Illinois

Joanna Marsh
Cy Lakes High School
Katy, Texas

Tara Mende
Pleasant Grove High School
Elk Grove, California

Kay Richards
Abilene High School
Abilene, Texas

Jim Satterfield
Jefferson County High School
Dandridge, Tennessee

*Energy and Agriculture* has been designed with key features to enhance your learning experience.

**KEY CONCEPTS**

*After reading this chapter, you should be able to:*

1. Explain the various methods of coal mining.
2. Identify two negative effects of coal mining.
3. Explain why coal needs to be processed and how it is processed.
4. Discuss the negative effects of coal processing.
5. Identify the two main methods for transporting coal.
6. Identify what percentage of our electricity is produced by burning coal.
7. Describe the process by which coal is used to produce electricity.
8. Define the term fluidized bed combustion.
9. Identify three coal combustion byproducts.
10. Discuss three pollution control techniques applied to coal combustion.
11. Identify how long coal reserves in the United States are predicted to last.
12. Explain why not all coal reserves can be mined.
13. Identify two technologies that can burn coal while producing less air pollution.
14. Define the term carbon sequestration.
15. Explain three methods of carbon sequestration.
16. Explain four atmospheric pollutants produced by burning coal.
17. Discuss the ways in which coal mining can affect aquatic ecosystems.

- *Key Concepts* highlight the import topics you should focus on learning in each chapter.

**TERMS TO KNOW**

| | | |
|---|---|---|
| surface mining | Rakine cycle | silviculture |
| overburden | fluidized bed combustion | photochemical smog |
| contour mining | coal reserves | acid precipitation |
| contours | syngas | silviculture |
| deep mining | carbon sequestration | |
| coal slurry | spoil pile | |

- *Terms to Know* highlight the vocabulary you should study to improve your understanding of the topics discussed in each chapter. Each term is defined in the margin to make review of the terms easy.

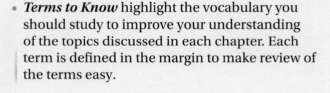

**CAREER CONNECTIONS**

## River Towboat/Tugboat Captain

River captains operate tugboats and towboats that deliver barges up and down rivers. Twenty percent of the coal transported in the United States each year is done via rivers. Fleets of towboats and tugboats steer barges that transport over 800 million tons of coal each year. Each barge can be as long as 1,000 feet and hold 20,000 tons of coal. There are more than 4,000 tugboats and towboats that direct 30,000 barges on our nation's rivers. Towboat captains must have experience with the operation and navigation of boats on inland waterways. Knowledge of technology is also essential because modern towboats are equipped with computer–aided navigation and controls. Shifts can last as short as 6 hours, or as long as 28 days depending on where the cargo is being delivered. Many towboat and tugboat captains get their start as deckhands and assistants, so a four-year college degree is not always necessary to become a boat captain.

- *Career Connections* are highlighted throughout the book to help you identify how these topics relate to work in the field. Areas of interest can be followed up with further research and investigation.

## Fun Fact

In 1865, German chemist Robert Bunsen perfected a device that produced a controllable flame by mixing natural gas with oxygen. This became known as the Bunsen burner and helped pave the way for natural gas to be used as a source of fuel for heating and cooking.

- ***Fun Facts*** are presented to highlight information of interest related to the topics covered in the reading.

## REVIEW OF KEY CONCEPTS

1. Natural gas is another important fossil fuel used in the United States that is made up of mostly methane ($CH_4$), ethane ($C_2H_6$), propane ($C_3H_8$), and butane ($C_4H_{10}$). Natural gas accounts for approximately 25% of our total energy use.

2. Natural gas extracted from the ground as part of the oil drilling process is known as conventional natural gas.

3. Natural gas associated with coal deposits, shale rock, or methane hydrates is called unconventional natural gas. The

types of unconventional natural gas are deep gas, tight gas, gas shale, and geo-pressurized gas.

4. The technology used to extract conventional natural gas is similar to that of oil well drilling, because the gas flows out of the ground under its own pressure.

5. Shale gas is natural gas trapped within sedimentary rocks known as black shale. In order to free up the gas, a drilling technique known as hydraulic fracturing or hydrofracking is used. A mixture of

- ***A Review of Key Concepts*** aids you in study of the important topics covered in the chapter.

## CHAPTER REVIEW

### Short Answer

1. What are the four main hydrocarbons that make up natural gas?

2. What is conventional natural gas?

3. Describe four sources of unconventional natural gas.

4. What are four ways natural gas is used?

5. At the present rate of use, how long are proven reserves of natural gas in the United States expected to last?

6. What are three negative effects natural gas has on the environment?

7. Describe how natural gas is used by production agriculture in the United States.

### Energy Math

1. How many cubic feet of natural gas does the residential sector in the United States use in one year?

2. Determine how much natural gas it takes to meet the demands of nitrogen fertilizer use in the United States for one year:

- ***Chapter Review*** provides a variety of questions and activities that test your understanding of the topics read in each chapter.

# Energy Use in the Pre-Industrial and Industrial Eras

---

## TOPICS TO BE PRESENTED IN THIS UNIT INCLUDE:

- Evolution of energy usage in the Pre-industrial Era
- The coal and oil eras of industrial energy use

## OVERVIEW

Since the dawn of humankind, people have searched for creative ways to harness the natural energy resources available on our planet. The use of fire to cook food, as well as provide warmth and protection, has been in use by humans and their ancestors for millions of years. Wood, plant material, and other forms of biomass were the fuels of choice for sustaining early fires. Eventually wood gave way to other forms of higher density fuels hidden within the Earth's crust. Coal soon took over as the major energy source to produce heat and began the technological revolution. The discovery of petroleum reserves in Europe and America soon helped oil join with coal as one of the world's top energy resources, especially with the advent of motorized transportation. As civilization entered the twentieth century, newer energy

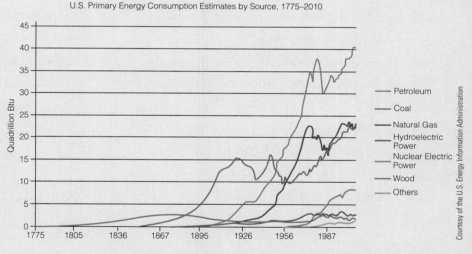

U.S. Primary Energy Consumption Estimates by Source, 1775–2010

technologies emerged and began to supply the growing need for electricity (Figure U-1). These technologies included nuclear fission, large-scale hydroelectric dams, and the development of renewable energy resources such as solar, wind, hydrogen, and geothermal. As civilization begins the twenty-first century, the need to supply the ever increasing energy demands of the world has created a realization to not only sustain our energy resources, but to also make them environmentally friendly.

## CHAPTER 1

# Pre-Industrial Energy Use

## KEY CONCEPTS

*After reading this chapter, you should be able to:*

1. Define the terms hominid, bipedalism, charcoal, and pyrolysis.

2. Identify the first human ancestor believed to use fire and eat cooked food.

3. Discuss the benefits of eating cooked food versus raw food.

4. Identify the region of the world where agriculture began, and how long ago it began.

5. Identify the first plants and animals used in early agricultural systems.

6. Explain what energy sources were used in early agricultural systems.

7. Identify the first light sources, besides fire, to be used by humans and how they were fueled.

8. Discuss the ways that humans first used heat energy.

9. Explain the process by which charcoal is made, and why it is used instead of wood.

10. Identify the products made by using charcoal as a heat source.

11. Describe the process of peat formation.

12. Discuss the different types of water-wheels used by early societies.

13. Explain how wind energy was captured and used by early societies.

## TERMS TO KNOW

| | | |
|---|---|---|
| anthropology | Fertile Crescent | pyrolysis |
| hominid | treadwheel/treadmill | lime |
| taxonomic | domesticated | ore |
| bipedalism | work animal | smelting |
| *Homo erectus* | sustainable | peat |
| absorb | energy crop | waterwheel |
| hearth | petroleum seep | windmill |
| agriculture | tallow | |
| civilization | charcoal | |

# INTRODUCTION

In his book, *Catching Fire*, Harvard professor of anthropology Richard Wrangham presents the idea that controlling fire and cooking food created an evolutionary advantage for the ancestors of modern humans. He makes a good hypothesis, for the ability to use fire as an external energy source for protection, warmth, and to prepare food must have paved the way for the development of our species. This most certainly changed the way we interacted with the natural world. The quest for controlling energy had begun, and when coupled with the advent of agriculture, would lead to the development of civilization.

## HUMAN ORIGINS AND ENERGY

The science of **anthropology** studies the evolutionary development of human beings and their culture. For more than a century, anthropologists have unearthed hundreds of fossils that helped determine the timeline of how our species evolved. Many of these discoveries were made in Africa, which is widely regarded as the birthplace of humanity.

Modern humans are what remain of a long line of now extinct **hominids**. The term *hominid* describes the **taxonomic** family of primates who share physical features including **bipedalism**.

The human quest for energy most likely began with the hominid species *Homo erectus*. *Homo erectus* lived between 1.8 million and 300,000 years ago. Physical evidence of the first use of fire by our human ancestors ranges between 400,000 to 250,000 years ago. This comes in the form of ancient fire pits discovered in China suggesting that it was most likely *Homo erectus* that began the controlled use of fire. *Homo erectus* probably used fire for protection, heat, making weapons, and also cooking.

The use of fire to cook food makes the intake of nutrients much more efficient. Before the use of fire to prepare food, hominids ate raw, uncooked food. The consumption of only raw food, whether meat or plant based, greatly reduces the ability to **absorb** nutrients into the body. Eating only raw food also requires a lot of time devoted to consuming it. Modern gorillas spend almost half their day just chewing their food! When cooked food is eaten, the eating process becomes more efficient, and less time is needed to eat. This is because heat generally breaks down the tough structures raw meat and plant material are made of, making it easier for the body to absorb. Cooking has other advantages as well. The heat sterilizes the food by killing off any microorganisms present on or in it, therefore making it safer to eat. By killing the microorganisms, the cooking process also helps preserve the food so it lasts longer and does not spoil. One last benefit of course is the significantly improved taste!

Besides finding physical evidence like old fire pits and **hearths** to determine when hominids first used fire to cook, changes in the anatomy of hominids helps determine when this may have occurred. A close ancestor to *Homo erectus* was the older hominid form known as *Homo habilis*, or

**anthropology**—the study of the evolutionary development of human beings and their culture

**hominid**—classification of primates sharing physical features; early ancestors of modern humans

**taxonomic**—a system of classifying living organisms

**bipedalism**—the ability to walk upright

*Homo erectus*—a hominid species; an early ancestor to modern day humans

**absorb**—to take in

**hearth**—foundations of a fireplace

"Handy Man." *Habilis* lived between 2.4 and 1.5 million years ago and is the first hominid species believed to use tools.

The differences in the anatomy between *Homo erectus* and *Homo habilis* suggest the use of fire had a large impact on the evolution of our ancestors (Figure 1-1). *Habilis* had larger, sharper teeth and a larger jaw suggesting it ate a diet of mostly tough, raw foods. The skulls of *erectus* contain smaller, smoother teeth and a reduced jaw better suited to chewing cooked food.

*Habilis* also has longer arms and fingers, indicating it spent much of its time up in the trees avoiding predators. The arms and fingers of *Homo erectus* are smaller and not well designed for tree climbing. If *Homo erectus* slept on the ground, then harnessing fire would certainly aid in its protection while it slept. Finally, changes in the digestive systems between *habilis* and *erectus* point to the possibility the diet of erectus was most likely prepared food. When food is cooked, it can be absorbed more easily into the body, therefore requiring a smaller digestive system. The large, flared out rib cages of *Homo habilis* point to the idea it had a larger digestive tract similar to that of modern apes (Figure 1-2). This is because *habilis* ate a diet of mostly raw food. A longer digestive system is needed for organisms eating uncooked plants and meat so there is enough time for the food to be broken down and absorbed. *Homo erectus* fossils show a reduction in rib cage size, and therefore a reduction in the size of the digestive system similar to that of modern humans. This is most likely because of a changing diet consisting of mostly cooked food.

Therefore, the evidence the regular use of an external energy resource, in this case fire, by our ancestors probably occurred sometime between the appearance of *Homo habilis* and *Homo erectus* about 1.5 million years ago. Although the exact point in time when the harnessing of fire began by our ancestors is unknown, the evidence suggests the use of an external energy source had a key role in the evolution of our species.

(A) *Homo Habilis*                    (B) *Homo Erectus*

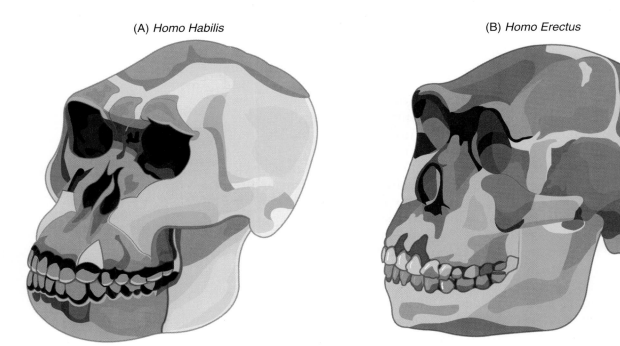

**FIGURE 1-1** Comparison of *Homo habilis* and *Homo erectus* skulls. (A) *Homo habilis*. (B) *Homo erectus*.

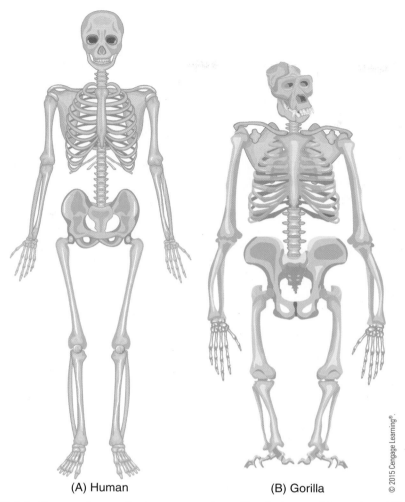

(A) Human        (B) Gorilla

© 2015 Cengage Learning®.

**FIGURE 1-2** Comparison of a human and gorilla skeleton. Note the differences in size of the rib cage.

## CAREER CONNECTIONS

### Physical Anthropologist

A physical anthropologist, also known as a biological anthropologist, is a type of scientist who studies the development of human beings and their ancestors. Physical anthropologists examine how humans have adapted to their environment, how these adaptations differ around the world, and how they have changed through time. This fascinating field is a specialized subject of anthropology; the general study of human behavior and culture. Physical anthropologists can work in a variety of jobs in the private sector, academic world, and for the government. They often study bones and artifacts to unlock the past history of human societies. Forensic anthropologists even assist in murder investigations. Typically a four-year college degree and a graduate degree is required to find a job as a physical anthropologist. To find out more about career opportunities in this field, visit the American Association of Physical Anthropologists at http://physanth.org/.

# THE AGRICULTURAL REVOLUTION

**agriculture**—living off the land by raising crops and livestock

**civilization**—the stage of human development that is considered most advanced

**Fertile Crescent**—region in the Southwest part of Asia; believed to be the location where agriculture began

After the harnessing of fire and its ability to cook food, the next big leap in the advancement of human use of energy was the birth of **agriculture**. The ability to stay in one place and produce your own food created a system of living that eventually lead to the development of **civilization**. Prior to this, humans were constantly on the move in search of food as hunter-gatherers. Although people might stay in one place seasonally, eventually they would have to leave to find new sources of food. With the development of agriculture, humans could now remain in one location and establish permanent communities. The quest to control the food supply through raising plants and animals in one place also began a more concentrated use of energy resources. Some historians classify agriculture itself as an energy resource because it fuels humanity by providing an abundance of calories and nutrients for us to consume. The rise of agriculture, also known as the agricultural revolution, began in the southwestern part of Asia known as the **Fertile Crescent** around 10,000 years ago (Figure 1-3).

This area today encompasses parts of Iran, Iraq, Turkey, Israel, Jordan, Lebanon, and Syria. The first crops grown here included barley and wheat. Goats, sheep, pigs, and eventually cattle where the first animals raised by humans in this location. Other parts of the world including Asia, Africa, Australia, and North, South and Central America began to practice agriculture between 4,000 and 8,500 years ago. Major crops grown in these regions were rice, corn, beans, sorghum, potatoes, squash, and sweet potatoes. At first, the main energy source to fuel the new practice of agriculture was human labor itself. It is estimated 95% of the food energy raised during these early times, went directly to feeding the farm labor force. The remaining 5%, although small, allowed for a portion of the population to practice other things besides supplying food. This lead to the birth of civilized society.

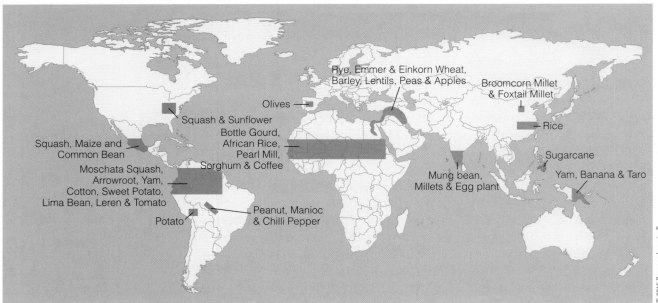

© 2015 Cengage Learning®

**FIGURE 1-3** The location of the Fertile Crescent indicating the first plants and domesticated animals used in agriculture.

FIGURE 1-4 A treadwheel powered by human energy.

One of the more interesting devices used to harness the muscle power of humans during early times was called the **treadwheel** or **treadmill**.

Today's modern treadmill currently used for exercise derived its name from these early human-powered machines. Treadwheels and mills used wooden conveyor belts, steps, or large wheels that people would walk and use the power of their legs to drive water lifting devices or mill grinding stones (Figure 1-4).

**treadwheel** or **treadmill**—device used to power grinding stones or to move water

## ANIMAL POWER

Eventually the use of livestock as an energy source became more widespread. This reduced the percentage of people who needed to work on the farm. Livestock was used to pull plows and power other agricultural equipment. The types of animals used as power sources in early civilizations varied according to the regional availability of strong mammals that could be tamed. These animals are also known as **domesticated** animals. Egyptian artifacts showing donkeys performing work around 3,000 years ago are believed to be the oldest artistic evidence of the use of animal power by humans. Other examples of animals, also known as beasts of burden or **work animals**, include cattle, horses, camels, and elephants. The incorporation of animals as energy sources was easy because they could be fed by the same crops they helped to produce. This new system of animal-powered agriculture was **sustainable** because some of the food energy harvested in the fields was fed to these animals. This was also one of the first uses of an **energy crop**. Although a portion of the crops grown were used to fuel work animals and not feed people, the work the animals performed replaced many laborers, further freeing up humans to do other things. For example, it took about 160 hours for one human to till a one acre field using a hand tool to prepare it for planting. This time is reduced to 40 hours by using cattle to till a one acre field pulling a plow. As a result of this development one person could do the work of many with the aid of animal power.

**domesticated**—to adapt to live in with and for the benefit of humans

**work animal**—an animal used to accomplish tasks such as plowing; also known as beasts of burden

**sustainable**—the ability to be maintained at a certain rate or level

**energy crop**—an agricultural product grown solely for use as an external energy source

## EARLY ENERGY USED FOR LIGHT

The use of wood fires to cast light at night and indoors has been used for thousands of years. Over time other sources of energy began to be used as a means to produce light. Oil lamps discovered in ancient caves have been dated to around 12,000 years ago. These early lamps most likely used animal fat as a fuel source. Other sources of fuel used in lamps included oils derived from plants, fruits, and nuts such as olive oil and castor oil. Fish oil, whale oil, and **petroleum seeps** were also used to fuel early oil lamps. Another form of early light included candles. The oldest candles discovered so far were found in China, and made from whale fat about 2,000 years ago. Other sources of fuel used for candles included animal fat, also known as **tallow**. Tallow is a solid at room temperature, and can be easily molded to make candles. The use of tallow candles continued until the 1800s. Candles were also made from beeswax and many different types of plants that could yield the waxy oil needed to produce them.

**petroleum seep**—an area where crude oil flows naturally out of the ground

**tallow**—substance made from animal fat

## EARLY HEAT ENERGY

The availability to generate heat energy by ancient humans was an important advancement. Not only did it allow people to keep warm and survive in colder climates, but it also became an important resource for the manufacturing process. We know using heat to cook food was a crucial step in our advancement, but it was also valuable for use in creating tools. For example, hardening wood by charring it in fire created stronger weapons. The eventual discovery of how fire renders metals like copper, bronze, and iron greatly changed the way humans made things.

The first major change to the traditional use of wood to provide heat energy came approximately 3,000 years ago with the use of **charcoal**. Charcoal is a dark gray-black residue of impure carbon produced by heating wood in a low oxygen environment. The process of making charcoal is known as **pyrolysis**, which literally means "fire - separate" in Greek. Although charcoal is made from wood, it burns at a much higher temperature than wood. An open wood fueled fire can burn at around 600°F (315°C), whereas a fire fueled by charcoal can reach a much higher temperature of 900°F (482°C), without the addition of forced air. Early production of charcoal involved stacking wood in large piles and covering the pile with soil (Figure 1-5). The wood was then set on fire and allowed to smolder for many days. The soil was removed and the charcoal was harvested from the pile.

**charcoal**—substance created from heating wood in a low oxygen environment; used as a heat fuel source

**pyrolysis**—method of using fire to break a substance into its component parts

The extreme heat generated from charcoal allowed for the production of iron, bricks, and **lime**. Lime is an important component in making paper, mortar, concrete, and a wide array of other products. Although charcoal production greatly improved the generation of high heat furnaces, it came at a cost. The production of 2 pounds of charcoal required 11 pounds of wood. This resulted in a great loss of forests in order to meet the demands of charcoal production. By far the biggest user of charcoal was for the production of iron. The high temperatures needed to process iron from iron **ore**, also known as **smelting**, could only be produced by charcoal. It is estimated it took about 440 pounds of wood to create the charcoal needed to make 2 pounds of iron. Although wood had been sought out as a fuel source for thousands of years, it was the advent of iron production and the use of charcoal that began to put a strain on forests to meet the energy demands of growing civilizations around the world.

**lime**—a common term for calcium oxide (CaO)

**ore**—a mineral used to make valuable metals

**smelting**—the process of melting a substance to promote a chemical change to create another more useful or valuable substance

**FIGURE 1-5** Charcoal mound.

## PEAT ENERGY

Another form of early energy was the use of **peat** as a fuel. Peat is the partially decomposed remains of plant material associated with swamps and bogs. Aquatic plants, trees, grasses, and shrubs that grow in wetland areas are prevented from decaying by the acidic conditions of the water. This forms a fibrous, spongy, low-density material called peat, that when dried, can be burned (Figure 1-6).

The combustion (burning) of peat produces a temperature much less than that of wood, making it only useful for generating low temperature heat. Peat can be found in many parts of the world, including the Netherlands, England, Ireland, Germany, Sweden, and Russia. It was first used as a fuel around 2,000 years ago. The first large-scale use of peat began in the 1500s by the Dutch to fuel their growing economy.

**peat**—partially decomposed remains of plant material that can be burned as an energy source when dried

## EARLY WATER ENERGY

The first use of water energy began around 2,000 years ago in Greece and shortly thereafter in China. The principal means to produce power from flowing water involved the **waterwheel**. A waterwheel is either a horizontal or vertical wheel mounted on an axle, that is then propelled by the power of moving water in a river or stream. Most early waterwheel-powered grinding stones in mills. The vertical axis waterwheel used two different devices to harness the power of water. These are known as the overshot waterwheel and undershot waterwheel (Figure 1-7). The overshot waterwheel uses the flow of water spilling down over the wheel, causing it turn. The undershot waterwheel uses the flow of water moving underneath it to move it.

Both wheels were extremely effective in producing power. Estimates show an early waterwheel could perform the same work as at least twenty people. Waterwheels ultimately harness the power of gravity, that causes water to flow from areas of high elevation to low elevation. Waterwheels became an important energy source for society for thousands of years and are still in use today.

**waterwheel**—a device use to produce power through the use of a wheel, axle, and flowing water

**FIGURE 1-6** Peat bog.

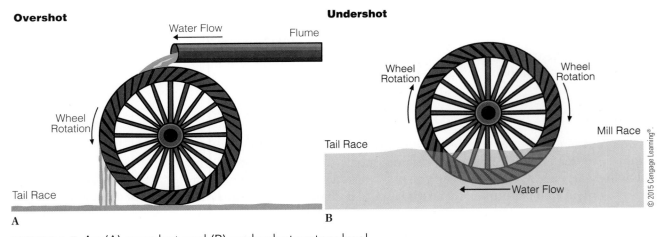

**FIGURE 1-7** An (A) overshot and (B) undershot waterwheel.

## EARLY WIND ENERGY

The use of wind energy began around 2,000 years ago; about the same time as waterwheel power. First documented in ancient Persia, the vertical axis **windmill** used a series of sails to capture the force of the wind, causing them to turn. The vertical shaft attached to the sails usually turned a grindstone used for milling or was used to pump water. The horizontal axis windmill was first used along the coastline of Greece. Cloth sails captured

**windmill**—a device used to create power by the use of wind

the force of the wind turning a horizontal shaft, that transferred the energy through a gear to a grindstone.

The horizontal axis windmill was much more efficient then the vertical windmill, and soon spread into Europe and China. It is estimated the Netherlands had more than 9,000 windmills in operation during the 1600s (Figure 1-8).

**FIGURE 1-8** Early use of windmills for milling and pumping water.

# REVIEW OF KEY CONCEPTS

1. A hominid is an early ancestor of the modern human. One of the characteristics of the hominid was bipedalism, the ability to walk upright.

2. *Homo erectus* is the earliest ancestor to modern humans that was believed to be the first to use fire and eat cooked food.

3. The benefits of cooked food include: easier absorption of nutrients, sterilization or reduction of bacteria and microorganisms found in raw foods, preservation of foods making them last longer, and the improved taste.

4. The birth of agriculture that began the agricultural revolution occurred approximately 10,000 years ago in the Fertile Crescent region of Southwestern Asia. Agriculture created a new relationship between humans and the environment, that increased the need for external energy resources.

5. The first crops grown in the Fertile Crescent included barley and wheat. Goats, sheep, pigs, and eventually cattle where the first animals raised by humans in this location.

6. The early use of work animals and the human-powered treadwheel improved the efficiency of agriculture in early agricultural systems.

7. Ways of generating artificial light were developed from animal fats and plant based oils used in oil lamps and candles made from waxes and tallow.

8. Humans first used heat energy for protection, cooking, and providing warmth.

9. Charcoal is a substance made from wood. Charcoal was produced from wood by pyrolysis; the act of exposing something to high heat in a low oxygen environment.

10. Charcoal was made by placing piles of wood under soil and heating the wood. Charcoal production began to put a great strain on forests and the wood they supplied. Charcoal burns at higher temperatures than wood.

11. Charcoal is used as a fuel source for generating high heat fires used to smelt iron and other metals.

12. Peat is the partially decomposed remains of plants deposited in bogs. The plant materials do not decompose fully because the acid environment of the swamps.

13. Two types of waterwheels were used as early sources of water energy; the overshot and undershot waterwheel. The overshot waterwheel used the flow of water spilling down over the wheel, causing it turn. The undershot waterwheel used the flow of water moving underneath it to turn the wheel.

14. A series of sails were attached to a vertical shaft to capture the wind to turn a grindstone used for milling or to pump water.

## CHAPTER REVIEW

### Short Answer

1. Describe two changes in hominid anatomy that most likely resulted from eating cooked food.

2. What are four advantages of eating cooked food?

3. List four sources of energy used by ancient agricultural societies.

4. Briefly discuss some positive and negative impacts that agriculture and the early use of energy resources had on the environment and human society.

### Energy Math

1. If it took approximately 160 hours for one person to prepare a one acre field for planting, but only 40 hours to plow the same field using an ox driven plow, how many humans does it take to do the same work as one ox?

2. How many pounds of wood are needed to smelt 1 pound of iron in a charcoal fueled furnace?

### Multiple Choice

1. The study of the evolution of human beings and their culture is known as:
   a. archaeology
   b. anthropology
   c. biology
   d. paleontology

2. Bipedalism is the ability to:
   a. eat food
   b. make fire
   c. walk on two legs
   d. climb trees

3. Which human ancestor is believed to be the first to use fire and cook with it?
   a. *Australopithecus*
   b. *Homo habilis*
   c. *Homo sapiens*
   d. *Homo erectus*

4. Which of the following parts of human anatomy are believed to have resulted from eating cooked food?
   a. Larger teeth
   b. Shorter arms
   c. Larger jaw
   d. Smaller ribcage

5. Approximately how long ago did the agricultural revolution begin?
   a. 1,000,000 years ago
   b. 250,000 years ago
   c. 10,000 years ago
   d. 2,000 years ago

6. The Fertile Crescent is the name given to the location where agriculture began, approximately where is it located?
   a. Africa
   b. Southwestern Asia
   c. Europe
   d. South America

7. In early agricultural societies, approximately how much energy harvested went to feed the labor force?
   a. 95%
   b. 75%
   c. 50%
   d. 5%

8. Work animals used in early agriculture could perform the work of approximately how many people?
   a. 2
   b. 4
   c. 10
   d. 20

9. After the use of fire, which light source is the oldest?
   a. Light bulb
   b. Candle
   c. Match
   d. Oil lamp

**10.** Which fuel source was the first to be used to generate high temperature heat?
a. Oil
b. Charcoal
c. Gas
d. Peat

**11.** Which fuel source came from bogs?
a. Oil
b. Charcoal
c. Gas
d. Peat

**12.** The process of heating a substance in a low oxygen environment is known as:
a. Pyrolysis
b. Combustion
c. Distillation
d. Smelting

## Matching

*Match the terms with the correct definitions*

a. anthropology
b. hominid
c. taxonomic
d. bipedalism
e. *Homo erectus*
f. absorb
g. hearth
h. agriculture

i. civilization
j. Fertile Crescent
k. treadwheel
l. domesticated
m. work animals
n. sustainable
o. energy crop
p. charcoal

q. pyrolysis
r. lime
s. ore
t. smelting
u. peat
v. waterwheel
w. windmill

**1.** ____ The use of two legs for walking.

**2.** ____ A device for transferring walking energy into mechanical energy.

**3.** ____ The study of the evolution of human beings and their culture.

**4.** ____ A mechanical device used to capture wind energy.

**5.** ____ To take in or soak up.

**6.** ____ The stage in human development that is considered advanced.

**7.** ____ A brown spongy material made of partially decomposed plant material.

**8.** ____ The floor of a fireplace.

**9.** ____ The process of extracting metal from ore by heating and melting.

**10.** ____ The family of primates that includes humans and their ancestors.

**11.** ____ The process of heating something in a low oxygen environment.

**12.** ____ A mechanical device used to capture the energy of flowing water.

**13.** ____ To tame or keep as a pet.

**14.** ____ A source of calcium in the form of calcium oxide (CaO).

**15.** ____ The scientific classification of a living thing.

**16.** ____ The ability to maintain at a certain level.

**17.** ____ The dark gray-black residue of impure carbon produced by heating wood in a low oxygen environment.

**18.** ____ The naturally occurring rock that contains valuable metals or minerals.

**19.** ____ The science of growing plants and animals for use as food or providing other products.

**20.** ____ The species classification of human beings.

**21.** ____ The region in Southwestern Asia where agriculture began.

**22.** ____ An agricultural crop raised for use as an energy source.

**23.** ____ Animals used as a source of power.

## CHAPTER 2

# The Coal Era

## KEY CONCEPTS

*After reading this chapter, you should be able to:*

1. Discuss the physical properties of coal.

2. Define the terms hydrocarbon and organic.

3. Explain how and when coal was formed.

4. Identify the four different types of coal and their unique characteristics.

5. Explain the reasons why people began using coal.

6. Discuss the problem humans faced that lead to the development of the steam engine.

7. Identify the basic function of a simple steam engine.

8. Explain the advantage of using steam turbines.

9. Define the term coke and explain the advantages of using it instead of coal.

10. Explain how coal gas is produced.

11. Discuss the advantages and disadvantages of using coal gas.

## TERMS TO KNOW

| | | |
|---|---|---|
| coal | bituminous coal | heat engine |
| hydrocarbon | sub-bituminous coal | turbine |
| organic | lignite coal | coke |
| anthracite coal | coal seam | coal gas |

# INTRODUCTION

As human civilization progressed, the use of plant and animal based energy resources gave way to fossil fuels. A fossil fuel is an energy resource formed from the remains of living things from millions of years ago. The dawn of the age of fossil fuel use created changes to human civilization that forever altered our planet. The era of industrial energy began, tapping into fuels created millions of years ago, long buried deep within the Earth.

The use of coal alone gave rise to the industrial revolution that started almost two hundred years ago. The invention of the steam engine and the coal needed to power it resulted in an accelerated advancement of technology at a rate never seen before. The link between steam power and coal was born, and launched civilization into its next great energy era—the coal era.

## COAL FORMATION

**Coal** is a black or brown, rock-like material made up of **hydrocarbons** that came from the partially decomposed remains of plants living millions of years ago. Technically, coal is not a rock, because rocks are formed from minerals, and minerals do not contain organic material. The term **organic**, when applied in physical science, refers to something that contains carbon. Because all living things contain carbon, organic usually describes a living thing. Much of the world's coal was formed during a period of time known as the Carboniferous Period. This geologic period existed between 362 and 290 million years ago. During that time on Earth, the climate was much warmer and wetter than it is today in many parts of the world.

During the Carboniferous Period, large areas of land now known as coal swamps were covered in lush vegetation consisting of ferns, club mosses, and horsetails that grew the size of trees (Figure 2-1). When these plants died, they fell into the shallow water and were partially decomposed. Over time these deposits were covered with sediments, compressed, and transformed into coal. Scientists believe the vast amounts of partially decomposed plant material deposited during this period existed because the specific bacteria needed to break down the woody material had not yet evolved. The great fluctuations of sea level during the Carboniferous Period may have also played a role in the preservation of this coal-forming material. The ebb and flow of seawater into the terrestrial coal swamps helped cover and preserve the organic material with sediments.

### Types of Coal

The varying amount of heat and pressure these organic remains were exposed to resulted in the formation of four types of coal (Figure 2-2). **Anthracite coal** is the black, shiny, dense coal formed from exposure to the greatest amount of heat and pressure. Also known as hard coal, anthracite is the cleanest burning coal, contains the least amount of impurities, and 86–96% carbon. Anthracite is used in industry and as a source of home

**coal**—a black or brown, rock-like material made up of hydrocarbons that came from the partially decomposed remains of plants that lived millions of years ago

**hydrocarbon**—a molecule made up of hydrogen and carbon atoms that is plentiful in plants

**organic**—something containing carbon

**anthracite coal**—black, shiny, and dense; highest amount of carbon, cleanest burning with lowest amount of impurities

**FIGURE 2-1** Carboniferous Period coal swamp.

**bituminous coal**—dark black, brittle, less dense; lower amount of carbon

**sub-bituminous coal**—dull black, low density, low carbon content, low heating value

**lignite coal**—brown, lowest density, least amount of carbon

heating fuel. **Bituminous coal** is a less dense, dark black coal that is brittle and contains approximately 45–86% carbon. **Sub-bituminous coal** is a dull black, low-density coal that formed much later than the Carboniferous Period; about 100 million years ago. This type of coal contains 35–45% carbon and has a much lower heating value than bituminous coal. Both of these coals are used for electric power generation, cement production, iron and steel manufacturing, and industrial processes. Finally, **lignite coal** is the brown colored, lowest density coal that was not exposed to heat and pressure and formed fairly recently in geologic time. Lignite contains 25–35% carbon and is mainly used for electric power generation.

## EARLY COAL USE

The first documented use of coal as a supplemental fuel occurred during the Bronze Age, about 5,000 years ago.

The Roman occupation of Britain was the first recorded use of coal as a fuel source. It was used in iron smelting, grain drying, and to produce heat. This began around the year 200 CE (Contemporary Era, formerly known as AD). Coal use began in China shortly after this time. The term "sea coal" was often used to identify the early use of coal in England. Sea coal refers to natural deposits of coal that washed up along the shoreline. This was gathered by people and used as a minor fuel source. The term sea coal also refers to the delivery method of coal. Beginning in the eleventh century, coal was often delivered by ships in England.

As the demand for charcoal increased during the middle ages in Europe, the loss of forests forced people to search for an alternative fuel to use in metal fabrication. This increased use lead to the widespread search for coal and the development of the coal mining industry. At first, early coal mines consisted

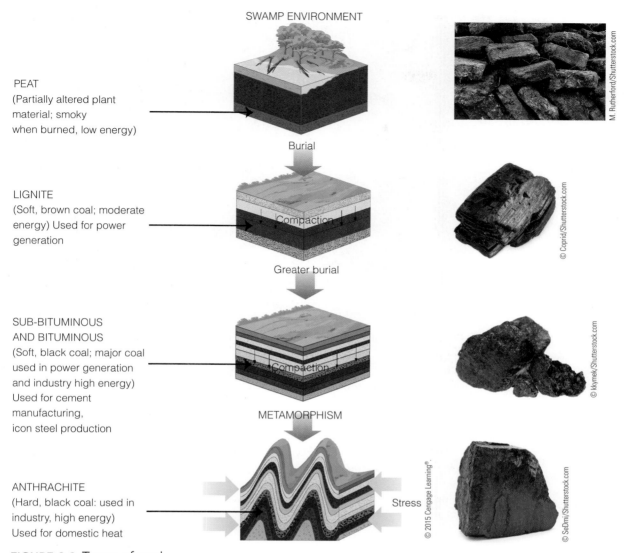

SWAMP ENVIRONMENT

PEAT
(Partially altered plant material; smoky when burned, low energy)

Burial

LIGNITE
(Soft, brown coal; moderate energy) Used for power generation

Compaction

Greater burial

SUB-BITUMINOUS AND BITUMINOUS
(Soft, black coal; major coal used in power generation and industry high energy) Used for cement manufacturing, icon steel production

Compaction

METAMORPHISM

ANTHRACHITE
(Hard, black coal: used in industry, high energy) Used for domestic heat

Stress

**FIGURE 2-2** Types of coal.

of open pits that were dug into existing outcrops of coal exposed at the surface. As the demand for coal grew in Europe, more complex, underground mines called deep shaft mines began to operate. Deep shaft mines involved digging narrow shafts into the **coal seams** (Figure 2-3). Some coal seams can reach hundreds, even thousands of feet into the ground. Deep shaft mining is one of the world's most dangerous operations. As the search for coal lead deep underground, new problems facing mining engineers emerged. One of these was the seeping of groundwater into the mineshafts. As miners dug down past the water table, groundwater flooded into the mines. The water table is the area beneath the ground where water completely fills pore spaces in rock and soil. The use of pumps to remove groundwater from these early coal mines initiated the search for an energy source to power them. The result of this search was the development of the modern steam engine.

**coal seam**—a narrow band of coal sandwiched between surrounding layers of rock

## COAL AND THE STEAM ENGINE

The steam engine is a **heat engine** that employs the power of pressurized steam that results when liquid water is heated to the boiling point. The force of the steam is used to perform work. Crude steam engines have existed

**heat engine**—a device that uses heat to produce mechanical power

**FIGURE 2-3** Old mine tunnel.

back to the time of ancient Greece when they were used to demonstrate the power of boiling water, but these devices were never put to practical use. One of the earliest steam engines is known as the Hero Engine, invented around 1 CE by Heron of Alexandria, a Greek mathematician who lived in Egypt (Figure 2-4).

In 1698, Thomas Savery produced a simple steam-powered vacuum pump that was used to remove water from English coal mines. The Savery

**FIGURE 2-4** Hero's steam engine.

pump was powered by burning coal. Shortly after, in 1712, Englishman Thomas Newcomen invented the first true steam engine to power a piston that was forced upwards by steam pressure. His device worked much better than Savery's and was soon applied to the coal mining industry. The Newcomen steam-driven water pump was used for the next seventy years. Although the Newcomen engine was widely successful in its ability to remove water from coal mines, it was extremely inefficient. In 1776, James Watt, a Scottish engineer, invented the first truly efficient steam engine known as the Boulton and Watt engine (Figure 2-5). Mathew Boulton was

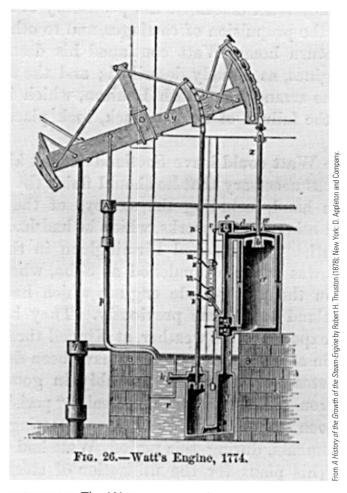

FIG. 26.—Watt's Engine, 1774.

From *A History of the Growth of the Steam-Engine* by Robert H. Thruston (1878); New York: D. Appleton and Company.

**FIGURE 2-5** The Watt steam engine.

## Fun Fact

Although James Watt is regarded as the inventor of the first practical steam engine that propelled the industrial revolution, he is mostly associated for the unit of power that was named after him; the watt. The watt is a unit of energy commonly used for electrical power and equates to one joule per second, or the product of amps multiplied by volts (see Chapter 16).

Watt's financial partner. Their new steam engine used 75% less coal than the Newcomen engine, and soon revolutionized the use of steam power. The Boulton and Watt engine could pump 1,700 gallons of water per minute and greatly improved the coal mining process.

These early steam engines were known as low-pressure steam engines because they injected large cylinders with steam under relatively low pressures needed to move the piston up and down. After 1800 when the patent protecting the design of Watt's engine expired, the rapid development of high-pressure steam engines occurred. These new devices operated at pressures one hundred times greater than Watt's engines. This high-pressure steam created much more power and consumed 66% less coal to fuel them. Soon these new steam engines replaced Watt's, and the power of steam was no longer limited to the coal mining industry, but spread quickly as a source of power for manufacturing. The development of high-pressure steam allowed steam engines to become much smaller. The standard Boulton and Watt engine was the size of a large house; the new high-pressure engines were greatly reduced in size. This widened the applications of these engines to include transportation. The first steam-powered rail locomotive engines were first used in mining operations. Eventually these steam rail locomotives transported people starting in 1810. Compared to horse drawn trains, steam locomotives could pull far heavier loads. Water transportation also benefited from the power of steam. Paddle wheeled steamers began to shuttle passengers and cargo up and down rivers in Europe and America. American inventor Thomas Fulton operated successful steamboat ferries along the Mississippi and Hudson rivers beginning in 1814. Unfortunately, these early steam engines reached a limit to their power output. Whether it was high- or low-pressure steam, these early engines shared one common flaw; their power came from the up and down motion of their pistons that was forced to move with steam pressure. Transferring this vertical motion into rotary circular motion was difficult and inefficient. During the end of nineteenth century, engineers realized that water **turbines**, developed to replace large inefficient waterwheels in France, could be applied to steam engine technology. The development and use of steam-driven turbines increased greatly with the invention of electrical generators that required rapid rotation (Figure 2-6). Standard, vertical motion steam engines could not move fast enough to power electrical generators, so engineers adapted water turbines for use in the steam turbine engine.

**turbine**—a circular wheel or rotor, that is forced to spin at high speed when water, air, steam or any moving fluid transfers its energy to a series of angled blades or vanes attached to a central pivot point

Steam turbine engines operate as high-pressure steam is forced through the turbine, causing it to rotate rapidly. In 1888, English engineer Robert Parsons developed a steam turbine engine that could rotate at 4,800 revolutions per minute. Soon, steam turbines powered by coal became the main power source for the production of electricity. The Parson designed steam turbine has since been the main source of electricity production in the United States for more than one hundred years. Steam turbine technology was also applied to water transportation. In 1897, Parsons developed a steam turbine engine for a 100-foot long boat he called the *Turbinia*. He demonstrated its power in 1897 when it raced along the water at 32 knots. During that time boats traveled at speeds much lower than 20 knots. Soon after, all new ships were constructed with steam turbine engines.

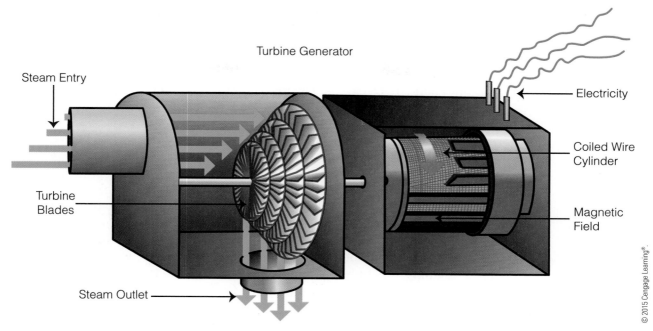

Turbine Generator

Steam Entry

Electricity

Turbine
Blades

Coiled Wire
Cylinder

Magnetic
Field

Steam Outlet

© 2015 Cengage Learning®.

**FIGURE 2-6** Steam turbine engine.

## COKE

**Coke** is a solid hydrocarbon made from coal that is similar to charcoal and produced by pyrolysis. The practice of making coke is the same as making charcoal. Coal is superheated in large vessels lacking oxygen. The process drives off many impurities present in coal, including sulfur. Once wood charcoal became scarce, coal was soon sought out as a replacement fuel. However, the use of coal presented problems for the metal refining industry. Coal contains sulfur that is released when burned. The sulfur caused problems in the iron making process, and could not be used to manufacture iron products. This is when the production of coke began. Besides preserving wood for other uses, the manufacturing and use of coke had another benefit as well. Because of charcoal's fragile makeup, the size of a charcoal-fueled iron producing blast furnace was limited to a height of about 24 feet. At heights greater than 24 feet, stacks of iron ore and charcoal would collapse under their own weight. Using coke to fuel iron blast furnaces enabled more iron to be produced in one burn. Stacks of coke fueled iron furnaces could now be built to as high as 78 feet. Coke became the fuel of choice over charcoal and coal because it was a cleaner burning fuel that produced high temperatures and did not require the use of wood.

**coke**—a solid hydrocarbon made from coal similar to charcoal that is produced by pyrolysis

## COAL GAS

The use of coal was not limited to the steam engine or to produce coke at the start of the industrial revolution. Coal could also be used to create a combustible gas, known as **coal gas**. Coal gas is composed of carbon monoxide and hydrogen. Coal gas today is also known as *syngas*, but in the nineteenth century it was called town gas because many towns in both Europe and North America supplied their citizens with this combustible gas. Coal gas is

**coal gas**—a combustible gas made of composed of carbon monoxide and hydrogen

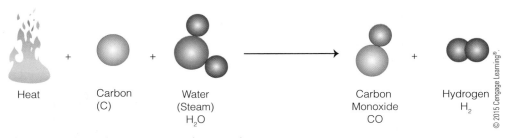

Heat + Carbon (C) + Water (Steam) H$_2$O → Carbon Monoxide CO + Hydrogen H$_2$

© 2015 Cengage Learning®.

**FIGURE 2-7** Coal gas reaction by pyrolysis.

produced by the pyrolysis of coal (Figure 2-7). This involves the heating of coal in a low oxygen environment. Coal gas was first made as a byproduct of coke production.

Soon people learned they could use coal gas to produce a fuel for lighting, cooking, and heating in homes. Prior to the use of coal gas, indoor lighting was only possible using oil lamps and candles. Coal gas had an advantage over these light sources because it produced a bright flame with no smoky residue. Soon coal gas manufacturing plants, known as coal gas works, began to spring up in many towns and cities across Europe and the United States. These facilities used coal to produce coal gas that was then delivered to homes and businesses by a network of gas pipes. Coal gas greatly improved the lives of people because they no longer needed to haul heavy wood or coal into their homes to use as fuel for heating, lighting, or cooking. Now the fuel was delivered to them by a network of pipes. Although coal gas improved the quality of life, it had major drawbacks. Not only was it a dangerous flammable gas, but it also contained carbon monoxide that was toxic to breathe. The production of coal gas was improved around 1850 when steam was injected into the furnace during the heating of the coal. This process enriched the resulting gas with combustible methane, hydrogen, and carbon monoxide. Coal gas was distributed to homes and factories in many cities in Europe and America beginning around 1815 and became the principle fuel for lighting and cooking for one hundred years. Because of the need for a central coal gas works and a network of pipes to deliver it, the use of coal gas was limited to cities. Rural communities still relied on traditional fuels like wood and coal to heat their homes, and fuel oil lamps and candles to produce light. Eventually coal gas use began to decline at the beginning of the twentieth century. This was the result of the more widespread use of electricity and the electric light, and the availability of methane gas. Methane, also known as natural gas, was much safer than coal gas because it did not contain toxic carbon monoxide (see Chapter 6).

## REVIEW OF KEY CONCEPTS

1. Coal is a hydrocarbon; it is formed from organic materials.

2. An organic material is one that contains carbon. A hydrocarbon contains both carbon and hydrogen atoms.

3. Coal is the remains of ancient plants living in swamps that existed on the Earth millions of years ago. These plants were buried in sediments, compressed, and transformed into coal. Much of the

world's coal deposits were formed during the Carboniferous Period that existed between 362–290 million years ago.

4. There are four types of coal that are classified by their carbon content and density. These are anthracite, bituminous, sub-bituminous, and lignite.

5. The use of coal as a principle fuel source began in Europe as a replacement for charcoal, used for metal manufacturing. Charcoal production required large amounts of wood that lead to the decline of forests. The use of coal to replace charcoal began the process of coal mining.

6. The development of deep coal mines required a means to remove groundwater that often flooded the mines and lead to the development of the steam engine. The first efficient steam engine to be used to power water pumps for coal mining was invented in 1776 by James Watt.

7. Early steam engines used low-pressure steam to power large pistons.

8. The invention of the steam turbine that used high-pressure steam greatly improved the power of coal-fired steam and was soon used to drive electrical generators.

9. Coal was also used to produce coke, the high carbon content fuel used for the manufacturing of iron and steel.

10. Coal gas was produced by the pyrolysis of coal. Pyrolysis involves the process of heating coal in the absence of oxygen. This resulted in the formation of solid coke and coal gas.

11. Coal gas produced a flame with little smoke and could be delivered via a pipeline. Coal gas could be dangerous to human health because it contained carbon monoxide, a deadly gas.

## CHAPTER REVIEW

### Short Answer

1. Briefly explain the process by which coal is believed to have formed.

2. What are the four types of coal?

3. How long ago did most of the earth's coal form, and during what geologic time period did this occur?

4. Describe two ways that coal is related to the steam engine.

5. Explain the process of producing coke, and discuss two advantages of using it instead of coal.

6. What is town gas and how was it used?

### Energy Math

1. If a Newcomen steam engine used 300 pounds of coal per day to pump water from a coal mine, how much coal would be used per day if it was replaced by a more efficient Watt steam engine which uses 66% less fuel?

2. Approximately how many times more iron could be produced by using coke as a fuel in a blast furnace instead of coal?

## Multiple Choice

1. The geologic period when much of the world's coal was formed is known as the:
   a. Cambrian Period
   b. Jurassic Period
   c. Carboniferous Period
   d. Cretaceous Period

2. Coal is formed mostly from the remains of:
   a. plants
   b. phytoplankton
   c. animals
   d. minerals

3. This type of coal is the cleanest burning, dense type of coal:
   a. lignite
   b. sub-bituminous
   c. bituminous
   d. anthracite

4. Coal was used as a replacement for:
   a. oil
   b. natural gas
   c. water power
   d. charcoal

5. The first efficient steam engine was invented by:
   a. James Watt
   b. Thomas Savery
   c. Thomas Newcomen
   d. Robert Parsons

6. The process by which coke and coal gas are made from coal is called:
   a. combustion
   b. distillation
   c. pyrolysis
   d. fractioning

7. During the Roman occupation of Britain coal was used for:
   a. iron smelting
   b. cooking
   c. heating homes
   d. weaponry

8. Name the individual who invented a steam-powered vacuum pump used to remove water from coal mines.
   a. Newcomen
   b. Boulton
   c. Watt
   d. Savery

9. What substance is released when coal is burned that causes problems in the production of iron?
   a. Carbon
   b. Hydrogen
   c. Sulfur
   d. Oxygen

10. What substance produced from the burning of coal gas was toxic to breathe?
    a. Oxygen
    b. Helium
    c. Carbon monoxide
    d. Hydrogen

# Matching

*Match the terms with the correct definitions*

a. coal
b. hydrocarbon
c. organic
d. anthracite coal

e. bituminous coal
 f. sub-bituminous coal
g. lignite coal
h. turbine

i. coke
j. coal gas
k. coal seam

1. _____ The black, shiny, cleanest burning, dense coal formed from exposure to the greatest amount of heat and pressure.

2. _____ A molecule made up of hydrogen and carbon atoms.

3. _____ Carbon containing.

4. _____ A dark black coal that is brittle and contains approximately 45–86% carbon.

5. _____ A black or brown, rock-like material made up of hydrocarbons that came from the partially decomposed remains of plants that lived millions of years ago.

6. _____ A circular wheel or rotor that is forced to spin at high speed.

7. _____ A dull black, low-density form of coal.

8. _____ The brown colored, lowest density coal that was not exposed to heat and pressure, and formed fairly recently in geologic time.

9. _____ A combustible gas that is composed of carbon monoxide and hydrogen.

10. _____ A solid hydrocarbon made from coal that is similar to charcoal that is produced by pyrolysis.

11. _____ A narrow band of coal sandwiched between surrounding layers of rock.

## CHAPTER 3

# The Oil Era

## KEY CONCEPTS

*After reading this chapter, you should be able to:*

1. Define the term petroleum.

2. Explain the process by which petroleum was formed and what the environment was like at the time.

3. Define the terms anoxic, aerobic, and oil seep.

4. Discuss how scientists use the Gingko leaf to infer what past climates were like.

5. Identify the type of rocks that form petroleum.

6. Define the term alkane and identify four examples of alkanes.

7. Explain the process of fractional distillation.

8. Identify the various products derived from fractional distillation in the relative order they condense out of the distillation tower.

9. Identify four early uses of petroleum by ancient people.

10. Discuss the factors that lead to the widespread use of kerosene.

11. Identify the world's largest producers of oil in the 1900s.

12. Explain the advantages liquid petroleum fuels had over coal and wood.

13. Discuss the contributions made by Nikolas Otto, Gottlieb Daimler, and Rudolph Diesel.

## TERMS TO KNOW

| | | |
|---|---|---|
| petroleum | bitumen | fractional distillation |
| phytoplankton | catagenesis | oil seep |
| anoxic | anticline | kerosene |
| kerogen | alkane | |
| black shale | natural gas | |

# INTRODUCTION

The discovery of readily available, abundant petroleum created another leap forward in our ability to provide power to improve the quality of life. Not only was oil an easily transportable fuel, it was the source of chemical compounds that could be used to produce numerous products that created the modern world in which we live.

## OIL FORMATION

**Petroleum**, also known as *crude oil*, is a dark brown, flammable liquid hydrocarbon that was formed millions of years ago from the remains of marine organisms. The term petroleum comes from the Latin words for "rock oil". The formation of crude oil is similar to that of coal, because both of these fossil fuels formed from the remains of photosynthesizing organisms deposited in aquatic sediments. Unlike coal which formed in a mostly freshwater environment, most of the world's crude oil originated from marine diatoms—a type of **phytoplankton** living in coastal oceans. Diatoms are a type of single-celled phytoplankton that builds a protective shell out of silicates, a mineral composed of silicon and oxygen. It is believed much of the world's oil was formed during the Mesozoic Era that began around 250 million years ago. This was also the time when dinosaurs first appeared on the Earth. During the early Mesozoic Era, the planet's continents where one large landmass known as Pangaea, and the climate was much warmer and wetter than it is today (Figure 3-1)

Together all of these factors played a role in oil formation. A large sea called the Tethys Sea, existed at this time and was located near the eastern edge of Pangaea, along the equator. This vast, warm, shallow marine

**petroleum**—a dark brown, flammable liquid hydrocarbon formed millions of years ago from the remains of marine organisms; also known as crude oil

**phytoplankton**—tiny organisms like algae that float freely in water and gain their energy from photosynthesis

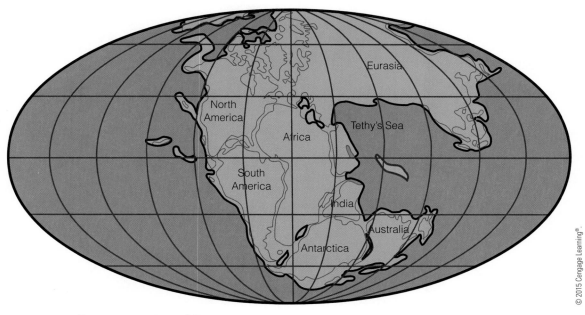

**FIGURE 3-1** Representation of Pangaea.

environment was home to an abundance of marine life, including phytoplankton. Today, phytoplankton is abundant along the coastlines of continents where runoff from the land provides vital nutrients the plankton need to flourish (Figure 3-2A and B). Back in the Mesozoic, phytoplankton would have been concentrated along the coast as well. The Tethys Sea was a perfect habitat for the growth of marine algae, because of its close proximity to Pangaea.

The warmer climate existing back then was because of a higher concentration of carbon dioxide in the atmosphere. The $CO_2$ absorbed heat given off from the planet's surface causing the temperature of the atmosphere to rise. The evidence to support higher levels of $CO_2$ in the atmosphere during the early Mesozoic can be found in the fossilized leaves of the Ginkgo tree. This ancient tree grew on Pangaea and is still alive today. Scientists revealed a relationship between the number of stoma on the leaves of the Ginkgo and the amount of atmospheric $CO_2$. Stomas are the tiny pores located on the underside of a leaf to allow gas exchange. During times when $CO_2$ is abundant, Ginkgo leaves have less stoma. When $CO_2$ concentrations decrease, there are more stomas distributed on the leaves. To support this theory, scientists grew Ginkgo trees in greenhouses that had different levels of $CO_2$ in them. The trees were exposed to lower $CO_2$ concentrations had a much higher number of stoma than the trees exposed to increased levels of $CO_2$. Fossils of Ginkgo leaves from the Middle and Late Jurassic and Early Cretaceous periods show a low number of stoma.

This suggests the levels of atmospheric $CO_2$ were much higher than today and therefore the planet was warmer. Because the climate was

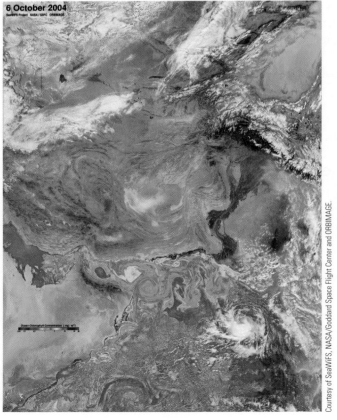

Courtesy of SeaWiFS, NASA/Goddard Space Flight Center and ORBIMAGE.

**FIGURE 3-2** Satellite images of phytoplankton bloom in the Arabian Sea.

warmer, it also tended to be wetter as well. Increased moisture was the result of higher evaporation rates that pumped water into the atmosphere forming more clouds and precipitation. The warmer and wetter climate of the early Mesozoic Era also played a role in the formation of oil. Higher rainfall rates lead to increased runoff from the land that feed nutrients into the coastal waters like the Tethys Sea. This caused an increase in the population of marine phytoplankton in the sea known as an algal bloom. These short-lived organisms reproduced rapidly, died, and sank to the bottom. This accumulation of dead algae led to the formation of a dead zone along the ocean floor known as an **anoxic** (lacking oxygen) zone. The lack of oxygen results in a complete stoppage of the decomposition of the dead phytoplankton by bacteria. This causes it to build up in the bottom sediments forming a rich layer of organic material. Over time the accumulation of this organic ooze is buried by marine sediments, exposed to increasing pressure, and is transformed into a substance called **kerogen** (Figure 3-3). The rocks produced from the deposition of phytoplankton in an anoxic environment are known as **black shale,** and is where crude oil originates. The kerogen is what gives the black shale its dark color (Figure 3-4).

The rocks that form petroleum are known as source rocks. Over time, the shale is buried deeper in the earth, and the kerogen within it is exposed to extreme heat and pressure. Eventually these forces cause the hydrocarbons

**anoxic**—lacking oxygen

**kerogen**—a compound made up of many different types of hydrocarbons that formed from organic remains

**black shale**—a sedimentary rock composed of a mixture of clay size sediments and kerogen

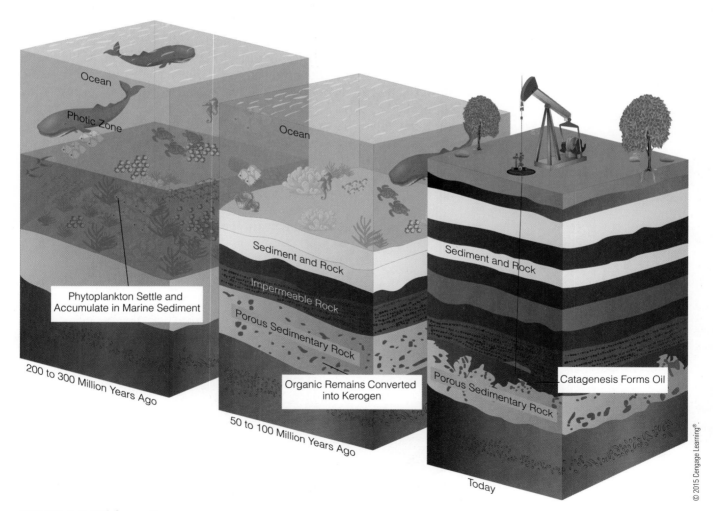

**FIGURE 3-3** Oil formation.

© 2015 Cengage Learning®.

FIGURE 3-4 Black shale.

**bitumen**—a thick, black, hydrocarbon compound much like tar

**catagenesis**—the process of heat and pressure breaking the long chains of hydrocarbon molecules in kerogen into smaller ones

**anticline**—a layer of folded rock in the shape of an arch

making up the kerogen to cook and fracture forming **bitumen** and petroleum. This is known as **catagenesis**. If the pressure on the black shale is too great, the kerogen within it is converted into lighter hydrocarbons like methane, also known as natural gas. If the pressure on the black shale is not great enough, sticky, black oil shale rock forms. This process occurred more than millions of years, forming great deposits of hydrocarbons that make up today's oil reserves. The tectonic forces that move the continents caused the rocks containing the black shale to buckle, collide, and be buried deep underground. The low density, liquid crude oil would then slowly seep upward into porous rock. Eventually it would reach a layer of impermeable folded rock called an **anticline** and pool up there (Figure 3-5).

The processes of oil formation that began during the Mesozoic Era can be seen today in areas like the Gulf of Mexico. The Mississippi River delivers large amounts of nutrients from the runoff that drains almost one third of the United States. This eventually dumps into the Gulf of Mexico where it feeds blooms of phytoplankton, creating an anoxic zone. The buildup of organic ooze today along the bottom of the Gulf of Mexico near the mouth of the Mississippi is similar to modern oil deposits that began as tiny plankton in the Tethys Sea 200 million years ago.

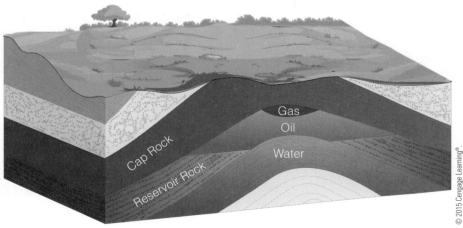

FIGURE 3-5 Diagram of an oil reservoir in an anticline.

# OIL CHEMISTRY

Crude oil is a liquid compound is made up of many types of hydrocarbons. Each hydrocarbon in oil has unique properties and uses. The chemistry of oil is called organic chemistry because it deals with all compounds that contain carbon. Carbon is unique among the other chemical elements because it bonds easily to other carbon atoms. Hydrocarbon compounds are divided up into categories based on the types of covalent bonds they form with other carbon atoms. A covalent bond is a link that is formed between two or more atoms that shares a pair of electrons. These can include single, double, and triple bonds. A single bond shares one pair of electrons, a double bond shares two pairs, and a triple bond shares three pairs of electrons. The hydrocarbons in petroleum are made up mostly of single-bonded carbon compounds called **alkanes**. These compounds form chains of molecules that follow the basic chemical formula: $C_nH_{2n+2}$ where $n$ represents the number of carbon atoms. To determine the number of hydrogen atoms in an alkane, you multiply the number of carbon atoms by 2 and then add 2. Petroleum is made of many mixtures of alkanes and fall into different categories based on the number of carbon atoms (Figure 3-6).

The simplest category of alkane present in petroleum is known as **natural gas** and includes methane ($CH_4$), ethane ($C_2H_6$), propane ($C_3H_8$), and butane ($C_4H_{10}$). Because propane and butane are larger molecules, they can remain in a liquid state under high pressure. This makes them useful as a portable fuel because they can be easily transported. Propane is often used in rural areas as a source of combustible gas for cooking and heating,

**alkane**—a type of carbon compound made of single-bonded carbons, also known as saturated hydrocarbons

**natural gas**—a flammable hydrocarbon consisting of mostly methane mixed with smaller amounts of propane and butane

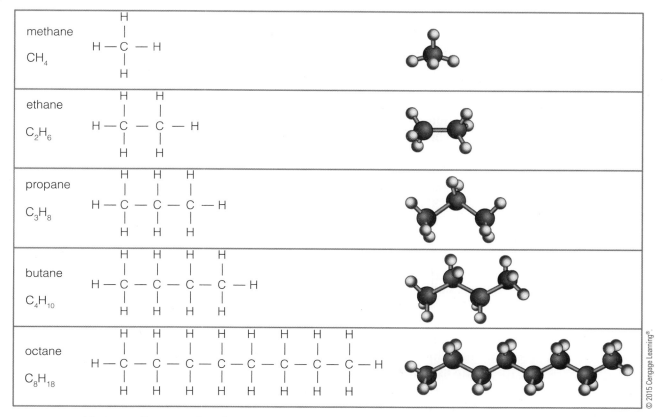

© 2015 Cengage Learning®.

**FIGURE 3-6** Various alkane molecules.

and butane is used in cigarette lighters. The next class of alkanes found in petroleum are referred to as naphtha that contain 5 to 10 carbons, and include pentane ($C_5H_{12}$), hexane ($C_6H_{14}$), heptane ($C_7H_{16}$), octane ($C_8H_{18}$), nonane ($C_9H_{20}$), and decane ($C_{10}H_{22}$). The heavier hydrocarbons with 10 to 20 carbon atoms are called paraffin oils and made up of large alkanes like dodecane ($C_{12}H_{26}$). Docosane ($C_{22}H_{46}$) is part of the gas oils within petroleum that include alkanes having between 20 and 70 carbon atoms. The largest hydrocarbons in petroleum are called asphaltenes. Asphaltenes are tiny solid particles of kerogen that contain 60 or more carbon atoms. Because of the wide variety of hydrocarbon molecules within petroleum, there are many uses for it as energy sources and as the raw material for industrial products.

## PETROLEUM REFINING

Crude oil's different assortment of hydrocarbons makes it difficult to use directly; therefore it must first be refined before it can be used. Crude oil may also contain many impurities such as sulfur, nitrogen, and various metals. The process of refining involves the separation of the different classes of hydrocarbons that are then treated to produce useable products and the removal of impurities. The first step in the refining process is called **fractional distillation**. Fractional distillation is the process where heat is used to separate liquids based on their different boiling point temperatures. The fractional distillation process of crude oil begins when oil is passed through a closed pipe that is superheated to a temperature of more than 800°F (427°C). The crude oil is then turned into a gas by the process of vaporization. The petroleum vapor is then passed into a distillation tower. As the heated vapor rises into this tall column, it begins to cool and condense. The different hydrocarbons that compose crude oil all condense into a liquid at different temperatures (Figure 3-7). The process of separating liquids based on their boiling points is known as *fractioning*. The different petroleum products separated by this process are referred to as *fractions*. Larger hydrocarbon molecules have higher condensation points, so they condense first and are collected in the tower, then removed.

The first hydrocarbons to be fractioned are the asphaltenes that condense at approximately 800°F (427°C). After collection, the asphaltenes can be used to make asphalt, charcoal, and coke. The next class of hydrocarbons to fracture is the gas oils that have a condensation point of approximately 70°F (371°C). Gas oils are then processed into lubricants and industrial fuels used in power plants and ships. As the vapor continues to cool and rise in the tower, at about 500°F (260°C) diesel fuel condenses. At about 400°F (204°C) kerosene is condensed. Kerosene can be used directly as a fuel for lamps or heating units, and can also make jet fuel. Heavy and light naphtha are condensed at temperatures between 275°F and 100°F (135°C and 38°C). These hydrocarbons are then used to make gasoline, and petrochemicals like plastics and resins. The lightest of the alkenes make it to the top of the tower and include gases like methane, ethane, and propane that are used as fuels for heating and cooking. Some of the heavier petroleum products like the asphaltenes and gas oils can be further processed to break down into lighter naphthas. This process is called cracking and involves exposing the hydrocarbons to intense heat and pressure, along with hydrogen gas and metal catalysts that break the larger molecules into smaller ones.

**fractional distillation**—the process by which heat is used to separate liquids based on their different boiling point temperatures

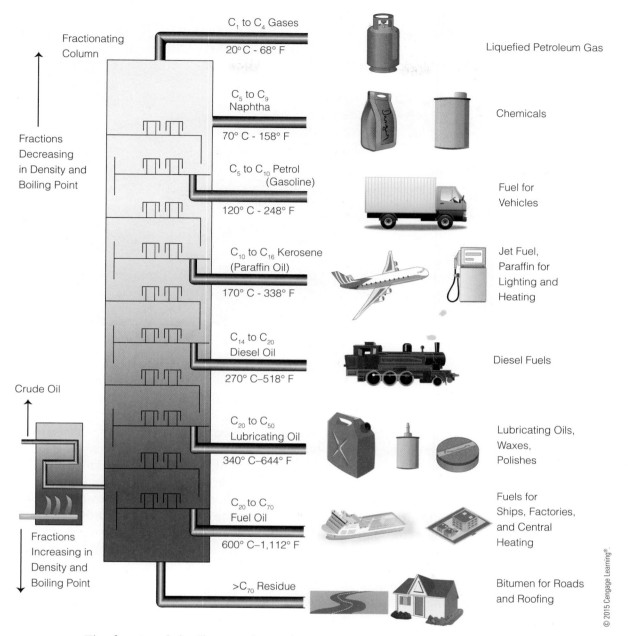

**FIGURE 3-7** The fractional distillation of petroleum products.

## EARLY USE OF OIL

In areas where crude oil deposits are located near the surface of the earth, oil can flow out under its own pressure forming surface deposits. This is known as an **oil seep** (Figure 3-8). Oil seeps occur all over the world and were the first crude oil resources used by humans. Today, oil seeps can still be found. The famous La Brea tar pits located in California are an example of an oil seep. These natural flows and pools of petroleum have been exploited by people for thousands of years. Oil seeps consist mainly of thick, sticky bitumen, also known as tar or asphalt.

Ancient societies harvested the asphalt for a variety of uses. The earliest evidence of the use of tar was found in Syria. Weapons made by Neanderthals living there 40,000 years ago used asphalt as a way to glue

**oil seep**—an area where crude oil flows naturally out of the ground

Courtesy of the State of California, Department of Conservation; Photographer Stephen P. Mulqueen

**FIGURE 3-8** Oil seep.

sharp points made of stone to wooden spear shafts. About 6,000 years ago, also in Syria, asphalt was applied as a sticky mortar used to join stones together in the construction of buildings, and as a way to waterproof small boats made from grasses. The asphalt used in ancient Syria is believed to be from an oil seep located 90 miles northwest of the modern city of Bagdad in Iraq. In ancient Egypt, bitumen was also used in the making of mummies. The word mummy itself is derived from the Persian term for bitumen wax, called "mum". This term was applied to mummies because of their dark appearance that was attributed to the bitumen used to preserve them. The use of bitumen was also applied to medicine in ancient times. People believed it possessed many healing properties and was used both inside and outside of the body. Natural springs of bitumen were used as religious sanctuaries where the flammable material was set on fire and allowed to burn indefinitely. In ancient Persia, oil seeps were converted into Fire Temples where eternal flames burned. These were called Ateshgahs, and can still be seen today in the modern day city of Baku in Azerbaijan, located near the rich oil fields along the Caspian Sea. In ancient Greece, the famous Oracle at Delphi temple was a place where the priests entered into trances that provided prophecies and advice to the people. It is now believed these trances were induced by the oracles breathing in hydrocarbon vapors coming from the bitumen containing rocks buried deep beneath the temple.

Another natural source of hydrocarbons seeping out of the ground is found in New York State, just south of Buffalo. Known as an eternal flame, this hydrocarbon seep is located behind a waterfall in the Chestnut Ridge County Park located in Erie County. The eternal flame is believed to have been lit hundreds or even thousands of years ago by Native Americans and still burns today. The flame is powered by natural gas seeping out from deep underground. The gas emitted from this seep contains the world's highest known percentages of propane and ethane.

The Byzantine Empire also used bitumen to create a deadly weapon known as Greek fire (Figure 3-9). This flammable mixture of bitumen and

Image from an illuminated manuscript, the Skylitzes manuscript in Madrid.

**FIGURE 3-9** Greek flamethrower.

sulfur was poured into clay pots that were set on fire and then thrown onto the decks of the ships of their enemies. The pots broke open and spilled the flammable mixture all over the ship, catching it on fire. The combustible mixture was nearly impossible to extinguish.

The Seneca Indians of Western New York skimmed oil from seeps and used it to repel insects and treat skin sores. They also traded oil with colonists. Today, NASA scientists using remote sensing technology estimate there are approximately 600 natural oil seeps on the bottom of the Gulf of Mexico. These alone pump more than 12 million gallons of crude oil into the Gulf every year! During the massive Deepwater Horizon oil spill in the Gulf of Mexico in 2010, a few of the gobs of crude oil washed onto beaches were mistakenly attributed to the oil spill when they were actually the result of natural oil seeps.

Although asphalt seeps were used widely by many cultures to create adhesives, as a means to waterproof boats, for roofing materials and as medicine, the first use of crude oil as a fuel can be traced back to more than 2,000 years ago in China. It was there that crude wells were first dug into oil seeps using bamboo pipes. This was then used to fuel lamps in some parts of China. The birth of the modern use of crude oil as a fuel source began in Europe and America around the 1850's. Crude oil back then was known as rock oil or carbon oil. The first modern era oil well was hand drilled in 1854 near Boryslov, Ukraine near the Carpathian Mountains, an oil rich region that gave birth to the **kerosene** industry. This region had been the source of crude oil for hundreds of years because of its many natural oil seeps. In 1858, the first crude oil distillery was constructed there, that at the time was governed by Austria. The use of kerosene in oil lamps provided a bright

**kerosene**—a liquid hydrocarbon compound produced from the distillation of crude oil that has a condensation point of around 400°F (204°C)

## Fun Fact

Because many Egyptian mummies were thought to contain bitumen, they were often ground up into dust, and made into a medicine! The use of mummy dust as a medicinal ingredient reached its peak in the Middle Ages in both Europe and Southwestern Asia. King Francis I of France often referred to as the Renaissance King, took a pinch of mummy dust every day to keep him healthy!

source of light, with virtually no smoky residue. Until the 1850s, the principal source of lamp fuel was in the form of whale oil, camphene, lard, and coal oil. Whale oil is waxy oil produced from the blubber, or fat of many species of whales. The highest quality whale oil came from the sperm whale. At its height, around 1845 the whaling industry produced 18 million gallons of whale oil. Camphene is a lamp oil mixture produced from alcohol, turpentine, and camphor. Both turpentine and camphor were produced from the resins of evergreen trees. Coal oil was a type of kerosene produced from coal. Lard is a solid substance at room temperature and made from animal fat. The arrival of kerosene to the lamp oil market provided an abundant, relatively cheap means to produce indoor light. Kerosene took over the lamp oil market as a result of the U.S. government applying a new tax on alcohol that drove up the price of camphene, in addition to the decline of whale oil production because of the over harvesting of whales. Shortly after the first kerosene distillery opened in the Ukraine, a vast oil field in Pennsylvania was tapped. In 1859, near Titusville, Pennsylvania, the first oil well was drilled in the United States (Figure 3-10).

This began the rapid rise of oil production in the United States. By 1901, the oil boom spread from Pennsylvania to East Texas and California. The United States became the biggest producer of oil in the world, followed by Azerbaijan and Austria's Carpathian oil. Most of the world's crude oil was refined into kerosene until a German inventor by the name

© Caitlin Mirra/Shutterstock.com

**FIGURE 3-10** Current site of the now historical oil well in Titusville, Pennsylvania.

of Nikolas Otto produced the world's first four-stroke internal combustion engine that used gasoline (see Chapter 17). At that time coal and wood were the primary fuels used to produce heat to drive steam engines. Steam engines are known as external combustion engines because the fuel used to power them is burned outside the engine. The ability to produce a liquid fuel in the form of kerosene allowed engineers to create new types of engines. Using liquid petroleum fuel rather than solid fuels like coal or wood has many advantages. Liquid fuels can be easily transported via tanks and pipes and burn much cleaner. They can also be used to create smaller, more powerful engines. In 1876 Otto produced his Otto Cycle Engine that used the power of exploding fuel inside the cylinders to create the pressure to drive pistons. By 1886, Gottlieb Daimler produced the first four-wheeled motor vehicle powered by a gasoline fueled Otto engine (Figure 3-11).

Gasoline quickly became the fuel of choice for internal combustion engines because it was easier to vaporize than kerosene. Daimler's car company went on to produce the world's first motorized truck. Amazingly, cars powered by Otto's internal combustion engine grew from the first one developed in 1886, to more than 1.7 million in the United States by 1914! This wave in the change of transportation rapidly altered the petroleum industry from its focus on providing lamp fuel in the form of kerosene, to producing gasoline for transportation. In 1893, Rudolph Diesel developed the diesel engine. The diesel engine operated differently than the four-stroke internal combustion engine and used the heavier alkane, *diesel*. The diesel engine was more fuel efficient and reliable, and eventually began replacing the steam engine. Soon, boats and trains were powered by diesel engines. In less than fifty years, the world rapidly changed from its heavy reliance on coal, to the new fuel of choice; petroleum.

FIGURE 3-11 First gasoline-powered automobile, as seen at the Essen Motor Show in Essen, Germany.

## REVIEW OF KEY CONCEPTS

1. Petroleum, or crude oil, is a liquid hydrocarbon formed from the remains of marine phytoplankton.

2. Much of the world's petroleum was created more than 200 million years ago when large amounts of marine phytoplankton lived, died, settled to the bottom of shallow seas, and formed a hydrocarbon-rich ooze. Eventually these hydrocarbons where buried in sediments and exposed to great amounts of heat and pressure. This extreme force transformed the plankton remains into sedimentary rocks called black shale that contain kerogen. Kerogen is a solid hydrocarbon compound. As these rocks were exposed to more heat and pressure, the kerogen was transformed into bitumen and petroleum. This process, called catogenesis, breaks the large hydrocarbon molecules into smaller ones. Bitumen is a solid hydrocarbon.

3. The anoxic means a lack of oxygen while the term aerobic means an oxygen environment.

4. Scientists discovered a relationship between the number of stoma on the leaves of the Ginkgo tree and the amount of atmospheric $CO_2$. During times when $CO_2$ is abundant, Ginkgo leaves have less stoma. When $CO_2$ concentrations decrease, more stomas are distributed on the leaves.

5. The rocks from which petroleum is derived are known as source rocks. Once petroleum is formed, it slowly rises upwards through porous rocks and can either flow out onto the surface as an oil seep, or pool under domed layers of impermeable rock known as anticlines.

6. Petroleum is made up of many hydrocarbon molecules called alkanes that are composed of different numbers of carbon atoms. There are five basic categories of alkanes that comprise petroleum. They include natural gas, naphtha, paraffin oils, gas oils, and asphaltenes. Natural gas has the least amount of carbons, and asphaltenes the most. Crude oil is refined to separate the different hydrocarbons and to remove impurities.

7. Fractional distillation is part of the refining process and uses heat to separate the various petroleum compounds.

8. The products of petroleum refining include natural gas, gasoline, diesel fuel, kerosene, jet fuel, asphalt, and many other hydrocarbons.

9. Humans used natural oil seeps as a resource for thousands of years. Early uses included using petroleum as an adhesive, a building material, weapons, and even a medicine.

10. The first modern use of petroleum began with the distillation of kerosene. Kerosene was widely used as a lamp fuel.

11. The first oil well in the United States was drilled in 1859 in Titusville, Pennsylvania.

12. Liquid fuels were more advantageous than coal or wood because they could be easily transported and delivered, burn cleaner, and can be used to run smaller, more powerful engines.

13. The development of the internal combustion engine in 1876 created a demand for gasoline that was refined from oil. Soon gasoline and diesel engines were used to drive vehicles on both land and water as the coal era declined and the oil era began to rise.

# CHAPTER REVIEW

## Short Answer

1. What is an oil seep?

2. Name some ways that ancient people used petroleum.

3. Which petroleum product was first used by people and for what purpose?

4. Why did kerosene become the favorite fuel of choice for indoor lamps?

5. Which three countries where the primary producers of oil around 1900?

6. What are some of the advantages liquid petroleum fuels have over solid fuels like coal and wood?

7. Who were Nikolas Otto, Gottlieb Daimler, and Rudolph Diesel?

8. Briefly explain the oil refining process.

## Energy Math

1. If all of the natural oil seeps at the bottom of the Gulf of Mexico produced the same amount of oil, how much oil would each seep spill into the Gulf?

2. Approximately 19 gallons of gasoline and 10 gallons of diesel fuel are made from one barrel of oil. If one barrel of oil equals 42 gallons, what percentage of one barrel of crude oil is made into gasoline and diesel fuel?

## Multiple Choice

1. Petroleum is formed from the remains of:
   a. plants
   b. phytoplankton
   c. animals
   d. minerals

2. The zone at the bottom of the ancient Tethys Sea where petroleum formed is known as:
   a. an aerobic zone
   b. an anoxic zone
   c. a benthic zone
   d. an aphotic zone

3. The source rock from which petroleum forms is called:
   a. black shale
   b. sandstone
   c. bitumen
   d. coal

4. The basic chemical formula for an alkane is:
   a. $C_nH_n$
   b. $CH$
   c. $C_2H_{n+2}$
   d. $C_nH_{2n+2}$

5. Which of the following hydrocarbons is in natural gas?
   a. Dodecane
   b. Methane
   c. Octane
   d. Heptane

6. Which of the following hydrocarbons is a naphtha?
   a. Docosane
   b. Butane
   c. Dodecane
   d. Octane

7. The process by which petroleum is refined includes:
   a. combustion
   b. fractional distillation
   c. pyrolysis
   d. catagenesis

8. The first modern use of petroleum included refining it into:
   a. kerosene
   b. gasoline
   c. diesel fuel
   d. natural gas

9. The first four-stroke internal combustion engine was invented by:
   a. Rudolph Diesel
   b. Gottlieb Daimler
   c. James Watt
   d. Nikolas Otto

## Matching

*Match the terms with the correct definitions*

a. petroleum
b. phytoplankton
c. anoxic
d. kerogen
e. black shale

f. bitumen
g. catagenesis
h. anticline
i. alkane
j. natural gas

k. fractional distillation
l. oil seep
m. kerosene

1. ____ The process of heat and pressure breaking the long chains of hydrocarbon molecules in kerogen into smaller ones.

2. ____ Lacking oxygen.

3. ____ A liquid hydrocarbon compound produced from the distillation of crude oil with a condensation point of around 400 degrees Fahrenheit.

4. ____ Tiny, free floating plants.

5. ____ A domed or arching structure of sedimentary rocks.

6. ____ A naturally occurring surface deposit of petroleum.

7. ____ The process by which heat is used to separate liquids based on their different boiling point temperatures.

8. ____ A flammable hydrocarbon consisting of mostly methane mixed with smaller amounts of propane and butane.

9. ____ Also known as crude oil, a dark brown, flammable liquid hydrocarbon that was formed millions of years ago from the remains of marine organisms.

10. ____ A sedimentary rock composed of a mixture of clay size sediments and kerogen.

11. ____ A thick, black, hydrocarbon compound much like tar.

12. ____ A compound made up of many types of hydrocarbons that formed from organic remains.

13. ____ A type of carbon compound made of single bonded carbons, also known as saturated hydrocarbons.

# UNIT II

# Contemporary Non-Renewable Energy Resources

---

## TOPICS TO BE PRESENTED IN THIS UNIT INCLUDE:

- Modern coal
- Petroleum
- Natural gas
- Nuclear power

## OVERVIEW

In his book, *Big Coal*, author Jeff Goodell writes, "We may not like to admit it, but our shiny iPod economy is propped up by dirty black rocks." This is indeed true, for 90% of the coal mined in the United States is used to produce approximately half of our electricity. But coal alone does not power our modern world. The use of petroleum as our main energy source surpassed that of coal in the middle of the twentieth century. Crude oil and its multitude of products account for 36% of our energy consumption that is mostly consumed for transportation. Natural gas is also a big player in our nation's energy supply making up 25% of our energy use, with new reserves being discovered every year. Coal produces 20% of our total energy use. Nuclear power also fills an important role in the production of electricity and accounts for 8% of our total energy. Together these four energy resources provide almost 90% of the energy produced in the United States, and are considered to be non-renewable energy resources. This means they cannot be replenished in a reasonable amount of time. This fact alone confronts us with three important questions about our nation's energy future: How long will these resources last? How do they impact the environment? What will replace them when they are exhausted?

## CHAPTER 4

# Modern Coal

## KEY CONCEPTS

*After reading this chapter, you should be able to:*

1. Explain the various methods of coal mining.

2. Identify two negative effects of coal mining.

3. Explain why coal needs to be processed and how it is processed.

4. Discuss the negative effects of coal processing.

5. Identify the two main methods for transporting coal.

6. Identify what percentage of our electricity is produced by burning coal.

7. Describe the process by which coal is used to produce electricity.

8. Define the term fluidized bed combustion.

9. Identify three coal combustion byproducts.

10. Discuss three pollution control techniques applied to coal combustion.

11. Identify how long coal reserves in the United States are predicted to last.

12. Explain why not all coal reserves can be mined.

13. Identify two technologies that can burn coal while producing less air pollution.

14. Define the term carbon sequestration.

15. Explain three methods of carbon sequestration.

16. Explain four atmospheric pollutants produced by burning coal.

17. Discuss the ways in which coal mining can affect aquatic ecosystems.

## TERMS TO KNOW

surface mining
overburden
contour mining
contour
deep mining

coal slurry
Rakine cycle
fluidized bed combustion
coal reserve
syngas

carbon sequestration
spoil pile
silviculture
photochemical smog
acid precipitation

# INTRODUCTION

Coal has dominated the energy landscape in the United States for more than one hundred years and will continue to play an important role in America's energy future. This resource is the most plentiful of the fossil fuels, but its use does not come without a cost to the environment. Learning how we get coal, transport it, use it today, how it affects the environment, and how it might be used in the future is crucial in understanding our nation's energy resources.

## COAL EXTRACTION

Coal is America's most plentiful non-renewable energy resource by far. Coal is a combustible black or dark brown sedimentary rock formed from the ancient remains of plants. Deposits of coal are called coal seams that can be anywhere from 20- to 300-feet thick and stretch over hundreds of square miles (Figure 4-1). Coal deposits also vary in the depth of their location and determine the type of mining operation used to extract it.

### Surface Mining

Coal seams located approximately 200 feet or less underground are extracted using surface mining, also known as opencast mining. **Surface mining** involves the removal of material covering a resource that is located close to the ground surface (see Figure 4-3). **Overburden** is the rocks and soil that cover over a resource.

**surface mining**—the removal of material covering a resource that is located close to the ground surface; also known as opencast mining

**overburden**—the rocks and soil that cover a resource

### Methods Used for Surface Mining

There are several ways surface mining can be done. An area surface mine is used to extract coal located in a relatively flat area. Extraction involves excavating trenches in the ground that are approximately 100 to 200 feet wide in order to extract the coal. Strip mining is another technique that involves the extraction of coal in long strips that follow the coal seams that are close to the surface. Excavation at these two types of surface mines follows the same general pattern.

- First, the site is surveyed for existing vegetation so that it can be replaced when the mining operation is complete.
- Next, bulldozers remove the vegetation and topsoil.
- Holes are then drilled into the remaining overburden, and explosives are placed in them.
- The explosives are ignited and break apart the overburden that is then removed by either large dragline buckets or front-end loaders.
- Finally the coal is removed.

Once the coal has been extracted from a surface mine, the coal mining company is responsible for replacing the overburden to its original contours. This process is known as surface mining land reclamation. The overburden is returned and is then covered by topsoil, and native vegetation is replanted.

**FIGURE 4-1** Coal seam.

**contour mining**—the over-burden is removed following the contours of the coal seam within the mountain

**contour**—the shape of a sur-face feature

In mountainous regions where coal seams are found located near the tops of a mountain, a method known as mountaintop removal is used (Figure 4-2). Here, the overburden, in this case the entire top of the moun-tain, is blasted and relocated. The overburden is then piled into an adjacent valley in a process called *valley fill*, and the coal is shoveled out. **Contour mining** is also used to extract coal in mountainous regions. Here, the overburden is removed following the **contours** of the coal seam within the mountain. The side of the mountain covering the coal is blasted away,

**FIGURE 4-2** Mountain top mining.

exposing a wall of coal. Often an auger is used to dig into the side of the mountain to access the coal inside. Once again, the overburden is replaced to restore the original contours of the mountain after the coal has been extracted.

Another type of surface mine is called open-pit mining. Open-pit mining involves the excavation of a deep pit to access a coal seam. This technique involves the excavation of a deep crater into the Earth.

All methods of surface mining use large amounts of explosives to both remove the overburden and break up the coal so it can be easily removed. Approximately 70% of industrial explosives in the United States are used to extract coal. Surface mining accounts for approximately 68% of the coal produced in the United States—more than 748.8 million tons in 2011.

## Deep Mining

Coal found deeper beneath the surface is extracted using **deep mining**. Deep mining is a method of extracting coal found 300 feet or more below the surface.

**deep mining**—a method of extracting coal that is usually found 300 feet or more below the surface

### Methods Used for Deep Mining

There are three types of deep mining operations: drift mining, slope mining, and shaft mining (Figure 4-3). These mines can be as deep as 1,000 feet beneath the surface. The deepest coal mine in the United States is in Alabama, reaching over 2,000 feet into the ground. The type of deep mining operation used is dependent on the location of the coal seam within the ground.

Drift mining involves the process of excavating a horizontal shaft into the side of a mountain to access the coal. Slope mining excavates a shaft that slopes downward into the ground to reach the coal seam. Shaft mining involves the digging of vertical shafts down into the ground to extract the coal.

Once the shafts arrive at the coal seam, two methods can then be used to excavate and remove the coal; these are known as the "room and pillar" method and "long wall" mining. The room and pillar method extracts the coal in the shape of large rooms approximately 20 to 30 feet wide. Between each room, a pillar about 100 square feet is left to support the overburden above the mine. These pillars prevent the collapse of the coal mine. Either

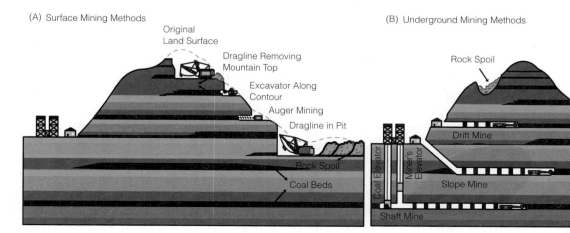

(A) Surface Mining Methods

Original Land Surface
Dragline Removing Mountain Top
Excavator Along Contour
Auger Mining
Dragline in Pit
Rock Spoil
Coal Beds

(B) Underground Mining Methods

Rock Spoil
Drift Mine
Coal Elevator
Miner's Elevator
Slope Mine
Shaft Mine

© 2015 Cengage Learning®.

**FIGURE 4-3** Various types of mining methods.

explosives or a machine called a continuous cutter is used to remove the coal. The coal is shuttled out of the mine by a conveyor belt. The ceilings of the large rooms are supported by drilling long bolts into the overburden. This helps prevent roof collapse. Once all of the coal has been reached, the pillars are then removed, one at a time working from the deepest room within the mine. As the pillars are removed, the rooms begin to collapse. Eventually all of the pillars are taken down, and the mine is abandoned. Long wall mining involves digging a long, wide shaft into the coal seam using a cutter. The shafts are typically 800-feet wide and thousands of feet long. The shaft is supported by hydraulic supports called shields that keep the overburden from collapsing on the mine. As the cutter moves into the seam, the roof shielding follows it along and the excavated mine sections collapse. The coal is continuously fed out of the mine via a conveyor system. Approximately 31% of all deep mining operations use the long wall method of coal extraction.

### Dangers of Deep Mining

Deep mining is probably the world's most dangerous profession. Coal miners are exposed to a number of hazardous working conditions that are not limited to mine collapses. On the job hazards include fires and explosions because of the flammable gases often associated with coal seams, exposure to coal dust that can lead to respiratory ailments, and exposure to the dangerous equipment used to remove the coal such as industrial explosives and cutting machinery.

## COAL PROCESSING

Once coal is extracted from the ground it is in a raw form known as *run of mine coal* that needs to be processed to remove any impurities present within the rock. Most processing of the coal takes place near or at the mine site. Processing includes washing the coal to remove rocks, soil, and metals mixed in the coal. It is estimated 20% to 50% of the raw coal material arriving from the mine is removed by the washing process. The more impurities removed from the coal, the higher its value in the coal market. During processing, coal is crushed, screened, and washed using a variety of methods. This usually includes the use of water to separate coal from impurities and large centrifuges to remove water from the coal after it is cleaned.

Sulfur removal is an important part of the washing process. Sulfur present in coal when burned is a criteria air pollutant (see the following section: Environmental Impacts of Coal). Sulfur found in coal usually comes in two forms; organic and pyritic. Pyritic sulfur is in the form of pyrite, also known as fool's gold, a mineral composed of iron and sulfur ($FeS_2$). Pyrite is often found within coal and can be washed out by crushing the coal and sending it to large water tanks. The coal floats at the top of the water while the denser pyrite sinks to the bottom. Organic sulfur is harder to remove and is often bonded within the coal itself. This type of sulfur can be removed when the coal is burned.

The byproduct of coal processing is a substance known as **coal slurry** or blackwater. Coal slurry is a mixture of soil, rocks, and other impurities removed from the coal and the slurry has the consistency of soft-serve ice

**coal slurry**—a mixture of soil, rocks, and other impurities removed from the coal with the consistency of soft-serve ice cream

cream. The coal slurry is usually stored in a slurry impoundment, an artificial pond often made from mine overburden. Valleys in mountainous regions are often damned using overburden to create an artificial reservoir to store the coal slurry. The impoundments holding the slurry are considered toxic because they can contain lead, mercury, arsenic, chromium and other heavy metals. The biggest threat of a coal slurry impoundment to the environment is a possible leak or spill. There have been more than forty reported spills of coal slurry in the United States since 2000. The majority of these spills occurred in West Virginia and Kentucky. The largest slurry spill took place in October, 2000 in Kentucky, when a slurry impoundment that was constructed over an abandoned mine collapsed. More than 250 million gallons of slurry spewed out of the old mine shafts and into 100 miles of streams and rivers, including the Big Sandy and Ohio rivers. Newer technologies are being investigated to replace the current coal washing practices that create slurry impoundments. These technologies include using dry separation methods to remove impurities from the coal that would eliminate the creation of slurry. Also, alternative storage of slurry in abandoned mines instead of in surface impoundments is being explored. Researchers are also looking into the possible use of coal slurry. Technologies are being developed to use the slurry as a co-firing fuel or as an additive to building materials. Co-firing means that the dried slurry is mixed in with coal as it is burned.

## TRANSPORTATION OF COAL

Once the coal has been cleaned it is ready to be transported to coal-fired electric plants. Much of the coal mined in the United States needs to be transported long distances. Some coal-fired electric plants are adjacent to coal mines, making them more cost effective. In this case coal is transported to the power plant by conveyor belts or trucks. Long distance transport of coal involves mostly the use of trains and barges. Approximately 71% of coal used in the United States is transported by trains. Coal shipped out of Wyoming uses approximately 80 trains a day containing more than one hundred cars each, stretching almost one mile long. Each car, known as a hopper, carries about 100 tons of coal. The long distance transport of coal affects its overall cost. Approximately 50% of the cost of coal is attributed to its long distance transportation. Shipping coal by barges is by far the most cost effective means of transport (Figure 4-4). Shipping costs via water routes cost about 30% less than train transport. Delivery of coal by barge is limited because not all coal destinations are located on waterways. Alternative means of transporting coal have also been explored. The use of a coal slurry pipeline was employed at a coal mine in Nevada to link it to a coal-fired electric generating station located 275 miles and sent down an 18-inch diameter pipeline. Once the coal arrived at the plant it had to be dewatered before away. The coal was pulverized, mixed with water, it was burned. This option tended to be water intensive and inefficient. Another alternative transport mechanism is called the coal log pipeline. This technique compresses coal into logs that are then injected into a pipeline and transported. The coal logs require less dewatering than slurry and use about 70% less water than slurry transport.

**FIGURE 4-4** Transporting coal by barge.

## CAREER CONNECTIONS

### River Towboat/Tugboat Captain

River captains operate tugboats and towboats that deliver barges up and down rivers. Twenty percent of the coal transported in the United States each year is done via rivers. Fleets of towboats and tugboats steer barges that transport more than 800 million tons of coal each year. Each barge can be as long as 1,000 feet and hold 20,000 tons of coal. There are more than 4,000 tugboats and towboats that direct 30,000 barges on our nation's rivers. Towboat captains must have experience with the operation and navigation of boats on inland waterways. Knowledge of technology is also essential because modern towboats are equipped with computer-aided navigation and controls. Shifts can last as short as 6 hours, or as long as 28 days depending on where the cargo is being delivered. Many towboat and tugboat captains get their start as deckhands and assistants, so a four-year college degree is not always necessary to become a boat captain.

## COAL USE

By far the main use of coal in the United States is for electricity production. Nearly 42% of our electricity is produced by burning coal, representing 93% of the total coal used in the United States. The other uses of coal include coke production for steel making and as a heat and steam source for many industries. By products derived from coal are also used to make plastics,

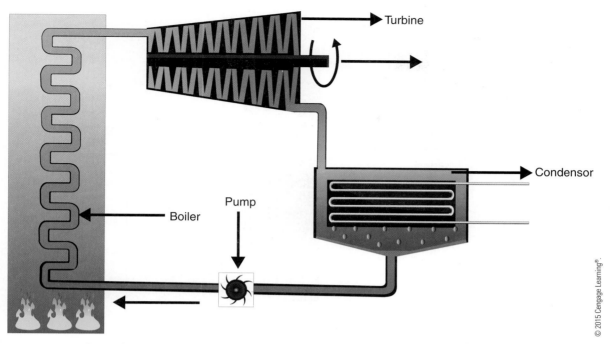

**FIGURE 4-5** The Rakine cycle.

© 2015 Cengage Learning®.

medicines, fibers, and fertilizers. Typically coal is used to heat water into steam, that is then forced to turn a turbine attached to an electrical generator. The steam then condenses and is pumped back into the combustion chamber to be re-heated. This process is known as the **Rakine cycle** (Figure 4-5). The Rakine cycle is a type of heat engine that uses heat to produce mechanical power. Typically the Rakine cycle consists of four stages. The first uses a pump to drive pressurized fluid, usually water, into a heated chamber. In the second step, the fluid is heated until vaporized. The third step involves the vapor being expanded into a turbine, that is then forced to turn and produce power. Last, the fluid enters a condenser that turns the vapor back into a liquid that is then pumped back into the heating chamber, thus completing the cycle. The use of the steam turbine to produce electricity has remained virtually unchanged for almost one hundred years. It is estimated that one pound of coal can produce enough energy to light a 100 watt light bulb for one hour. Because of this high energy density, we rely on coal as an energy resource.

Contemporary coal-fired electric plants pulverize the coal into a fine dust that is mixed with air and then injected into the furnace where it is burned. This is known as pulverized coal burning. Almost every coal-fired electrical generating plant in the United States uses this method of coal firing. One method of burning pulverized coal is called **fluidized bed combustion** (Figure 4-6). Fluidized bed combustion involves injecting coal dust into the boiler at high pressures where the dust burns in a flowing red hot, fluid-like mixture. The temperatures of around 1,500°F (815°C) generated by this technique are then used to heat water into high-pressure steam that is forced to turn a turbine electrical generator.

The efficiency of pulverized coal firing can range from 33% to 40%, meaning 33%–40% of the heat produced by the coal is used to produce electricity. The rest is lost in the heat transfer process that ends up in the atmosphere.

**Rakine cycle**—a type of heat engine that uses heat to produce mechanical power

**fluidized bed combustion**—the process of injecting coal dust into the boiler at high pressures, where the dust burns in a flowing red hot, fluid-like mixture

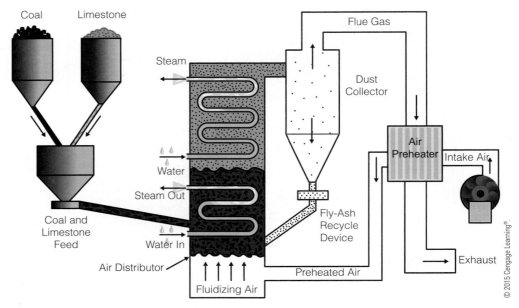

**FIGURE 4-6** Fluidized bed combustion.

One advantage of fluidized bed combustion is it allows potential air pollutants to be removed from the coal. Powdered limestone or dolomite, both calcium containing rocks, mixed into the combustion chamber when the coal is burned helps to remove sulfur from the exhaust gases. Another advantage of fluidized bed combustion is the possibility of using co-firing techniques. Co-firing is the process of adding other fuels, like biomass with the coal to burn (see Chapter 19). Fluidized bed combustion also lowers nitrogen oxide ($NO_x$) emissions. In old coal-fired boilers, coal was burned at high temperatures of around 3,000°F (1,649°C). In these high temperature conditions, nitrogen oxides form. The use of low temperature fluidized bed combustion does not produce temperatures high enough for excess $NO_x$ formation. Nitrogen oxides are a criteria air pollutant (see section on coal pollutants).

## COAL BYPRODUCTS

The products of burning coal create coal combustion byproducts (CCBs) that include bottom ash, fly ash, and FGD gypsum. Bottom ash is the residue that falls to the bottom of the combustion chamber and is usually composed of sand to gravel sized particles. Fly ash is composed of tiny dust sized particles captured from the exhaust systems by exhaust filtration units. The fly ash is composed of mostly silicates and makes up about 75% of the CCBs. Both bottom ash and fly ash can be used as additives to building materials like concrete or for roadway construction. Another product of pulverized coal is FGD gypsum, also known as fluidized gas desulfurization gypsum, which is produced as a byproduct of fluidized bed combustion. FGD gypsum is a fine powder composed of calcium sulfate, commonly known as gypsum; a product of mixing powdered limestone or dolomite with coal dust in the combustion chamber. FGD gypsum can be used to make building materials such as wallboard.

# COAL RESERVES

Coal is our nations' most plentiful fossil-fuel resource. The amount of coal left in the ground to be mined is known as our **coal reserves**. The Energy Information Administration, part of the U.S. Department of Energy, determines the estimates of coal reserves. This information is known as the demonstrated reserve base (DRB), includes the approximate amount of coal reserves that could potentially be mined using current technologies. As of January, 1, 2011, the DRB was estimated at 484.5 billion tons. This amount of coal however is not all readily available. This is due to many issues. For example, coal may be found in areas where it cannot be mined because the land is currently being used for something else. Also, the land on which coal reserves exist may be owned by someone who does not want to mine the coal. Other factors that make coal reserves inaccessible include the ability to get at the coal due to the local geology, land use patterns, and possible environmental impact of the extraction process. Together all of these factors reduce the estimates of the coal reserve base to approximately 261 billion tons, or about 50% of the DRB.

So how long will this last? In 2012 the United States used about 890 million tons of coal, a decline from the highest annual consumption of more than 1.1 billion tons in 2007 (Figure 4-7). If that rate stays the same, then our estimated reserves will last about 292 years. This assumes there will be no increase in the use of coal. Estimating an increase in coal consumption as a result of population growth and the decline in the availability of petroleum brings the reserves down to approximately 146 years. Whatever the number, coal will continue to be a major source of fuel for at least the next one to two hundred years.

**coal reserve**—the amount of coal that is left in the ground to still be mined

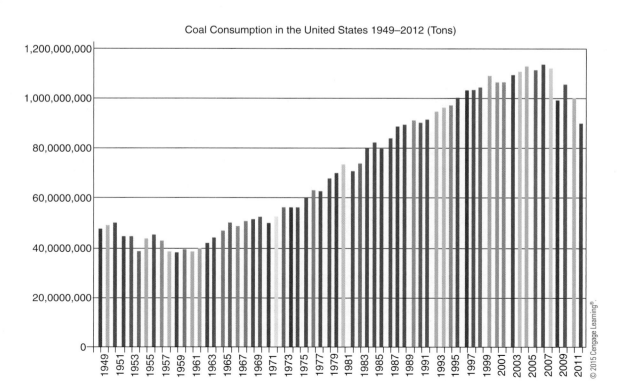

FIGURE 4-7 U.S. coal consumption.

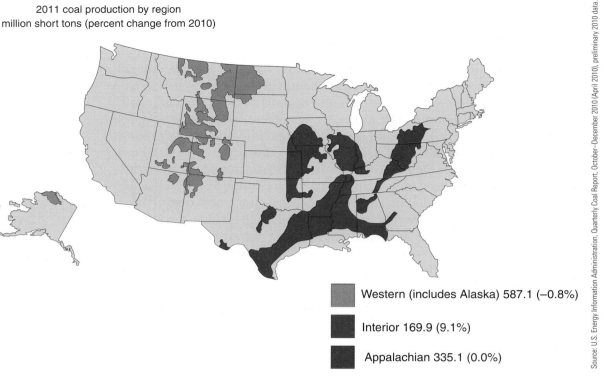

2011 coal production by region
million short tons (percent change from 2010)

Western (includes Alaska) 587.1 (−0.8%)

Interior 169.9 (9.1%)

Appalachian 335.1 (0.0%)

Source: U.S. Energy Information Administration, Quarterly Coal Report, October–December 2010 (April 2010), preliminary 2010 data.

**FIGURE 4-8** Coal-producing regions.

Coal reserves are divided into three main categories based on the geographic region where they can be found in the United States (Figure 4-8). These include the Appalachian, Interior, and Western regions. The Appalachian coal reserves exist in Pennsylvania, Ohio, West Virginia, Kentucky, Tennessee, Alabama, and Georgia. These reserves consist of mainly bituminous coal, with a smaller percentage of anthracite that makes up about 1.5% of the reserve base. The Interior Region encompasses Illinois, Michigan, Indiana, Iowa, Missouri, Kansas, Nebraska, Oklahoma, Arkansas, and Texas. This area holds mostly bituminous coal, with deposits of lignite found in the eastern part of Texas. Together the Appalachian and Interior regions account for nearly 50% of all the bituminous coal reserves for the United States. The Western region includes North Dakota, South Dakota, Montana, Wyoming, Colorado, Utah, New Mexico, and Arizona. The north slope of Alaska also holds coal reserves. The Western Region contains mostly sub-bituminous and lignite with smaller amounts of bituminous. Thirty-seven percent of U.S. coal reserves are made up of sub-bituminous western coal.

## NEW COAL TECHNOLOGIES

New types of coal combustion technologies are being developed to reduce the amount of pollutants that coal introduces into the environment. One of the most popular is coal gasification. This is the conversion of coal into a combustible gas known as **syngas**. Syngas is a flammable gas made up of mostly hydrogen and carbon monoxide. It can be produced by injecting coal dust into a gasifier. A gasifier is a heated chamber containing a low percentage of oxygen, filled with high-pressure steam and coal dust. The coal

**syngas**—a flammable gas that is made up of mostly hydrogen and carbon monoxide

goes through a chemical conversion where the hydrocarbon molecules are broken apart into syngas. This gas can be used to produce liquid transportation fuels similar to petroleum, industrial chemicals, or burned directly to power a turbine to generate electricity. Syngas was a fuel used during the nineteenth century to light lamps (see Chapter 3).

Coal gasification creates a cleaner burning fuel than untreated coal because the process removes impurities like sulfur and nitrogen. The development of a new electrical power plant known as an integrated gasification combined cycle system uses coal based syngas as its fuel (Figure 4-9). In this technology, syngas is used to power a turbine much like an aircraft engine that turns an electrical turbine. The hot exhaust gas that comes out of the turbine is then used to heat water that produces steam to turn another electrical turbine. This combined system of capturing the waste heat is much more efficient than current coal-fired power plants. Oxy-combustion is another method used to make coal burn cleanly. This combustion technology begins with the use of an air separation unit that separates oxygen from nitrogen in the air. The oxygen is then pumped into the combustion chamber along with pulverized coal where it is burned. This reduces $NO_x$ emissions, and also purifies the flue gases to be composed of mostly carbon dioxide and water vapor. The carbon dioxide can then be captured and stored.

Another potential alternative use for coal is to convert it into a liquid fuel much like gasoline or diesel fuel. This process is called coal liquefaction, or Coal to Liquids (CTL). There are two liquid transportation fuels can be produced from coal. The first is called indirect coal liquefaction. This method uses heat and pressure to form syngas from pulverized coal. The syngas is then passed through a series of catalysts to remove impurities like sulfur,

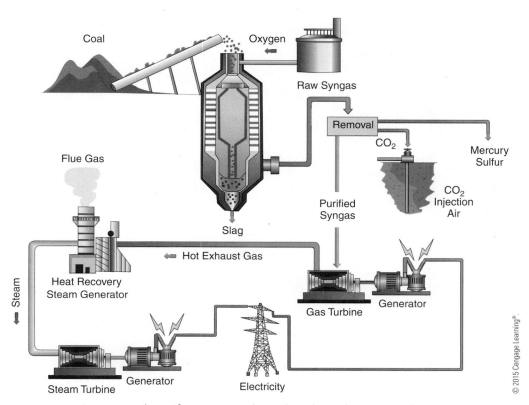

**FIGURE 4-9** Integrated gasification combined cycle with oxy-combustion.

and is converted into a liquid fuel similar to petroleum. This is also known as the Fischer-Tropsch process. During World War II, Germany used this method to produce fuel to power their military vehicles. Today, the U.S. Department of Defense has tested diesel and jet fuels derived from indirect coal liquefaction. These fuels are known as F-T fuels (Fischer-Tropsch). They have found these fuels are much cleaner burning than conventional fuels derived from petroleum. Another CTL method is known as direct liquefaction. This involves mixing pulverized coal with oil and hydrogen gas in a high-pressure environment. The result is a liquid fuel.

## CARBON SEQUESTRATION

One of the biggest drawbacks to burning coal as an energy source is that it produces carbon dioxide gas as a byproduct of combustion. As a result of burning large amounts of fossil fuels like coal and petroleum for the past 60 to 100 years, the levels of carbon dioxide have been increasing in the Earth's atmosphere. Scientists fear this rise in atmospheric $CO_2$ is causing the Earth's climate to warm. Increased levels of atmospheric $CO_2$ absorb the heat radiated from the Earth's surface, thereby increasing global temperatures. This is commonly called the greenhouse effect because the Earth's atmosphere and a greenhouse are heated in similar ways by sunlight. (Figure 4-10). The greenhouse effect occurs when short wave electromagnetic energy in the form of visible light emitted from the sun is transmitted through the Earth's atmosphere. This light energy is then absorbed by the

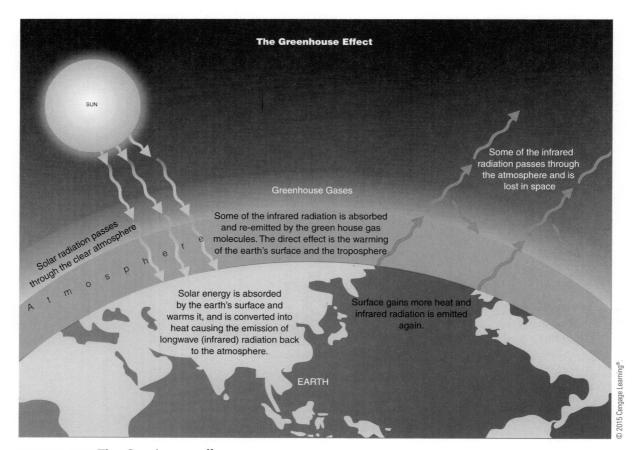

FIGURE 4-10 The Greehouse effect.

Earth's surface. Eventually this energy is radiated back from the surface into the atmosphere as longer wavelength infrared electromagnetic radiation. This infrared radiation is then absorbed by greenhouse gases like $CO_2$ and causes the temperature of the earth's surface to increase.

Global levels of atmospheric carbon dioxide have risen by more than 25% since 1958. Scientists agree the cause of this rise in atmospheric $CO_2$ is the result of burning fossil fuels like coal, petroleum, and natural gas. In 2012, the burning of these fossil fuels in the United States alone put 5,282 million metric tons of $CO_2$ into the atmosphere. Because of the concern that excess $CO_2$ is causing the Earth's climate to warm, technologies are being developed to prevent carbon dioxide from entering the atmosphere when coal is burned to produce electricity. This is called **carbon sequestration**. Carbon sequestration is the process by which carbon dioxide gas produced from combustion is captured and stored to prevent it from going into the atmosphere. Carbon sequestration involves two main processes, carbon capture and carbon storage. Exhaust gases from coal-fired power plants consist of a mixture of gases that include water vapor, nitrogen oxides, and carbon dioxide. In a typical coal-fired power plant, $CO_2$ makes up about 10–12% of the flue gas. In order to sequester the carbon, the $CO_2$ must be separated from the other flue gases. This can be done in a variety of ways. The most cost effective is to use a solvent to absorb the $CO_2$. In this method, flue gases are bubbled through a solvent that removes the $CO_2$. The solvent is then heated to release the $CO_2$ that is then captured. Other methods of carbon capture include the use of membranes or filters to separate gases. The use of oxy-combustion, where nitrogen gas is removed from air prior to burning coal can also help capture carbon. This process results in a flue gas that contains mostly $CO_2$ and water vapor. The water vapor can then be condensed and the carbon dioxide stored.

Once the carbon dioxide has been captured, it has to be stored indefinitely so it cannot get into the atmosphere. Carbon storage can be done in a number of ways (Figure 4-11). Terrestrial storage involves the process of injecting carbon dioxide into the Earth. In this storage technique, $CO_2$ can be stored in depleted oil wells, unusable coal deposits, and deep saline water formations. These three methods are possible because all of these potential storage sinks are widely available in the United States. The injected carbon dioxide will then be absorbed into these geologic formations. Current research is on-going to determine if these methods will be safe, and provide long-term storage of $CO_2$ for thousands of years or more. Injection into old oil wells has an added benefit that the $CO_2$ helps to free up oil that is still left in the rock but is too difficult to remove. This is known as enhanced oil recovery. Introducing $CO_2$ into old wells can free up oil that can then be pumped from the ground. Similarly, injecting $CO_2$ into deep coal seams may also be a way to produce natural gas. The $CO_2$ is more soluble than methane stored in coal seams, and helps drive off the gas that can then be collected. Currently there is a carbon sequestration project in use in Norway. A natural gas facility in the North Sea pumps $CO_2$ into a saline water enriched sand deposit that is located more than 2,600 feet below the seafloor. This injection well holds about 11 million tons of $CO_2$. Another carbon storage method involves injecting $CO_2$ into the deep ocean. If $CO_2$ is deposited at depths greater than 9,800 feet, it becomes denser than water and sink to the bottom of the ocean. Although this method has potential, concerns have been raised about the impact excess $CO_2$ might have on the ocean ecosystem. There is also the possibility ocean storage might alter the pH of the ocean

**carbon sequestration**—the process by which carbon dioxide gas produced from combustion is captured and stored to prevent it from going into the atmosphere

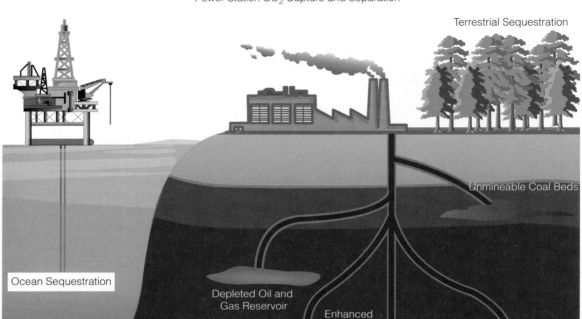

Power Station CO$_2$ Capture and Separation

**FIGURE 4-11** Methods of carbon sequestration.

by making it more acidic, because when carbon dioxide mixes with water it forms carbonic acid. There are ways to address these concerns that can be incorporated into ocean storage technology. Mixing the CO$_2$ with powdered limestone may buffer the solution, preventing it from lowering the pH of the water. Using biomass to trap CO$_2$ is also being explored. This involves setting aside forestland for use as a carbon capture source. Matching the amount of carbon that is emitted from a power plant to the appropriate amount of forest it would take to absorb the excess CO$_2$ could be a way to balance some carbon emissions. The use of single-celled algae to remove CO$_2$ from exhaust gases is under development. Passing exhaust gasses through photobioreactors filled with algae, can remove CO$_2$. A photobioreactor is a transparent tank or tube that uses light energy and photosynthesis to grow algae. In this case the source of the CO$_2$ for the algae in the reactor would be from exhaust gases. Also, the use of chemical scrubbers that react with CO$_2$ to form minerals is being researched as a means to remove carbon dioxide from the atmosphere. Magnesium and Calcium silicate minerals can absorb CO$_2$ when mixed forming a solid carbonate rock. Another method is passing CO$_2$ gas through water solutions containing calcium hydroxide and sodium hydroxide can convert it into solid carbon containing minerals.

All carbon sequestration technologies will undoubtedly raise the cost of electricity. It is estimated that rates would rise from 1 to 4 cents per kilowatt hour to employ this technology. This equates to a $10 to $20 increase in the price of the average American family's electric bill. The process of carbon capture is also energy intensive. It will use approximately 15% of the power produced at a coal plant. This will then cause an increase in the amount of fuel burned to provide the same amount of electrical output. Planting forests and reclaiming marginal lands can also help sequester CO$_2$. It is

## TABLE 4-1 Carbon Sequestration Methods for Agriculture and Forestry

| Agricultural Practices | Typical Definition and Some Examples |
| --- | --- |
| Conservation or riparian buffers | Grasses or trees planted along streams and croplands to prevent soil erosion and nutrient runoff into waterways. |
| Conservation tillage on croplands | Typically defined as any tillage and planting system in which 30% or more of the crop residue remains on the soil after planting. This disturbs the soil less, and therefore allows soil carbon to accumulate. There are different kinds of conservation tillage systems, including no till, ridge till, minimum till and mulch till. |
| Grazing land management | Modification to grazing practices that produce beef and dairy products that lead to net greenhouse gas reductions (e.g., rotational grazing). |
| Biofuel substitution | Displacement of fossil fuels with biomass (e.g., agricultural and forestry vastes, or crops and trees grown for biomass purposes) in energy production, or in the production of energy-intensive products like steel. |
| **Forestry Practices** | **Typical definition and some examples** |
| Afforestation | Tree planting on lands previously not in forestry (e.g., conversion of marginal cropland to trees). |
| Reforestation | Tree planting on lands that in the more recent past were In forestry, excluding the planting of trees immediately after harvest (e.g., restoring trees on severely burned lands that will demonstrably not regenerate without intervention). |
| Forest preservation or avoided deforestation | Protection of forests that are threatened by logging or clearing. |
| Forest management | Modification to forestry practices that produce wood products to enhance sequestration over time (e.g., lengthening the harvest. Regeneration cycle, adopting low impact logging). |

© 2015 Cengage Learning®.

estimated one acre of growing forest can capture between 2 to 9 tons of $CO_2$ per year. This would require millions of acres of forest just to absorb $CO_2$ from electrical power generation, which is not practical, although setting aside forest land could offset some carbon emissions. There are also several agricultural practices that can be employed to reduce carbon (Table 4-1). These include conservation tillage, grazing land management, the use of vegetative buffers, and the use of renewable biofuels to replace liquid transportation fuels.

Another idea to sequester carbon is known as ocean fertilization. This process proposes to seed the ocean with iron in order to increase the number of marine phytoplankton that would then take in $CO_2$ as part of the photosynthesis reaction. When spent, the algae would sink to the bottom where the $CO_2$ would be stored in ocean sediments. The problem with this technique is how the excess algae in the world's oceans might affect marine ecosystems. Probably the easiest method of lowering carbon emissions is to employ energy conservation and increase energy efficiency. Any practice that can be used to reduce the amount of electricity being used equates to less coal being burned, and therefore less $CO_2$ emitted.

# ENVIRONMENTAL IMPACTS OF COAL

The use of coal impacts the environment in two major ways. The first involves the mining process itself, and the second surrounds the emission of air pollutants produced when the coal is burned. The mining of coal is a destructive and dangerous process. It greatly disrupts the natural areas where the coal is buried. Surface mines expose rocks and sediments to precipitation that leads to excessive runoff and sediment pollution in surrounding surface waters. Although efforts are made by coal mining companies to lessen their impact on the environment, surface mining is one of the most disruptive processes of energy resource extraction. Reclaimed land on old valley fills and **spoil piles** have often compacted soil, and can have a negative impact on the growth of trees replanted in the area. Also, many streams located near the reclaimed lands are found to have elevated levels of sodium, selenium, sulfate, and zinc in the water. This negatively affects fish and macroinvertebrates living in these waters. The reclamation of surface mining lands can be improved by applying the agricultural sciences such as soil management, fertilizer application, and **silviculture** to restore the land to its former state.

The loss of terrestrial ecosystems like forests and grasslands affects the biodiversity in the region as well. The large areas of land blasted away to expose coal seams at surface mines also creates potential sources of sediment runoff that pollute surface water. The runoff from spoil piles can introduce sediments into water that increase the turbidity of the water. Turbidity is a measure of how clear the water is. Increased turbidity of water because of sediment pollution blocks photosynthesis that lowers the productivity of an aquatic ecosystem. The productivity of an ecosystem is a measure of how much energy is fixed by photosynthesis. This forms the base of the food chain. Sediments can also cover the bottoms of streams and lakes that can kill fish eggs and benthic macroinvertabrates living there. Benthic macroinvertebrates are organisms with no backbones, like aquatic insects, crustaceans, and mollusks that live on the bottom of an aquatic ecosystem.

The air pollutants produced by burning coal include nitrogen oxides ($NO_x$), sulfur oxides ($SO_x$), particulate matter (PM), carbon dioxide ($CO_2$), carbon monoxide (CO), mercury (Hg), lead (Pb), and other toxic chemicals. Nitrogen oxides form when atmospheric nitrogen ($N_2$) combines with oxygen in the combustion chamber. The atmospheric form of nitrogen is known as diatomic. Diatomic means the molecule is made up of 2 atoms. The high temperatures created by the combustion of coal causes $N_2$ to bond with oxygen and form $NO_x$. Nitrogen oxides, volatile organic compounds (VOCs), and sunlight lead to the formation of **photochemical smog**. (Figure 4-12). Photochemical smog is a type of air pollution that forms respiratory irritants like ozone gas ($O_3$) in the presence of sunlight. This is what gives urban areas a brownish colored haze on clear, sunny days. The VOCs that lead to the formation of smog often come in the form of evaporated fuels. **Acid precipitation** is a form of rain, snow, or fog that has a pH of lower than 5.6. In this case it forms when $NO_x$ reacts with water in the atmosphere to form nitric acid ($HNO_3$). Sulfur oxides form when the sulfur present in coal is burned. It then bonds with oxygen to form sulfur compounds like sulfur dioxide ($SO_2$). Sulfur oxides can also form acid precipitation. In this case the $SO_x$ reacts with water in the atmosphere to form sulfuric acid ($H_2SO_4$).

**spoil pile**—the stored remains of rocks and soil leftover from the mining process

**silviculture**—the art and science of growing trees

**photochemical smog**—a type of air pollution that forms respiratory irritants in the presence of sunlight

**acid precipitation**—a form of rain, snow or fog that has a pH of lower than 5.6

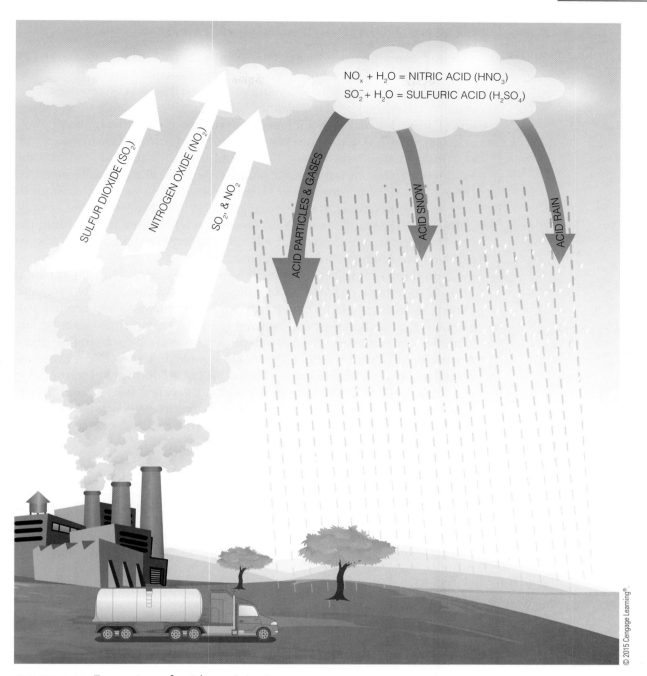

**FIGURE 4-12** Formation of acid precipitation.

Particulate matter is a criteria air pollution that originates from all fossil-fuel combustion. This pollutant consists of tiny solids of ash that can cause or exacerbate respiratory ailments, like asthma and bronchitis. Carbon dioxide is the natural byproduct of any type of combustion. $CO_2$ emitted from coal power plants contributes to elevated atmospheric $CO_2$ concentrations that contribute to global warming and climate change. Mercury can also be found in coal and is introduced into the environment when coal is burned. The mercury can then be transported in the atmosphere and deposited on land and in water when it mixes with precipitation. Once in the environment, mercury incorporates into the food chain where it accumulates to toxic levels in the bodies of organisms. This process is known as bioaccumulation.

## COAL AIR POLLUTION CONTROL

The pollutants produced from coal combustion—nitrogen oxides ($NO_x$), sulfur oxides ($SO_x$), carbon dioxide ($CO_2$), particulate matter (PM), carbon monoxide (CO), lead (Pb), and mercury (Hg)—can be removed from exhaust gases, making coal cleaner burning. All coal plants in use today employ a variety of control measures to reduce these atmospheric pollutants (Figure 4-13). Exhaust scrubbers come in four main forms: wet, dry, electrostatic, and baghouses. Wet scrubbers use water sprayed into the exhaust chamber to remove particulates, gases, and toxins like mercury. Often a wet scrubber uses plain water to remove unwanted exhaust products, and sometimes chemicals are added to the water to target specific pollutants. Limestone powder can be mixed with water, injected into the combustion chamber, and the precipitant collected by directing the exhaust to a baghouse to remove sulfur and particulates.

Dry scrubbers inject dry powders into the combustion chamber or flue gas to remove pollutants like sulfur oxides. Sodium bicarbonate, also known as baking soda, is commonly applied to flue gases to remove sulfur. Activated carbon dust can be used to remove toxins like mercury. Dry scrubbers can also use catalysts to remove nitrogen oxides from flue gases. A catalyst is a chemical or element that speeds up the rate of a chemical reaction. The removal of nitrogen oxides produced during coal combustion can be reduced by employing a new technique called staged combustion. This involves burning the coal with a high fuel to oxygen ratio, therefore preventing nitrogen oxides from forming. This method can reduce $NO_x$ emissions by up to 90%.

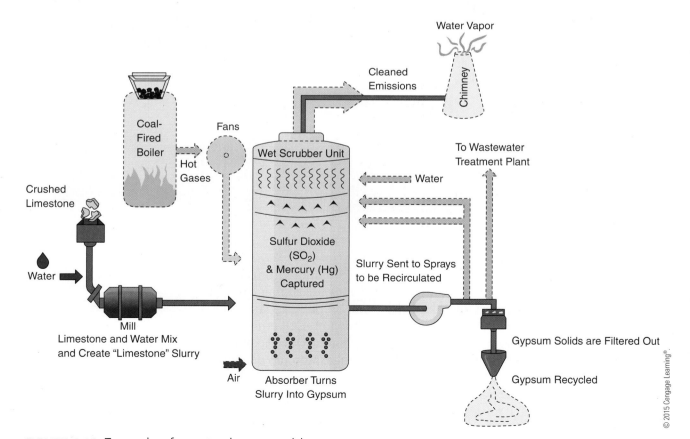

**FIGURE 4-13** Example of a wet exhaust scrubber.

Electrostatic precipitators are another control technique that uses large metal plates that are exposed to exhaust gases. An electrical current is passed through the plates giving them a negative electrostatic charge that attracts particulates in the flue gas. The tiny particles stick to the plates and are removed.

Finally, a baghouse filter system can also be used to remove pollutants. The exhaust gas is passed through large cloth bags that filter out unwanted dust, much like a giant vacuum cleaner.

## REVIEW OF KEY CONCEPTS

1. Coal alone produces about 42% of our electricity, making it extremely important to our economy and way of life. Coal is mined by employing both surface mining and deep mining techniques. Surface mining removes coal no deeper than 300 feet below the surface and deep mining removes coal deeper than 300 feet. There are a variety of methods employed for both surface and deep mining.

2. There are many hazards involved in mining. These consist of adverse environmental impacts, fire and explosions, and exposure to hazardous chemicals and pollutants.

3. Once coal is extracted from the ground it needs to be processed to remove impurities. This usually involves crushing the coal into dust and washing it with water.

4. The products of coal processing include coal slurry, a mixture of rock material and heavy metals removed from the coal. Slurry is stored in surface and subsurface impoundments and can be a risk to the environment if not contained properly.

5. Coal is mainly transported by train and barge. If the coal plant is close to the mining site conveyor belts and trucks may be used.

6. 93% of coal used in the United States is burned in combustion chambers to produce steam used to turn an electrical turbine. Most coal-fired power plants today use a form of fluidized bed combustion. Coal dust is injected into a combustion chamber with oxygen to produce heat that boils water.

7. Coal is used to heat water into steam that turns a turbine attached to an electrical generator.

8. Fluidized bed combustion is a method of burning pulverized coal. It allows a reduction in air pollutants released during the burning process and can be used in co-firing with low Btu fuels.

9. Coal combustion byproducts consist of bottom ash, fly ash, and FGD gypsum.

10. To reduce or eliminate air pollutants produced by burning coal, wet and dry control equipment is used.

11. U.S. coal reserves are expected to last 261 years.

12. Not all coal reserves are able to be mined. Some land on which a coal seam exists may already be in use, land owners may not permit mining of the land, and the location and geography may prohibit the mining process due to poor environmental impacts of doing so.

13. New forms of advanced coal technologies are being developed to make coal less polluting and more efficient; these include Integrated Gasification Combined Cycle and Oxy-combustion.

14. Carbon sequestration is a technique that removes and stores the carbon dioxide emitted from coal-fired power plants combat the greenhouse effect.

15. Methods of carbon sequestration include carbon capture and carbon storage.

16. Coal combustion can potentially produce a variety of atmospheric pollutants like sulfur dioxide ($SO_2$), nitrogen oxides ($NO_x$), carbon dioxide ($CO_2$), particulate matter (PM), carbon monoxide (CO), organic compounds (OC), lead (Pb), and mercury (Hg).

17. Streams or riverbeds located near mines may experience increased levels of sodium, selenium, sulfate, and zinc which has an adverse impact on aquatic ecosystems.

## CHAPTER REVIEW

### Short Answer

1. Explain four methods of coal mining.

2. What are two dangers associated with mining coal?

3. Describe why coal is processed and one negative effect of coal processing.

4. Describe the process of fluidized bed combustion.

5. What are three coal combustion by products?

6. List four coal combustion pollution control methods.

7. Describe two clean–burning coal technologies.

8. What are three methods of carbon sequestration?

9. Explain five ways burning or mining coal can be harmful to the environment.

10. Describe three agricultural practices that can sequester carbon dioxide.

### Energy Math

1. What percentage of coal was used to produce electricity in the United States in 2009 if the total use was 1,000,424 thousand tons, and 936,536 thousand tons was used to produce electricity?

2. 1,293 million tons of carbon dioxide was produced by coal-fired electrical power plants in 2009. How many acres of forests would be needed to absorb this carbon if one acre takes in approximately 6 tons of carbon per year?

### Multiple Choice

1. Which type of coal mining technique is most destructive to the environment?
   a. Shaft mining
   b. Surface mining
   c. Longwall mining
   d. Deep mining

2. Coal processing produces:
   a. carbon dioxide
   b. sulfur dioxide
   c. lignite
   d. coal slurry

3. Which type of coal combustion technology is used mostly in the United States?
   a. Fluidized bed combustion
   b. Coal liquefaction
   c. Integrated gasification
   d. Syngas

4. Approximately how much of our electricity is produced by coal?
   a. 5%
   b. 25%
   c. 40%
   d. 70%

5. What is the by product of using a wet scrubber during coal combustion?
   a. Nitrogen dioxide
   b. Sulfur dioxide
   c. Calcium sulfate
   d. Water

6. Coal reserves in the United States are expected to last:
   a. 40–50 years
   b. 75–100 years
   c. 150–200 years
   d. 1,000 years

7. Which type of coal makes up most of our reserves?
   a. Lignite
   b. Sub-bituminous
   c. Bituminous
   d. Anthracite

8. Which product of coal combustion is associated with acid precipitation?
   a. Sulfur dioxide
   b. Carbon dioxide
   c. Volatile organic compounds
   d. Particulate matter

## Matching

*Match the terms with the correct definitions*

a. surface mining
b. overburden
c. spoil piles
d. deep mining

e. coal slurry
f. fluidized bed combustion
g. coal reserves
h. syngas

i. carbon sequestration
j. photochemical smog
k. acid precipitation

1. _____ The removal of material covering a resource that is located close to the ground surface.

2. _____ Rocks and soil that cover over a resource.

3. _____ The process by which carbon dioxide gas produced from combustion is captured and stored to prevent it from going into the atmosphere.

4. _____ The stored remains of rocks and soil leftover from the mining process.

5. _____ A method of extracting coal usually found 300 feet or more below the surface.

6. _____ A flammable gas made up of mostly hydrogen and carbon monoxide.

7. _____ A mixture of soil, rocks, and other impurities removed from coal.

8. _____ A method of injecting coal dust into the boiler at high pressures, where the dust burns in a flowing red hot, fluid-like mixture.

9. _____ A type of air pollution that forms respiratory irritants in the presence of sunlight.

10. _____ The amount of coal left in the ground to still be mined.

11. _____ A form of rain, snow, or fog that has a pH of lower than 5.6.

## CHAPTER 5

# Petroleum

## KEY CONCEPTS

*After reading this chapter, you should be able to:*

1. Define the term oil trap.

2. Explain the methods used by geologists to locate oil.

3. Describe the process of rotational drilling.

4. Explain the difference between primary oil recovery and secondary oil recovery.

5. Discuss two types of unconventional sources of oil.

6. Identify approximately how much oil by percentage is imported into the United States each year.

7. Identify how much by percentage of oil in the United States is used for transportation.

8. List ten products derived from petroleum.

9. Discuss five potential environmental consequences of locating, extracting, transporting, and burning petroleum.

10. Identify about how many gallons of oil is needed to feed one American for one year.

11. Identify which agricultural product requires the least amount of energy to produce, and which requires a great amount of energy.

## TERMS TO KNOW

| | | |
|---|---|---|
| oil traps | secondary oil recovery | tar sands |
| seismic waves | enhanced oil recovery | oil shale |
| primary oil recovery | | |

## INTRODUCTION

No other fossil fuel has a larger impact on our economy and our daily lives than petroleum. The amount of oil the United States consumes is almost impossible to believe. In just one day we consume more than 789 million gallons! More than one-third of our total energy use is in the form of petroleum, and 71% of all the oil we consume, is in the form of transportation fuel. Our modern society is oil based. How do we supply this oil? How long can it continue and what affect does its use have on the environment? These are the questions that inevitably need to be answered in order to understand the true nature of modern oil and its impact on society.

## CONVENTIONAL CRUDE OIL EXTRACTION

Petroleum, also known as crude oil, is a dark brown, flammable liquid hydrocarbon formed millions of years ago from the remains of marine organisms (see Chapter 3). Conventional crude oil exists within the earth in naturally occurring deposits called **oil traps**. Oil traps are reservoirs of petroleum found within porous sedimentary rocks also known as reservoir rocks (Figure 5-1). Porous means the rocks contain tiny openings that allow the oil to accumulate in them, much like a sponge. Sandstone is an example of a reservoir rock. Oil traps are often associated with in specific geologic formations called geologic traps. Geologic traps are arrangements of rocks within the Earth's crust that can store large amounts of crude oil. Typically, reservoir rocks containing the oil are capped by a layer of impermeable non-porous rock known as trap rocks. These rocks prevent the oil from moving toward the surface.

**oil traps**—reservoirs of petroleum found within porous sedimentary rocks also known as reservoir rocks

### Oil Exploration

Because of their distinctive structure, petroleum geologists use a variety of techniques to identify where these formations might be found deep within the Earth. Geologists who search for oil must have a good knowledge of the geologic history of the region they are exploring. They also need to know the types of rocks and geologic formations that make up the area. On land this can be fairly easy, because you can see the rocks are at the surface that may identify if the area fits the geologic conditions that house oil traps. The search for oil deposits is not limited to land however. There are many areas below the sea surface where oil traps can be found, but this makes the search for oil more challenging.

One of the most successful methods used to locate oil is to conduct a seismic survey, using **seismic waves** to map the subsurface geology. Seismic waves are powerful shock waves that travel through the Earth's crust, much like ripples traveling across a pond when you throw a rock in the water. Geologists generate these waves on land by using explosives or a specialized truck called a Vibroseis (pronounced "vibrosize") that lowers a heavy, vibrating metal plate on the ground to generate seismic waves. These waves travel into the earth and reflect or refract off of the different type of

**seismic wave**—a powerful shock wave that travels through the Earth's crust, much like ripples traveling across a pond when you throw a rock in the water

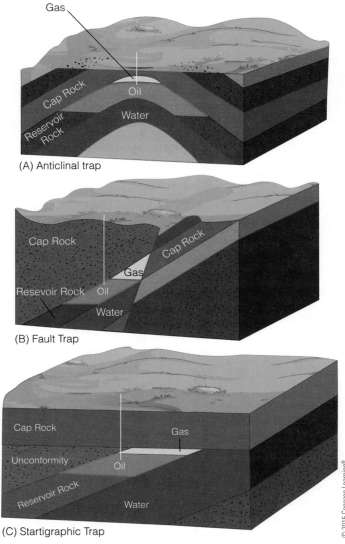

**(A) Anticlinal trap**

**(B) Fault Trap**

**(C) Startigraphic Trap**

© 2015 Cengage Learning®.

**FIGURE 5-1** Oil trap rock formations.

## CAREER CONNECTIONS

### Petroleum Geologist

A petroleum geologist helps oil companies locate oil deposits. Geologists use data gathered from the field on geology of a specific region to pinpoint areas that may contain oil. Using core samples, seismic data, and field investigations, petroleum geologists compile evidence to advise oil engineers an area could potentially produce oil.

Petroleum geologists require a four year college degree in applied geology or petroleum geology. Often a Master's degree is also required. Most petroleum geologists are employed in the private sector by oil companies.

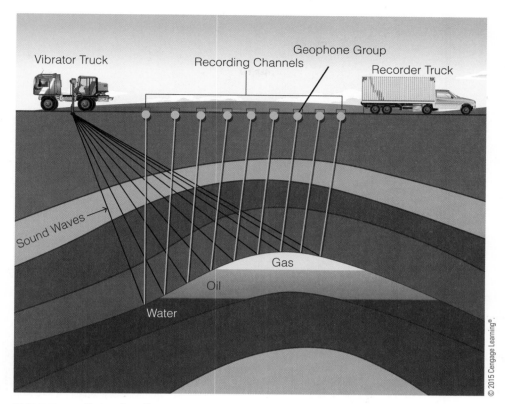

**FIGURE 5-2** Seismic oil survey.

rocks present in the area (Figure 5-2). These seismic reflections are then received by a network of special receivers called geophones. The reflection data is analyzed by a computer to reveal a two or three dimensional model of the underlying geology of the region. Geologists can then identify the potential location of geologic trap formations. When seismic surveys are conducted at sea, a ship tows a network of hydrophones through the water column, and air guns are set off in the water to create the shock waves.

Once a survey has revealed a potential oil trap, the next step in oil exploration is to drill a test well into the rocks. Most oil drilling wells use a method known as rotational drilling. Rotational drilling uses a sharp drill bit housed at the end of a long, steel pipe. This pipe is then turned by a motor called a rotary table that powers the drill and bores a hole into the earth. All drilling rigs are comprised of four main components: power supply, rotating system, hoisting system, and circulating system (Figure 5-3). These together are housed within a derrick. A derrick is a tower-like structure constructed at the drill site.

Some exploratory wells can also use a specialized truck that houses the drilling equipment. In this case, the operation is portable. As the drill bit bores into the ground, a series of pipes are fit together to form one long connection to the drill bit. The diameter of the bore hole can vary but is usually 21 inches. As the drill bit moves deeper into the earth, the circulating system pumps drilling mud into the pipe at high pressure. The drilling mud is a fine clay–water mixture that lubricates the drill bit and also removes the rock material that is cut out during the drilling process. The drilling mud is pumped at a constant high pressure so oil or natural gas does not burst up through the pipe. Because oil traps are buried deep within the earth, they are exposed to extreme pressure caused by the weight of the overlying

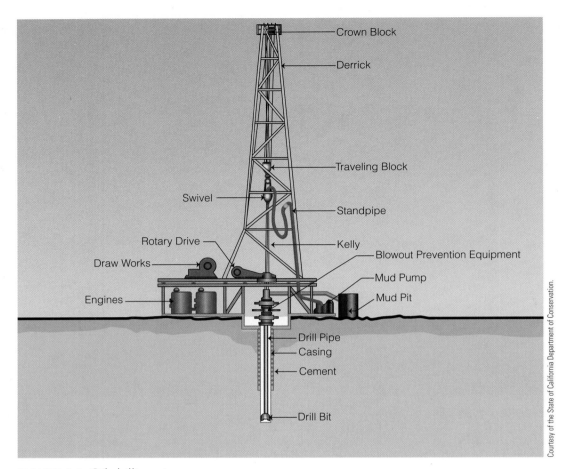

**FIGURE 5-3** Oil drilling rig.

rock. Oil wells are typically located at 5,000 to 6,000 feet below the surface. When the drill pipe enters the reservoir, the oil can flow out rapidly and travel up the pipeline dangerously. This is known as a blowout. The drilling mud seals the drill pipe, and prevents a blowout. As an extra precaution, drilling rigs also use an emergency device called a blowout preventer. This rapidly seals the pipe, and cuts off the flow to stop a blowout. If the test well confirms the presence of a significant oil trap, then a permanent drilling rig will be erected, and extraction of the oil begins.

## Drilling for Extraction

To access the oil, three types of drilling methods can be used: straight-hole, horizontal, or directional drilling. Straight-hole drilling bores straight down into the oil trap. In this case the derrick needs to be directly above the oil reservoir. Horizontal and directional drilling provides greater flexibility in accessing the oil, because the drill bit can be directed into the oil trap from many different angles (Figure 5-4). This allows the derrick to be located adjacent to the reservoir.

When the oil well is first drilled into the reservoir and oil begins to flow, it is called **primary oil recovery**. Primary oil recovery is the process where oil flows out under its own pressure to the surface to be collected. The oil is then collected in tanks at the drill site and transported by pipelines, trains,

**primary oil recovery**—the process by which oil flows out under its own pressure to the surface to be collected

© 2015 Cengage Learning®.

**FIGURE 5-4** Directional drilling.

or ships. Oil that is recovered can be classified as light or heavy, referring to how easily it flows, or as sweet or sour that is relative to its sulfur content. Light sweet crude oil flows freely like water and has low sulfur content. Eventually the oil from a trap will stop flowing when the pressure within the reservoir has been released. A typical oil well removes approximately 25% of the oil within the trap using primary recovery. The rest of the oil remains trapped within the reservoir rocks, because the pressure no longer exists to push it out to the surface.

When this occurs, **secondary oil recovery** begins. Secondary oil recovery is the process used to extract oil by injecting water into the well to wash more of the oil out. This method typically frees up another 5%–10% of the oil within the ground. In the past, after secondary recovery the oil well would be abandoned.

Today, oil drilling is using a process known as **enhanced oil recovery** to access the oil still trapped after secondary recovery within old reservoirs. Enhanced oil recovery uses steam or gases like carbon dioxide, methane, or nitrogen to drive oil out of the ground. It is estimated enhanced oil recovery could free up 30%–60% of oil trapped within old wells. This would significantly increase U.S. oil reserves. Another positive aspect of enhanced recovery is it can be joined with carbon sequestration (see Chapter 4). The large volumes of $CO_2$ required for enhanced oil recovery can be supplied by capturing the carbon dioxide emitted from coal-fired power plants. The $CO_2$ can be pumped into the ground to drive the remaining oil out, and then stored indefinitely within the reservoir rocks.

**secondary oil recovery—** the process by which oil is extracted by injecting water down into the well to wash more of the oil out

**enhanced oil recovery—** uses steam or gases like carbon dioxide, methane, or nitrogen to drive oil out of the ground

## CAREER CONNECTIONS

### Petroleum Engineer

A petroleum engineer locates oil deposits both on and offshore. They also plan the best method to extract oil from the deposit and manage the drilling operations. Petroleum engineers need to have a background in math, chemistry, and physics. In college, courses in engineering, petroleum geology, and engineering are required. Petroleum engineers work mostly outdoors and use a variety of techniques to locate oil. They also design the best drilling systems needed to extract oil from a specific deposit. Today, petroleum engineers work not only on discovering and extracting new oil deposits, but how to remove oil using enhanced recovery techniques.

# UNCONVENTIONAL OIL EXTRACTION

**tar sands**—rocks composed of sand, clay, water, and bitumen

**oil shale**—a black, sedimentary rock composed of clay-sized particles mixed with kerogen

Unconventional oil is petroleum that can be extracted from rocks. Unlike crude oil, these hydrocarbon deposits exist as solids in sedimentary rocks known as **tar sands** and **oil shale**. Tar sands are rocks composed of sand, clay, water, and bitumen. Bitumen is a solid hydrocarbon. Tar sands do not produce liquid oil like that of oil traps; instead the rocks must be mined using surface or underground mining techniques. Once the rocks are pulled from the earth they must be processed in order to extract the bitumen. The extraction process involves mixing the pulverized rock with hot water and sodium hydroxide, then agitating the mixture in a device much like a giant washing machine. The bitumen is separated from the sand and floats to the surface where it can be skimmed off and collected. The separation process uses a large amount of water. Approximately one hundred gallons of water is needed to extract one barrel (42 gallons) of oil. Also, about two tons of tar sands are needed to produce just one barrel of oil. Once the bitumen is separated and collected it needs to be processed so it becomes more like a fluid instead of a semi-solid. The remaining sand and clay is returned to the mine were it was extracted and the land is reclaimed. Tar sands may also be mined in situ, defined as "in place", by drilling wells into the rocks and heating them to free up the bitumen that is then extracted. So far, this method is energy intensive and inefficient. The world's largest deposits of tar sands exist in Canada (Figure 5-5). It is estimated there are approximately 170 billion barrels of oil locked up in Canada's tar sands. The Alberta tar sands is the third largest oil reserve in the world, with Saudi Arabia and Venezuela being number one and two respectively. Environmental concerns surrounding tar sand oil include the huge supply of water resources needed for extraction, land disruption as a result of the mining process, and water pollution produced from the spent water needed to process the sands. Currently in Northern Alberta, Canada there is 130 square kilometers of polluted tailing ponds. These wastewater ponds are the result of tar sand mining, and contain toxins like arsenic and naphthalene.

Another unconventional oil source is oil shale. Oil shale is a black, sedimentary rock composed of clay-sized particles mixed with kerogen (Figure 5-6). Kerogen is a thick sticky hydrocarbon from which petroleum is derived. Extracting oil from oil shale involves a similar process used for

**FIGURE 5-5** Oil sand mining and processing.

**FIGURE 5-6** Oil shale.

Courtesy of the U.S. Department of the Interior, Bureau of Land Management, Minerals, Realty, and Resource Protection Directorate and the Argonne National Laboratory.

tar sands. The shale is mined, crushed, and goes through a heating process known as retorting. Retorting exposes the shale to extreme temperatures that drive the kerogen out from the rock particles, and then collected. After the retort process, the kerogen needs to be further processed to make it flow like petroleum. Currently oil shale processing is not cost effective. The average cost of oil production from oil shale is approximately $60 per barrel of oil produced. The average cost of producing one barrel of conventional oil in the United States and Canada is about $36, and below $20 in the Middle East. These costs include locating the oil deposits and then extracting them. There is also an in situ process being developed to extract oil from oil shale. This is called the In Situ Conversion Process, that uses electric heaters placed deep in the earth to heat the oil shale to temperatures between 700°F–900°F. This process usually takes two to four years to heat the shale to the point that the oil begins to flow, and then pumped to the surface. Another large source of unconventional oil lies within the Bakken Shale. This 200,000 square mile oil shale deposit is located beneath North Dakota, Montana, and parts of Canada. The use of hydraulic fracturing, a process involving the injection of high-pressure fluid into the ground to crack the rock, has increased oil production there. In November of 2013, 976,000 barrels of crude oil was pumped from the Bakken Shale each day.

## OIL RESERVES

In 2011, the world consumed 87,605,300 barrels of oil per day. One barrel of oil contains 42 gallons of petroleum. This is a staggering amount of fuel. The United States is the world's largest consumer of oil, using 18,554,600 barrels a day in 2012, and accounts for nearly 22% of the world's total oil use. This equates to one American using approximately 2.5 gallons of oil each day. About 45% of the oil used in the United States is imported from other countries (Figure 5-7).

Much of the world's oil is controlled by The Organization of Petroleum Exporting Countries, also known as OPEC. OPEC is an organization of 12 nations that coordinate policies and pricing of oil. The most productive oil fields in the United States are in the Gulf of Mexico, Texas, and Alaska.

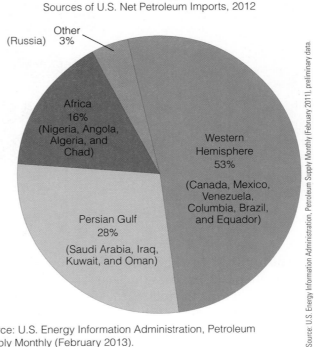

Sources of U.S. Net Petroleum Imports, 2012

(Russia) Other 3%

Africa 16% (Nigeria, Angola, Algeria, and Chad)

Western Hemisphere 53% (Canada, Mexico, Venezuela, Columbia, Brazil, and Equador)

Persian Gulf 28% (Saudi Arabia, Iraq, Kuwait, and Oman)

Source: U.S. Energy Information Administration, Petroleum Supply Monthly (February 2011), preliminary data.

Source: U.S. Energy Information Administration, Petroleum Supply Monthly (February 2013).

**FIGURE 5-7** Oil Imports.

Because the United States uses so much oil, it is important to know how much oil is left to be extracted. The answer to this question relies in estimating the world's oil reserves. Oil reserves are reasonable estimates on the amount of oil still available for extraction. These estimates, also known as proved reserves, are based on the amount of oil that exists in known reserves located by seismic studies, test drilling, and geologic field work. The U.S. Energy Information Administration estimates the proven oil reserves in the United States are approximately 23,267 million barrels. This does not include unconventional oil resources like oil shale and tar sands. As for the world's total oil reserves, these estimates are difficult to calculate, but are predicted to be about 1,473 billion barrels. This number may include unconventional reserves like tar sands that are not usually included as proven reserves in the United States. Whatever data you use, because oil is a non-renewable resource it will eventually be exhausted. Economists have predictions of the world running out of oil anywhere between 40 and 100 years at present rates of consumption. Using oil more efficiently can extend our reserves, but eventually we will have to seek another form of fuel.

Out of all the products made from petroleum (see Chapter 3) by far the biggest user is for transportation fuel. Approximately 71% of our petroleum use in the United States is for transportation. This use is usually in the form of diesel fuel and gasoline used to power cars and trucks. The U.S. Military is the largest buyer of fuel in the world, consuming almost 8 billion gallons of fuel each year. We can greatly extend our oil reserves if we can improve the efficiency of our use of petroleum fuels. Because the United States is so dependent on oil for its economy, the government has established a large stockpile of oil known as the Strategic Petroleum Reserve. This oil reserve was created by an act of Congress in 1977 to assure the United States would have access to oil in times of emergency. In 2011, the strategic reserve holds 695.95 million barrels of oil. Most of this oil is stored in abandoned salt mines in Louisiana and Texas. The U.S. government purchases oil to fill the reserve, and can sell the petroleum to oil companies in times of low supply.

## MODERN USE OF OIL

Oil is used for a variety of industries. Before oil can be used, it must first be refined. Refining is a process by which the different hydrocarbons that make up crude oil are separated. This is accomplished by using fractional distillation. This begins by superheating crude oil to turn it into a vapor. The petroleum vapor is then passed into a condensation tower where the different hydrocarbon molecules are separated according to their individual condensation temperatures. A more detailed explanation of the oil refining and fractional distillation processes are explained in Chapter 3. Currently the United States has 148 oil refineries in operation. The largest capacity refinery is located in Baytown, Texas and can refine more than 560,000 barrels of oil per day. Together, the total refining capacity of the United States is approximately 17,736,370 barrels per day, which is about 1.4 million barrels under the nation's annual consumption rate. The products of the refining process include liquefied natural gas, gasoline, diesel fuel, aviation fuel, kerosene, fuel oil, lubricants, waxes, petroleum coke, asphalt, road oil, and petrochemical feed stocks used for the manufacturing of synthetic rubber, plastics, and chemicals (Figure 5-8). Typically, almost half of a barrel of oil

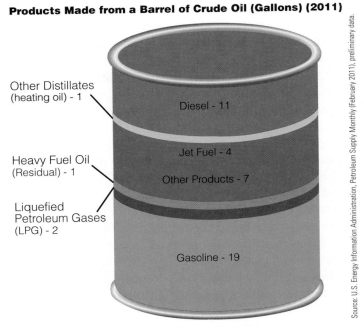

**Products Made from a Barrel of Crude Oil (Gallons) (2011)**

Other Distillates (heating oil) - 1

Diesel - 11

Jet Fuel - 4

Heavy Fuel Oil (Residual) - 1

Other Products - 7

Liquefied Petroleum Gases (LPG) - 2

Gasoline - 19

Source: U.S. Energy Information Administration, Petroleum Supply Monthly (February 2011), preliminary data.

**FIGURE 5-8** Products produced from one barrel of oil.

goes into making gasoline, with another 26% being made into diesel fuel. Transportation in the United States in 2011 used approximately 10.1 million barrels of oil per day.

## ENVIRONMENTAL IMPACTS OF PETROLEUM

The search for oil, its extraction, transportation, and use has many negative effects on the environment. Locating oil deposits can lead to deforestation, water pollution, and land disruption because of the use of heavy equipment required for the search for oil. Often new roads need to be constructed to gain access to the exploration area. If these areas are in remote locations, fragile habitats can be disturbed. In Alaska, millions of gallons of water are used to create ice roads that cover fragile tundra biomes to allow vehicles to access remote areas. The drilling of test wells and extraction of oil can also harm the environment. This often comes in the form of the discharge of drill cuttings. Drill cuttings are the crushed rock, clay, and drilling mud mixed with water and chemicals pumped in and out of the borehole and disposed either on land, or in the case of off shore drilling, at sea. Drill cuttings can contain radioactive materials, hydrocarbons, heavy metals and other toxins that negatively affect living things. It is estimated that offshore drilling in the North Sea has resulted in the build-up of more than 1.5 million tons of drill cuttings on the sea floor. This material has been shown to have devastating effects on the aquatic food chains that exist there. On land, drill cuttings are often discharged at the surface in impoundments or injected into the ground. The water used for petroleum exploration and extraction is known as Produced Water that contains numerous contaminants. These toxins pollute both surface and groundwater. The threat to human life is also a negative effect of oil drilling. Deaths related to oil and gas exploration and extraction are the highest in any industrial occupation. Fires, explosions, blowouts, and accidents at drilling rigs are the primary dangers faced by oil

Mobile, AL

New Orleans, LA

Oil Slick

Courtesy of NASA/MODIS Rapid Response Team.

**FIGURE 5-9** Satellite photo of the Deep Water Horizon oil spill in the Gulf of Mexico.

rig workers. The May 2010 explosion on the Deep Water Horizon offshore platform in the Gulf of Mexico killed 11 people (Figure 5-9).

Land-based oil drilling operations often occur in regions of the world that have political instability. One example is Nigeria. The oil extraction occurring around the Niger Delta in Africa illustrates the dangers involved with extracting oil in regions with unstable governments. The lack of government oversight can lead to severe environmental impacts. The area is strewn with rusting pipelines and open gas flares that burn 24 hours a day, filling the air with pungent fumes. Gas flares are created when natural gas produced from an oil well is burned off into the atmosphere. The unequal distribution of wealth in the area has resulted in the formation of angry militant groups who fight against the injustices created by the oil industry there. A report produced from the Shell Oil Corporation in 1997 reported 130 oil spills in the Niger Delta. The spills were caused by human error, sabotage, and equipment failure. In the city of Warri, Nigeria in 1998, a devastating fire caused by an exploding oil pipeline killed 700 people.

Another example of environmental disturbance created by the search for oil occurred in South America. In the fragile tropical rainforest ecosystem of the Amazon in Ecuador, the oil industry created more than 600 waste pits that have been leaking into surface and groundwater, causing severe contamination. The Ecuadorian government signed an agreement with Texaco Corporation to have these sites cleaned up. Today there still are more than one hundred waste pits still harming the environment there.

The transportation of oil is probably the most visible form of harm to the environment. The use of pipelines, ships, trains, and trucks to transport petroleum products leads to harmful oil spills. The largest oil spill in history occurred in 1991 during the First Gulf War. The Iraqi military detonated explosives at more than 730 oil wells on land and in the Persian Gulf.

The resulting disaster created fires that burned for months and spilled more than one billion gallons of oil into the deserts of Kuwait. The ruptured wells also pumped nearly 460 million gallons of crude oil into the Gulf of Arabia. This disaster was a deliberate act of war and not an accident. The world's largest accidental oil spill occurred in April of 2010 in the Gulf of Mexico. As a result of a violent blowout and subsequent explosion, the Deepwater Horizon drilling platform sank. This caused more than 205 million gallons of oil to spill into the Gulf of Mexico off the coast of Louisiana. It took 87 days to contain the leak. The oil damaged the Gulf's aquatic and coastline ecosystems, resulting in widespread contamination and die-off of birds, mammals, fish, and other aquatic organisms. The Gulf of Mexico region has experienced the highest number of oil spills in the world with more than 200. Before the Deepwater Horizon spill, the most widely publicized oil spill was the 1989 Exxon Valdez tanker accident in Alaska. The ruptured tanker spilled more than 10 million gallons of oil into Prince William Sound. The main threat to aquatic ecosystems after an oil spill usually involves direct contact with the oil itself, or ingestion of oil by wildlife. This causes sickness and death for fish, mammals, birds, and invertebrates. Plant life is coated with oil and damaged or killed. On land, oil spills often contaminate soil and groundwater. The use of petroleum products has other adverse effects on the environment as well. When petroleum-based hydrocarbon fuels are burned, toxins can be produced as a product of combustion. Carbon monoxide (CO), nitrogen oxides ($NO_x$), carbon dioxide ($CO_2$), particulate matter (PM), and volatile organic compounds (VOC's) are examples of air contaminants from the combustion of petroleum. Carbon monoxide is colorless, odorless toxic gas produced by incomplete combustion. CO is usually associated with starting a cold engine. The inhalation of CO causes sickness and death. Nitrogen oxides ($NO_x$) contribute the formation of ozone and acid precipitation (see Chapter 3). Particulate matter (PM) is a form of air pollution that originates from burning oil products such as diesel fuel. PM pollutants consist of tiny solids of ash that cause respiratory ailments, like asthma and bronchitis. Carbon dioxide is the natural byproduct of any type of combustion. $CO_2$ emitted from burning petroleum-based fuels, contributes to elevated atmospheric $CO_2$ concentrations that contributes to global warming and climate change. Volatile organic compounds (VOC's) include evaporated fuels like gasoline that form smog, and are a respiratory irritant.

## OIL AND AGRICULTURE

There is no doubt about the important link between petroleum and agricultural productivity in the United States. Agriculture on average ranks third in total end-use of diesel fuel and fuel oil, and farms consume more than 3.0 billion gallons of these fuels annually. The majority of this petroleum is used to power farm machinery, irrigation pumps, and agricultural facilities. Estimates are it takes 100 gallons of oil for American farmers to raise crops on 1 acre of land. This equates to 400 gallons of oil to feed one American for one year. This amount of oil represents 19% of the total oil use in the United States. An overview of the use of petroleum products used by agriculture goes as follows: 31% is used for fertilizer production, 19% to power field machinery, 16% for transportation, 13% to power irrigation systems, 8% for raising livestock, 5% to dry crops, 5% for the production of pesticides, and 8% for other activities or practices related to agriculture. Economists often

| Energy Input/Output Ratios for U.S. Agricultural Products (Kcal) | | | |
|---|---|---|---|
| Lamb | 57 : 1 | Rice | 1 : 2 |
| Grainfed Beef Cattle | 40 : 1 | Sorghum | 1 : 2 |
| Eggs | 39 : 1 | Soy Beans | 1 : 3 |
| Foraging Beef Cattle | 20 : 1 | Dry Beans | 1 : 2 |
| Swine | 14 : 1 | Peanuts | 1 : 1.4 |
| Catfish | 34 : 1 | Apples | 1 : 0.6 |
| Dairy | 14 : 1 | Oranges | 1 : 1 |
| Turkeys | 10 : 1 | Potatoes | 1 : 1.3 |
| Broilers | 4 : 1 | Corn Silage | 1 : 4 |
| Corn | 1 : 4 | Tomatoes | 1 : 0.26 |
| Wheat | 1 : 2 | Hay | 1 : 5 |
| Oats | 1 : 5 | Alfalfa | 1 : 6 |

© 2015 Cengage Learning®.

**FIGURE 5-10** Agricultural products energy input/output ratios.

use energy input/output ratios to determine the efficiency of a system (Figure 5-10). The ratio reveals how productive a system is and how much energy is needed to produce a product. The input/output ratio for fossil fuels in the form of oil and food production in the form of calories in the United States varies considering the type of crop being grown. Corn is by far the largest crop produced in the United States. In 2012, American farmers produced 10.8 billion bushels of corn. The energy input/output ratio for corn is about 1:4. This means it takes 1.0 kilocalorie of energy to produce 4.0 kilocalories of stored energy in corn. A kilocalorie is equal to 1,000 calories, and is the amount of energy required to raise the temperature of 1.0 liter of water, by 1°C. Typically, an average American consumes about 3,600 Kcal each day. Soybeans, the next largest crop grown in the United States has a ratio of 1:3. Other agricultural practices are much less efficient. For example, raising beef cattle has an input/output ratio of 40:1. This means it takes 40 kcal of oil to produce 1 kcal of energy stored in beef protein. This poor energy efficiency is because beef cattle are fed grains that also require energy inputs. Estimates show that the beef production energy ratio could be cut in half (20:1) by having cattle feed on forage in pastures instead of supplementing their diet with grains. The most efficient meat production technique that uses the least amount of oil energy is chicken farming. Raising chickens has an input/output ratio of 4:1. Most all grain production has a positive energy balance, meaning more energy in the form of food calories comes out than are put in from fossil fuels. Meat production carries a negative energy balance. Combined, the average energy input to output ratio for the entire food system in the United States is about 4:1. This shows how dependent our agricultural systems are on energy. Cornell University Professor of Ecology, David Pimental reports if all the people of the world grew and ate food like Americans, we would exhaust the world's oil reserves within 7 years. Clearly, increased efficiency in agriculture and the need for alternatives to petroleum is needed to sustain our food supply.

## REVIEW OF KEY CONCEPTS

1. Petroleum, also known as crude oil, is a form of fossil fuel originating from the remains of marine phytoplankton that lived in the oceans millions of years ago. Oil is located deep within the earth in large reservoirs called oil traps. These are usually associated with porous rock–like sandstone in geologic features called anticlines.

2. Petroleum deposits are located by performing seismic surveys. Once an oil deposit is discovered, drills are used to bore down into the oil trap and extract the oil.

3. Rotational drilling uses a sharp drill bit housed at the end of a long, steel pipe. This pipe is turned by a motor called a rotary table that powers the drill and bores a hole into the earth. All drilling rigs are comprised of four main components: power supply, rotating system, hoisting system, and circulating system. These together are housed within a derrick. A derrick is a tower-like structure constructed at the drill site.

4. Primary oil recovery is the process that a well is drilled into the oil trap and the petroleum flows out under its own pressure. Secondary oil recovery uses steam or brine pumped into the well to free up any oil that was not released during primary recovery. Enhanced oil recovery involves the use of carbon dioxide or nitrogen gas injected into the ground to loosen any remaining oil.

5. There are also sources of oil known as unconventional oil resources. These include tar sands and oil shale that bind hydrocarbon compounds within the rocks.

6. Approximately 45% of the oil used in the United States is imported from other countries. It is estimated the world's total oil reserves will last another 40–100 years.

7. Crude oil is used mainly for transportation fuel in the United States that is mainly in the form of gasoline and diesel fuel. Approximately 71% of oil in the United States is used for transportation.

8. Once oil is extracted it is made into useable fuels and industrial feedstocks by the process known as refining. Besides fuels, petroleum is also used to manufacture chemicals, fertilizers, plastics, rubber, lubricants, and many other products.

9. Burning petroleum-based fuels emits atmospheric pollutants like carbon monoxide (CO), nitrogen oxides ($NO_x$), carbon dioxide ($CO_2$), particulate matter (PM), and volatile organic compounds (VOCs). Other environmental risks that occur as a result of our use of petroleum include oil spills. In 2010, the largest oil spill accident occurred in the Gulf of Mexico, off the coast of Louisiana. This accident spilled more than 205 million gallons of crude oil into the Gulf.

10. Approximately 20% of all the oil used in the United States goes to grow food. Oil not only fuels the American farm, but is important for fertilizer and pesticide production. It requires approximately 400 gallons of oil to feed one American for one year.

11. An energy input-output ratio is used to show the efficiency of crop production and shows how much energy is required to produce crops. Corn is one of the most efficient crops in the United States with a ratio of 1:4. One of the most inefficient agricultural products is beef cattle with a ratio of 40:1.

## CHAPTER REVIEW

### Short Answer

1. What is an oil trap and what type of rocks are associated with oil traps?

2. What are two methods geologists use to locate oil?

3. Describe the differences between primary, secondary, and enhanced oil recovery.

4. What are two unconventional sources of oil?

5. How is most of the oil in the United States used? Approximately what percentage of oil is used for this purpose annually?

6. List ten products derived from petroleum.

7. Describe three negative effects petroleum has on the environment.

8. Explain what an energy input/output ratio is and what crop in the United States has the most efficient energy input/output ratio.

### Energy Math

1. If the total U.S. oil use in 2011 was 18,554,600 barrels of oil, how many gallons of oil was used for transportation in 2011?

2. Determine the energy ratio for potato production in the United States if the input is 17.5 million kcal/ha, and the output is 23.3 million kcal/ha?

### Multiple Choice

1. Oil that flows out of the ground under its own pressure through a drilling rig is known as:
   a. enhanced recovery
   b. secondary oil recovery
   c. primary oil recovery
   d. unconventional oil

2. Tar sand and oil shale are known as:
   a. enhanced oil
   b. secondary oil
   c. primary oil
   d. unconventional oil

3. Approximately what percentage of oil is used for transportation in the United States?
   a. 5%
   b. 25%
   c. 45%
   d. 70%

4. Approximately how much oil is imported into the United States each year?
   a. 5%
   b. 25%
   c. 50%
   d. 70%

5. About how much oil is needed to feed an average American each year?
   a. 50 gallons
   b. 100 gallons
   c. 200 gallons
   d. 400 gallons

6. Approximately how long are the world's oil reserves expected to last at present rates of consumption?
   a. 1,000 years
   b. 400–500 years
   c. 40–100 years
   d. 25 years

7. Which agricultural product grown in the United States is most energy efficient?
   a. Beef cattle
   b. Corn
   c. Chickens
   d. Soybeans

## Matching

*Match the terms with the correct definitions*

a. oil traps
b. seismic waves
c. primary oil recovery

d. secondary oil recovery
e. enhanced oil recovery

f. tar sands
g. oil shale

1. _____ A black, sedimentary rock composed of clay-sized particles mixed with kerogen.

2. _____ Rocks composed of sand, clay, water, and bitumen.

3. _____ The use of steam or gases like carbon dioxide, methane, or nitrogen to drive oil out of the ground.

4. _____ The process by which oil will flow out under its own pressure to the surface to be collected.

5. _____ The process where oil is extracted by injecting water into the well to wash more of the oil out.

6. _____ Powerful shock waves that travel through the Earth's crust that are used to locate oil.

7. _____ Reservoirs of petroleum found within porous sedimentary rocks.

## Short Answer

1. Explain what is meant by the following statement: "When you are eating food, you are eating oil."

2. Describe the difference between conventional and unconventional oil.

## CHAPTER 6

# Natural Gas

## KEY CONCEPTS

*After reading this chapter, you should be able to:*

1. Identify the three main gases that make up natural gas.

2. Define the term conventional natural gas.

3. List four sources of unconventional natural gas.

4. Describe the technology used to extract conventional natural gas.

5. Explain the process of hydraulic fracturing.

6. Identify five ways natural gas is used in the United States.

7. Explain how long proven natural gas reserves in the United States are expected to last.

8. Explain two negative effects natural gas has on the environment.

9. Discuss how natural gas is used by agriculture.

## TERMS TO KNOW

natural gas

unconventional natural gas

methane hydrate

hydraulic fracturing

anhydrous ammonia

# INTRODUCTION

Natural gas is the cleanest burning of the fossil fuels and supplies about 25% of the energy consumed in the United States. The United States supplies 90% of this fuel, making it an important domestic energy resource used primarily for residential heating and cooking and to generate electricity. Natural gas also has an important role in our nation's agricultural production. Many nitrogen containing fertilizers, like anhydrous ammonia, are made from natural gas feedstocks, many of which are imported from other countries.

## NATURAL GAS

**Natural gas** is a naturally occurring, colorless, odorless, combustible hydrocarbon gas. Natural gas is mainly composed of methane ($CH_4$), which typically makes up about 70–90% of its volume. Other hydrocarbon gases like ethane ($C_2H_6$), propane ($C_3H_8$), and butane ($C_4H_{10}$) can occupy up to 20% of the volume of natural gas. Trace gases like carbon dioxide, nitrogen, helium, and hydrogen sulfide can be found in natural gas as well. These impurities are usually removed during processing. There are two main natural gas deposits that are classified according to their accessibility: conventional and unconventional. Conventional natural gas is usually associated with oil deposits and forms under the same conditions that lead to petroleum formation. Unconventional natural gas is not as easily accessible and involves more energy intensive technologies to extract it. The type of methane associated with conventional and some unconventional sources of natural gas is known as thermogenic methane. Thermogenic methane was formed by the accumulation of organic sediments deposited in water millions of years ago. Over time, these deposits were buried deep within the earth and exposed to extreme heat and pressure. This process breaks the long chained organic molecules and converts them into simpler hydrocarbons like methane. It is believed that lower heat conditions within the earth converted organic sediments into liquid petroleum, while higher temperatures and pressures lead to natural gas formation. The gas accumulates in porous rock formations and can be trapped under layers of impermeable rocks.

**natural gas**—a naturally occurring, colorless, odorless, combustible hydrocarbon gas

### Conventional Natural Gas

Conventional natural gas deposits are often found within oil reservoirs. The lower density hydrocarbon gases like methane accumulate along the top of an oil trap. When oil drilling first began, it was common practice to burn off this "waste" natural gas that usually flowed out of the well before the oil. This process is known as gas flaring (Figure 6-1). Today, this gas is considered a valuable resource and is collected.

Another way methane is formed is known as biogenic methane. In this case, methane is formed as a result of a biological process involving methanobacteria. Methanobacteria is a class of anaerobic bacteria that produce methane gas in an environment without oxygen. These anaerobic bacteria break down organic material and form methane as a byproduct.

**FIGURE 6-1** Natural gas flaring from an oil well.

Most biogenic methane finds its way into the atmosphere, although there are some sources that can be tapped for use as a fuel source. Old landfills are a potential source of biogenic methane, where the gas can be collected and used.

### Extraction of Conventional Natural Gas

The technology for natural gas extraction is similar to that of oil drilling and is often done as a part of the oil extraction process. The first natural gas well in the United States built exclusively for gas extraction was created by William Hart in Fredonia, New York in 1821. During that time natural gas was mainly used as a fuel for street lights. Conventional natural gas collected along with liquid petroleum is known as wet gas or oil well gas. This accounts for approximately 21% of the gas produced in the United States in 2011. This type of gas is mixed with condensed hydrocarbons that make it "wet". These condensates must be separated from the methane to create dry natural gas. A gas well is a drilling operation used for natural gas extraction and not petroleum production (Figure 6-2). Today, because of new technologies developed for the petroleum industry, gas wells can be drilled either vertically or horizontally into the ground. Gas wells mainly extract natural gas produced by large deposits of sedimentary rock, such as black shales,

## Fun Fact

In 1865, German chemist Robert Bunsen perfected a device that produced a controllable flame by mixing natural gas with oxygen. This became known as the Bunsen burner and helped pave the way for natural gas to be used as a source of fuel for heating and cooking.

**FIGURE 6-2** Natural gas well.

that contain high percentages of organic material. These are also known as gas-bearing shales. Gas wells are responsible for nearly 73% of natural gas production during 2011 in the United States.

## Unconventional Natural Gas

Natural gas not associated with oil deposits, or that is not easily extracted from depths close to the surface, is known as **unconventional natural gas**. Unconventional sources include deep gas that is located more than 15,000 feet beneath the surface. At these extreme depths, high temperatures and pressures lead to the formation of nearly pure methane. Currently, accessing this natural gas reserve is not cost effective. Another form of unconventional natural gas is known as tight gas. Tight gas is methane trapped within rocks that have low permeability. Permeability is the ability for fluids to move through rocks in tiny pore spaces or cracks. Tight gas can be extracted by introducing chemicals into the rocks to dissolve them or create fractures within them. Another type of unconventional gas is located within rock formations known as gas shales. Gas shales are fine-grained, sedimentary rocks that contain high amounts of organic material. Trapped within this type of rock are large amounts of natural gas.

A fourth type of unconventional natural gas is called geopressurized gas. These extremely deep deposits of methane are located more than

**unconventional natural gas**—natural gas that is not associated with oil deposits, or that is not easily extracted from depths close to the surface

**methane hydrate**—a unique compound that consists of a methane molecule encased in a crystal lattice of ice

10,000 feet beneath the surface. Most of these gas deposits are located along the Gulf of Mexico and believed to hold large amounts of natural gas. Coal methane is a form of unconventional gas associated with coal seams. Natural gas is often found within coal deposits and has a potential for being collected and used. Traditionally, coal methane was a hazard to coal miners because of its flammability and potential for asphyxiation. **Methane hydrates** are also a form of unconventional natural gas. Methane hydrate is a unique compound that consists of a methane molecule encased in a crystal lattice of ice. This frozen form of methane forms deep on the bottom of the ocean in organic sediments where cold temperatures and high pressures encase the methane in ice (Figure 6-3). Methane hydrates are also found in the tundra of the arctic. The potential reserves of methane hydrates are vast, although the technology to access it is still under development.

### Extracting Unconventional Natural Gas

**hydraulic fracturing (fracking)**—the process of injecting high pressure water mixed with sand and chemicals into the ground to fracture rock, which is often associated with gas and oil recovery

The difficulty with extracting gas from shale is similar to that of tight gas. The shale must be fractured to allow the gas to flow and be collected. Horizontal drilling technology is often used to exploit shale gas. One of the largest potential gas shale deposits lies underneath Pennsylvania and Western New York. Known as the Marcellus Shale deposit, this black shale formation is between 50 and 200 feet thick, and located at depths between 4,000 and 8,500 feet (Figure 6-4). It is estimated to hold significant gas reserves. In order to produce gas from this shale deposit, a process known as **hydraulic fracturing** is used. Hydraulic fracturing, also known as

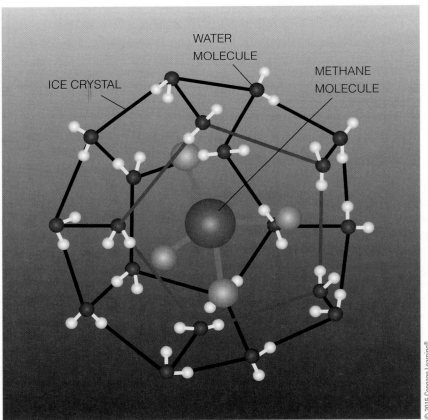

**FIGURE 6-3** Diagram of a methane hydrate.

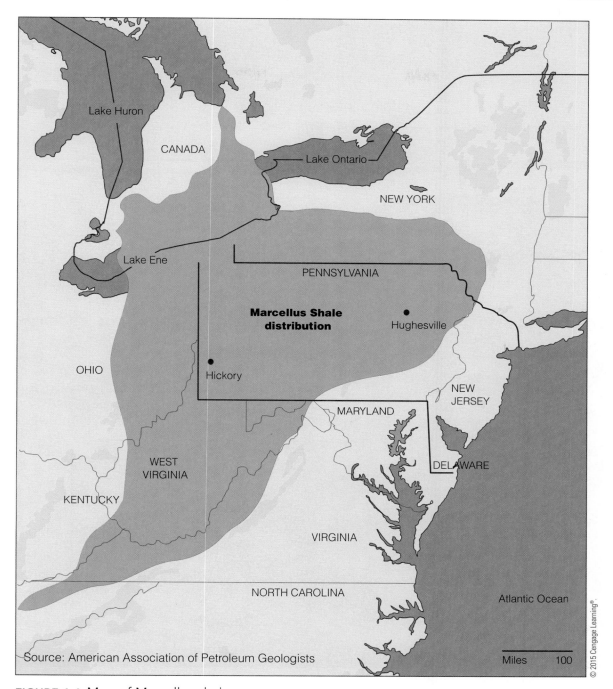

**FIGURE 6-4** Map of Marcellus shale.

fracking, involves the process of injecting water mixed with chemicals and sand to fracture the rock and keep the cracks open to allow gas to flow out and be collected by a gas well (Figure 6-5).

## NATURAL GAS USE

Natural gas is the cleanest burning of the fossil fuels. When natural gas undergoes combustion, mostly water vapor and carbon dioxide are produced. The lack of impurities within natural gas makes this possible. Natural gas combustion produces about 45% less carbon dioxide compared

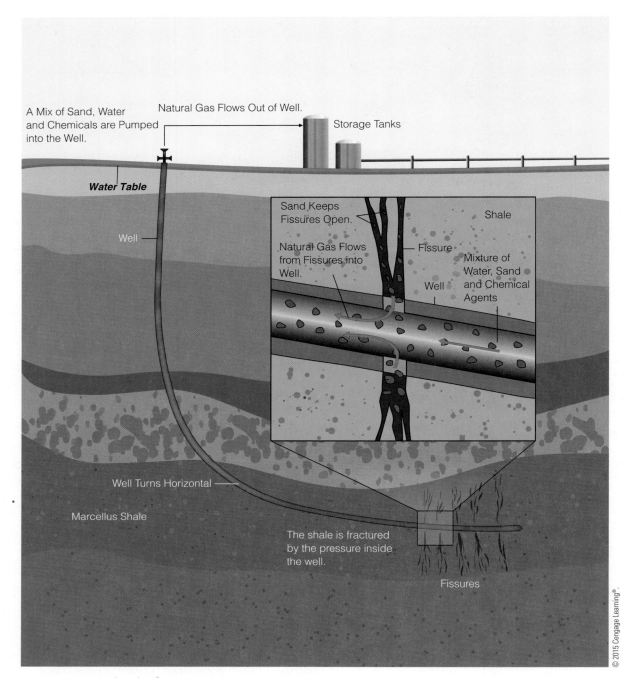

A Mix of Sand, Water and Chemicals are Pumped into the Well.

Natural Gas Flows Out of Well.

Storage Tanks

*Water Table*

Well

Sand Keeps Fissures Open.

Natural Gas Flows from Fissures into Well.

Fissure

Shale

Mixture of Water, Sand and Chemical Agents

Well

Well Turns Horizontal

Marcellus Shale

The shale is fractured by the pressure inside the well.

Fissures

**FIGURE 6-5** Hydraulic fracturing.

to burning petroleum-based fuels. Today, 25% of the total energy used in the United States is provided by natural gas and represents approximately 25.5 trillion cubic feet of natural gas. Unlike petroleum that is used mostly for transportation fuel, or coal that generates half of our Nation's electricity, natural gas is equally divided by its end-use. Approximately 36% is used to generate electricity, 28% for industry, 16% residential, and 14% for commercial use (Figure 6-6). Less than 1% is used for transportation. More than half of the homes in America use natural gas for heating or cooking. Natural gas is also an important energy source for the production of many products like paints, explosives, plastics, medicines, and fertilizers.

**Natural Gas Use, 2011**

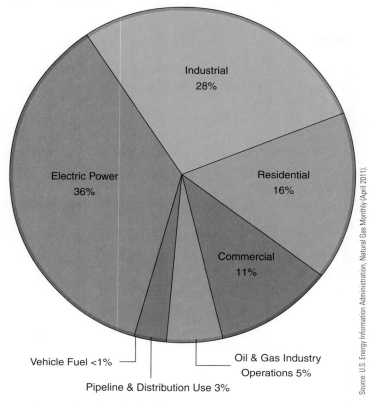

FIGURE 6-6 Natural gas usage 2011.

# TRANSPORTATION OF NATURAL GAS

Because natural gas has a condensation point of −260°F (−162°C), the temperature at which it becomes a liquid, 95% of natural gas used in the United States is delivered via pipeline in a gaseous form. Today, there are more than 1.2 million miles of natural gas pipelines in the United States that delivers this fuel to homes, businesses, and industry. This vast network of pipes, known as the natural gas transportation network, delivers gas throughout the United States. Typically, the gas is pressurized by compressor pumps or turbines spaced every 40 to 100 miles along the pipeline. The inside pressure of the gas pipelines ranges from 200–1,300 pounds per square inch (psi). The gas is delivered to local utility companies that reduce the pressure to 3 psi that is the standard in-home pressure for natural gas. Gas companies also add mercaptan, a rotten egg smelling substance that gives an odor to the gas to help detect leaks. Without the addition of mercaptan, colorless and odorless natural gas leaks would be impossible to detect, creating a dangerous situation. Natural gas is stored in large tanks as a liquid within depleted petroleum and gas deposits, or in old salt mines. Before natural gas can be used it must go through processing to remove any impurities within the gas. Processing facilities typically separate the methane from the ethane, propane, and butane, while also removing water, hydrogen sulfide, or any other impurities. The process of removing sulfur is called "gas sweetening".

**CAREER CONNECTIONS**

## Natural Gas Processing Technicians

The extraction, processing, and delivery of natural gas offer many career opportunities. Natural Gas Wellhead Pumpers operate the equipment at the gas well that pumps the gas from the ground. Natural Gas Treaters and Compressors run the machinery that removes impurities and water from natural gas, and readies it for delivery by pressurizing it to levels needed for long distance transport via pipelines. Gas Plant Operators run the machinery that liquefies natural gas, and Gas Service Technicians install and repair gas service lines. All of these careers require a mechanical background, and fits well for someone who likes to work with tools and complex machinery. Many of the jobs within the natural gas industry involve work outdoors. Although not a requirement, a two- or four-year degree at a technical college provides a good background for this type of career.

## NATURAL GAS RESERVES

Natural gas reserves are the estimates of how much natural gas is left within the earth that is technologically feasible to extract with known technology. Like oil, the Energy Information Administration classifies these reserves as proven reserves. As of December 31, 2010 the proven reserves for natural gas in the United States was determined to be 304.6 trillion cubic feet. This estimate includes accessible tight gas, coal methane, and shale gas. In 2012, the United States consumed 25.5 trillion cubic feet of natural gas. At present rates of consumption, this would supply the country with gas for almost 12 years; that is not a good, long-term supply. However there are undiscovered technologically recoverable natural gas resources that could extend our reserves for a little more than 80 years. These include potential gas reservoirs that have not yet been discovered or are currently not accessible. Many researchers believe the potential to harvest methane hydrates could provide us with natural gas for more than one thousand years! There is also great potential for using biogenic natural gas that could be considered a renewable resource. Imports represent 7% of our natural gas supply that comes mostly from Canada via pipeline. Liquefied natural gas is also imported into the United States from other countries like Trinidad, Tobago, Qatar, Yemen, and Egypt, representing 1% of our supply. The world reserves of natural gas are estimated at 6,707 trillion cubic feet. These reserves exist mostly in Africa and the Middle East.

## ENVIRONMENTAL IMPACTS OF NATURAL GAS

Although natural gas is the cleanest burning of the fossil fuels; burning it also produces carbon dioxide, a greenhouse gas that contributes to global warming. Natural gas itself, when not burned is composed of mostly methane that is also greenhouse gas. Methane can absorb 20 times more heat than carbon dioxide. Approximately 41% of U.S. methane emissions come from the natural gas industry. Agriculture produces approximately 30%, and waste management about 28% of atmospheric methane emitted from the United States. Extracting methane from the earth can also have environmental impacts. The drilling process can disrupt ecosystems as a result of clearing and constructing roads, while potentially causing surface water pollution. The use of hydraulic fracturing, or fracking, to exploit shale gas uses large amounts of water, mixed with additives and then injected deep into the ground. A typical gas well that uses fracking to free up natural gas in shale, uses and average of 4 million gallons per week of water mixed with chemical additives. Some of the additives include hydrochloric acid, biocides (bacterial growth inhibitors), salt, and ethylene glycol (antifreeze). The main concerns of the practice of fracking involve groundwater and surface water contamination. Treated water used for fracking that flows out of the well is known as flowback water. This contains the chemical additives necessary for the fracking process, along with dissolved salts and naturally occurring radioactive materials (NORMs) that exist with the shale. Flowback water is either stored in collection ponds on site or in tanks. The potential for flowback ponds to leak into water supplies is a concern. Once the flowback water has been collected it is then sent to wastewater treatment facilities located near the drill site. Many wastewater treatment plants are not designed to remove the impurities associated with fracking fluids, and have caused concerns about how to properly treat the flowback water. At many wells, millions of gallons of flowback water can be produced. A large concern about hydraulic fracking is the potential for groundwater contamination. Although drilling operations typically inject water thousands of feet below the surface, there is anxiety that aquifers could become contaminated leading to the pollution of drinking water. Although fracking has been used in the United States since the 1960s, during the autumn of 2010 in New York State, concerns about fracking and water pollution caused the legislature to put a ban on fracking operations used to extract natural gas from the Marcellus shale. This ban was imposed so an environmental impact study by the New York Department of Environmental Conservation could be performed.

## NATURAL GAS AND AGRICULTURE

Nitrogen fertilizer is extremely important to the production of crops in the United States and around the world. The main ingredient needed for nitrogen fertilizer is ammonia ($NH_3$). The process by which ammonia is produced is known as the Haber-Bosch process. This process involves exposing natural gas ($CH_4$) to high-pressure steam that combines atmospheric nitrogen with hydrogen in natural gas to form ammonia (Figure 6-7). This process is known as steam reforming. It takes 33,500 cubic feet of natural gas to make 1 ton of **anhydrous ammonia.**

Anhydrous ammonia is a form of ammonia that does not contain water. This ammonia can be used as a source of nitrogen for fertilizer, or to produce other forms of nitrogen fertilizers like urea, ammonium nitrate, and ammonium sulfate. In 2010, the United States used more than 12.4 million tons of nitrogen fertilizer. Forty percent of this nitrogen fertilizer was used to grow corn. Approximately 90% of the cost of nitrogen fertilizer production is based in the cost of natural gas, therefore the rise in cost of nitrogen fertilizers is closely linked to the cost of natural gas. Fifty-four percent of the nitrogen fertilizer used in the United States in 2011 was imported. Precise use of nitrogen fertilizers for agricultural production can not only

**anhydrous ammonia**—a form of ammonia that does not contain water. This ammonia can be used as a source of nitrogen for fertilizer, or to produce other forms of nitrogen fertilizers like urea, ammonium nitrate, and ammonium sulfate

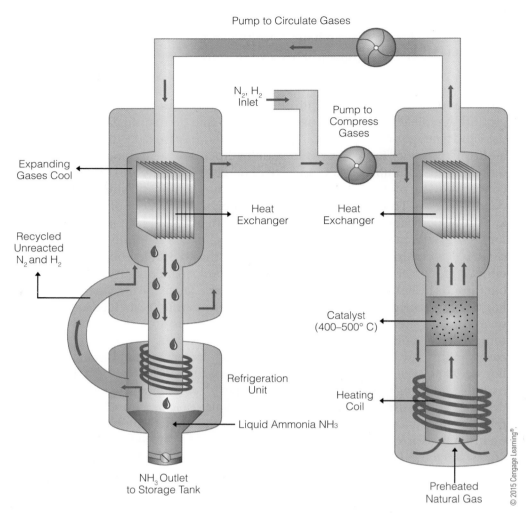

**FIGURE 6-7** Diagram of the Haber-Bosch process.

reduce the cost of growing food, but can reduce the use of natural gas and lessen the impact nitrogen has on water quality. Nitrogen fertilizers that get into surface and groundwater can lead to cultural eutrophication and algal blooms. The nitrogen that washes into water from farm fields causes the rapid growth of aquatic plants and algae that disrupt aquatic ecosystems. The algae goes through its short life cycle, dies, then sinks to the bottom of the water body and is decomposed by aerobic bacteria. These bacteria require oxygen as part of the decomposition process. This high level of decaying material removes oxygen from the water and creates dead zones, also known as hypoxic zones. Hypoxic means low oxygen, meaning the levels of oxygen within the water are 2 parts per million (ppm) or less. This has a severe effect on aquatic organisms that require oxygen for respiration. Estimates state that more than 1.7 million tons of nitrogen fertilizer is washed into the Gulf of Mexico each year from the Mississippi River and its tributaries that drains more than two-thirds of the United States. This nitrogen runoff has created a dead zone along the coast of the Gulf of Mexico near Louisiana and Texas that stretches over 7,700 square miles, an area about the size of Massachusetts (Figure 6-8).

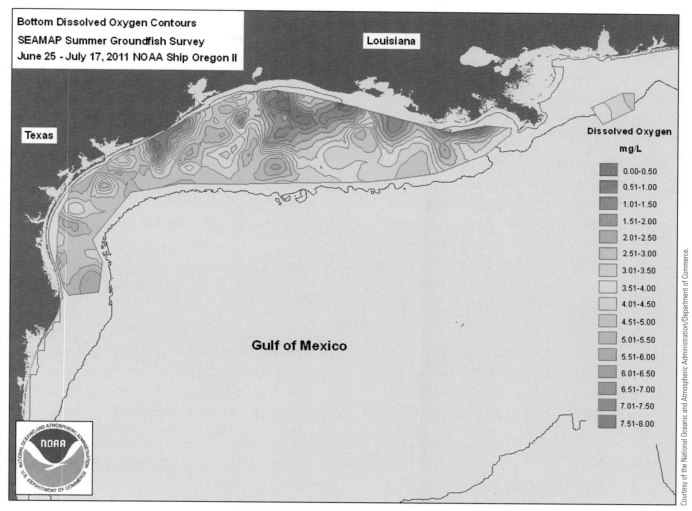

Courtesy of the National Oceanic and Atmospheric Administration/Department of Commerce.

**FIGURE 6-8** Hypoxic zones in the Gulf of Mexico.

## REVIEW OF KEY CONCEPTS

1. Natural gas is another important fossil fuel used in the United States that is made up of mostly methane ($CH_4$), ethane ($C_2H_6$), propane ($C_3H_8$), and butane ($C_4H_{10}$). Natural gas accounts for approximately 25% of our total energy use.

2. Natural gas extracted from the ground as part of the oil drilling process is known as conventional natural gas.

3. Natural gas associated with coal deposits, shale rock, or methane hydrates is called unconventional natural gas. The types of unconventional natural gas are deep gas, tight gas, gas shale, and geopressurized gas.

4. The technology used to extract conventional natural gas is similar to that of oil well drilling, because the gas flows out of the ground under its own pressure.

5. Shale gas is natural gas trapped within sedimentary rocks known as black shale. In order to free up the gas, a drilling technique known as hydraulic fracturing or hydrofracking is used. A mixture of

chemicals, sand, and water called fracking fluid, is injected into the ground at high pressure. This causes the shale to crack and allows the gas to flow into a collection well.

6. Natural gas is the cleanest burning of the fossil fuels, producing only carbon dioxide and water vapor when burned. Natural gas is used to produce electricity for industrial, commercial, and residential use. It is used in homes as a source for heating and cooking. Natural gas is a fuel source used to make paints, explosives, plastics, medicines, and fertilizers.

7. Approximately 80% of our natural gas comes from within the United States, and our reserves are estimated to last about 90 years. Natural gas is difficult to transport by truck or train, therefore most

of the gas in the United States is transported via pipeline.

8. Although natural gas is the cleanest burning fossil fuel, it still has a negative effect on the environment. Like all combustion, burning natural gas produces carbon dioxide, a greenhouse gas. Also, unconventional natural gas extraction techniques like hydrofracking can pollute surface and groundwater.

9. Agriculture relies heavily on natural gas because it is used for the production of nitrogen fertilizers. The process of converting methane into nitrogen fertilizers like anhydrous ammonia, urea, and ammonia nitrite is called the Haber-Bosch process. Approximately 90% of the cost of nitrogen fertilizer used by the American farmer is attributed to the cost of natural gas.

## CHAPTER REVIEW

### Short Answer

1. What are the four main hydrocarbons that make up natural gas?

2. What is conventional natural gas?

3. Describe four sources of unconventional natural gas.

4. What are four ways natural gas is used?

5. At the present rate of use, how long are proven reserves of natural gas in the United States expected to last?

6. What are three negative effects natural gas has on the environment?

7. Describe how natural gas is used by production agriculture in the United States.

### Energy Math

1. How many cubic feet of natural gas does the residential sector in the United States use in one year?

2. Determine how much natural gas it takes to meet the demands of nitrogen fertilizer use in the United States for one year:

## Multiple Choice

1. Which gas makes up almost 90% of natural gas?
   a. $CH_4$
   b. $C_2H_6$
   c. $C_3H_8$
   d. $C_4H_{10}$

2. What type of natural gas is associated with hydrofracking?
   a. Deep gas
   b. Tight gas
   c. Shale gas
   d. Methane hydrates

3. Which type of unconventional source of natural gas may extend reserves for hundreds of years?
   a. Deep gas
   b. Tight gas
   c. Shale gas
   d. Methane hydrates

4. What is the primary use of natural gas in production agriculture?
   a. Tractor fuel
   b. Nitrogen fertilizer production
   c. Barn heating
   d. Pesticide manufacturing

## Matching

*Match the terms with the correct definitions*

a. natural gas
b. unconventional natural gas
c. hydraulic fracturing
d. methane hydrate
e. anhydrous ammonia

1. _____ A naturally occurring, colorless, odorless, combustible hydrocarbon gas.

2. _____ Natural gas not associated with oil deposits, or not easily extracted from depths close to the surface.

3. _____ The process of injecting water mixed with chemicals and sand to fracture rock to allow natural gas to flow out.

4. _____ A form of nitrogen fertilizer produced from natural gas.

5. _____ A unique compound that consists of a methane molecule encased in a crystal lattice of ice.

## CHAPTER 7

# Nuclear Power

## KEY CONCEPTS

*After reading this chapter, you should be able to:*

1. Describe the difference between nuclear fission and nuclear fusion.

2. Explain the processes involved with extracting and refining uranium.

3. Describe the basic process by which nuclear power is used to generate electricity.

4. Explain the difference between light water and heavy water.

5. Explain the advantages of nuclear power.

6. List five negative effects that the use of nuclear power can have on the environment.

## TERMS TO KNOW

nucleon

nuclear fission

nuclear fusion

light water reactor (LWR)

heavy water reactor (HWR)

# INTRODUCTION

Nuclear power fills an important role in the production of electricity in the United States generating 19% of our present needs, and is fifth behind oil, coal, and natural gas for our total energy use. Because the concern that burning of fossil fuels has on our climate, many proponents of nuclear power predict an increase in its use because it does not contribute to global warming directly. The nuclear disaster in Japan as a result of the devastating tsunami in 2011 has focused attention on the potential dangers the use of nuclear power has. There is no question with the increased demand for energy and the reduction in the availability of fossil fuels, nuclear power will continue to be an important source of power long into the future.

# NUCLEAR REACTIONS

What exactly is nuclear power? The answer lies in understanding the structure of all matter on earth. The fundamental unit of all matter is the atom. An atom is composed of a positively charged nucleus surrounded by a negatively charged electron. The nucleus is made up of sub-atomic particles called **nucleons**. There are two types of nucleons, a positively charged proton, and a neutrally charged neutron. The number of protons within a nucleus is matched by the number of electrons surrounding it, making the atom stable. The simplest atom is hydrogen composed of one proton and one electron. The number of protons within a nucleus is known as the atomic number. The addition of neutrons to a nucleus changes the physical properties of an atom creating what is known as an isotope. For example, the addition of one neutron to a hydrogen atom creates an isotope of hydrogen called deuterium. An isotope is an atom with the same number of protons, but a different amount of neutrons. The number of protons and neutrons in an atom's nucleus is known as its atomic mass. Hydrogen has an atomic mass of 1 (actually 1.00794), because it has only one proton in its nucleus. The addition of a neutron to the nucleus of hydrogen increases its atomic mass to about 2 and forms deuterium. As neutrons are added to a nucleus, new atoms are formed. The periodic table organizes the different elements according to their atomic numbers. The force that holds the nucleus of an atom together is called the strong nuclear force. The discovery in the early 1900s that the nucleus of large atoms like uranium, that contains 92 protons, could be split apart to produce energy began the science of nuclear power. As a result of research done by physicists, scientists learned it is possible to use neutrons to break apart the nucleus of a large atom to form new elements and release a great amount of energy. This work simultaneously led to the development of nuclear power and nuclear weapons. The process of using neutrons to split apart an atom is known as **nuclear fission** (Figure 7-1). The element commonly used for nuclear fission is uranium 235. This is an isotope of uranium 238 that has an atomic number of 92, and

**nucleon**—the sub-atomic particles that make up an atom's nucleus, which include protons and neutrons

**nuclear fission**—the process of using nucleons to split apart an atom

**Nuclear fission**

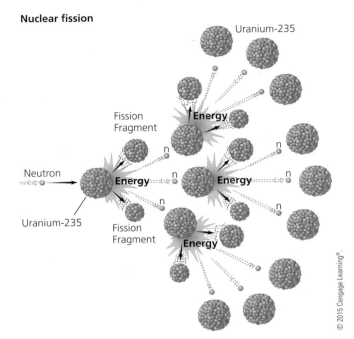

**FIGURE 7-1** Chain reaction associated with nuclear fission.

an atomic mass of 238. Uranium 235 is less stable, and can be split apart by firing a neutron at it that breaks apart the atom to form barium, krypton, and a great amount of energy. This basic nuclear reaction also produces three neutrons that then break apart other uranium nuclei to start what is known as a chain reaction.

Enrico Fermi, a physicist working at the University of Chicago, helped pioneer the work on nuclear fission by producing the first controlled chain reaction in 1942. When the chain reaction is in a steady state of energy production, it is known as being in a critical condition. Sustaining a critical chain reaction can then be used to generate a controlled source of heat that in turn produces steam to drive an electrical generator. This is fundamentally how a nuclear powered electric plant operates. One drawback of nuclear fission is the chain reaction not only produces heat, but radiation as well. The fission process emits harmful beta and gamma forms of radiation that has negative effects on living things.

**nuclear fusion**—the process of fusing two atoms together to produce a new element and energy

Another form of nuclear energy is called **nuclear fusion**. Nuclear fusion works opposite of nuclear fission. Instead of splitting the atom apart, two atoms are fused together to form a new element and a great amount of energy. This process was discovered by Cornell University physicist Hans Bethe in 1938. He attempted to unlock the mechanism the same way the sun generates energy. The process of hydrogen fusion that powers the sun, begins when two atoms of hydrogen are fused together to form the element helium and a great amount of energy. The possibility of using nuclear fusion as a power source on Earth is still being researched and is more technically challenging than nuclear fission. Currently research is focused on two methods to sustain a fusion reaction. One is called magnetic fusion that uses a strong magnetic field to force the hydrogen isotopes of deuterium and tritium together to begin the fusion reaction. The other form of fusion involves bombarding deuterium/tritium pellets with particle beams or lasers to start the fusion reaction.

# NUCLEAR FUEL

Nuclear fission requires the uranium 235 isotope in large enough quantities to sustain a chain reaction to continually produce heat. Uranium is most abundant on earth as uranium 238 and is the 48th most common element to exist in the Earth's crust (Figure 7-2). As a comparison, the amount of uranium estimated in the Earth's crust is 100 times that of silver. Uranium is a dense element that is approximately 19 times denser than water. For example, one gallon of water weighs 8.34 pounds. A gallon of uranium would weigh 150 pounds! About 99.3% of uranium found in rock is $U^{238}$ and the remaining 0.7% is $U^{235}$ which is required for nuclear power. Uranium usually occurs in rocks in the form of uranium oxide that makes up minerals like uraninite, also known as pitchblende because of its black color, and also carnotite.

Uranium ore is found all over the world and mined using surface mining techniques (see Chapter 4), or using in situ leach mining. The term in situ, means "in place" and is the least destructive form of mining to the landscape. In situ leach mining for uranium involves the injection of water into the ground that is mixed with an acid or base solution. The solution dissolves the uranium oxide and is then pumped out of the ground. This method accounts for approximately 26% of the uranium mined around the world and is widely used in the United States. In situ leach mining can only be applied where the uranium ore is located within porous rocks like sandstone. Once the uranium ore is extracted, it must be processed to isolate the uranium. This process is known as milling and involves the ore being crushed into a fine powder and washed with acid to leach out uranium oxide $U_3O_8$. This product is known as yellowcake because of its yellow color (Figure 7-3).

© Kletr/Shutterstock.com

**FIGURE 7-2** Uranium ore.

**FIGURE 7-3** Yellowcake.

In the United States, uranium ore typically holds between 0.3%–0.05% uranium oxide. In 2011, mines in the United States produced 4.1 million pounds of uranium oxide. The remaining rock material is left over as processing waste called mine tailings. Once the uranium oxide is extracted it is then converted into uranium hexafluoride ($UF_6$) gas, so the $U^{235}$ can be separated from the $U^{238}$. This process is known as enrichment and is complex (Figure 7-4). The enrichment process increases the proportion of $U^{235}$ needed to sustain a fission chain reaction. Typically enriched uranium contains 2%–5% $U^{235}$ compared to the natural proportion of 0.7%. Once the enrichment is complete, the uranium hexafluoride gas is converted into uranium oxide and made into fuel pellets. Each fuel pellet is about the width of a pencil and about 1 inch long. These tiny uranium pellets have the equivalent energy of about 150 gallons of oil or almost one ton of coal!

In 2012, more than 58 million pounds of uranium oxide was purchased in the United States for use in nuclear power plants. Approximately 83% of this was imported from other countries like Canada, Australia, Russia, Africa, and Britain. The remaining 17% was mined in the Southwestern United States. The Energy Information Administration estimates the total U.S. uranium oxide reserves to be more than 1,227 million pounds (Figure 7-5). At current rates of consumption this is expected to last 23 years. Sixty percent of these reserves exist in Wyoming and New Mexico. These low estimates are solely based on U.S. reserves, and does not take into account the world's remaining uranium that could extend reserves to a few hundred years.

Although nuclear fusion is still in the research phase and not expected to be a viable form of energy for another 30–50 years, it is still worth discussing the fuel source for this type of nuclear energy. Both methods of fusion currently under research use isotopes of hydrogen in the form of deuterium and tritium. Luckily both of these elements are present in large quantities in the ocean. Typically, one in every 6,700 hydrogen molecules that make up seawater is deuterium. This provides an almost inexhaustible amount of fuel for hydrogen fusion.

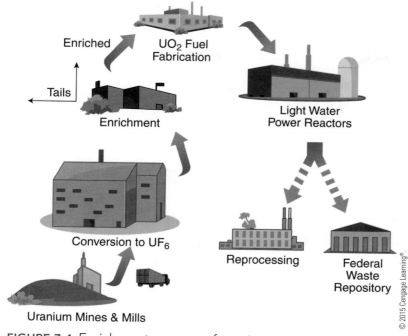

**FIGURE 7-4** Enrichment process of uranium.

**FIGURE 7-5** Map of uranium reserves.

## NUCLEAR POWER PLANTS

Currently there are 436 nuclear power plants in operation around the world. Sixty-five of these are located in the United States and contain a total of 104 reactors (Figure 7-6).

Although there are a variety of ways to harness nuclear power to generate electricity, most of the nuclear power plants in the United States are classified as **light water reactors (LWR)**. Light water refers to the use

**light water reactor**—a nuclear reactor run with the use of regular water ($H_2O$), made up of 2 hydrogen atoms with an atomic mass of approximately 2 and an oxygen with a mass of 16

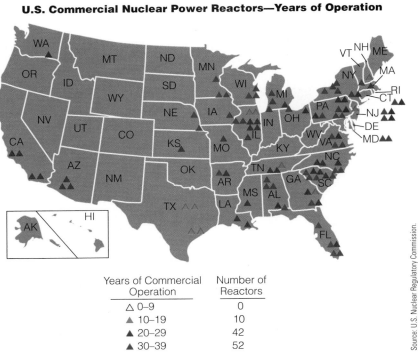

**U.S. Commercial Nuclear Power Reactors—Years of Operation**

| Years of Commercial Operation | Number of Reactors |
| --- | --- |
| △ 0–9 | 0 |
| ▲ 10–19 | 10 |
| ▲ 20–29 | 42 |
| ▲ 30–39 | 52 |

Source: U.S. Nuclear Regulatory Commission.

**FIGURE 7-6** Map of U.S. nuclear power plants.

**heavy water reactor**—a nuclear reactor run with the use of water that contains 2 deuterium isotopes of hydrogen bonded to one oxygen molecule ($D_2O$)

of regular water in the reactor ($H_2O$), that is, water made up of 2 hydrogen atoms with an atomic mass of approximately 2 and an oxygen with a mass of 16 (actually 15.9994). Together the atomic mass of regular water is 18. A **heavy water reactor (HWR)**, which is much rarer uses heavy water; water that contains 2 deuterium isotopes of hydrogen bonded to one oxygen molecule ($D_2O$). The atomic mass of this molecule of water is increased to approximately 20. Remember that deuterium has an atomic mass of 2, double that of regular hydrogen, and is why it is called "heavy" water. Approximately 78% of the world's nuclear power plants use light water reactors. The advantage to using heavy water is that the reactor does not need to employ enriched uranium as its fuel. Light water reactors require enriched uranium. Most all nuclear reactors use the same method to sustain high heat used to convert water to steam that powers an electrical turbine. Typically the uranium fuel pellets are placed in long cylindrical rods that are about 12 feet in length, then bundled together in groups of 270 to form a fuel assembly (Figure 7-7). A nuclear power plant uses hundreds of fuel assemblies.

The uranium in the rods is housed in the reactor core where they begin to undergo a fission chain reaction that generates heat. The rate of reaction can be controlled by introducing a neutron absorbing material into the reactor core to control the rate of the chain reaction; therefore controlling the temperature of the reactor core. The difference in reactor design usually involves the method of cooling the reactor and transferring its heat. Most light water reactors are classified as either Pressurized Water Reactors (PWR) or Boiling Water Reactors (BWR). A Pressurized Water Reactor uses high-pressure water to cool the reactor core and transfer the heat. The water is kept at a high pressure, about 3,200 pounds per square inch that prevents it from vaporizing. The reactor core heats the water to 500°F (260°C), then

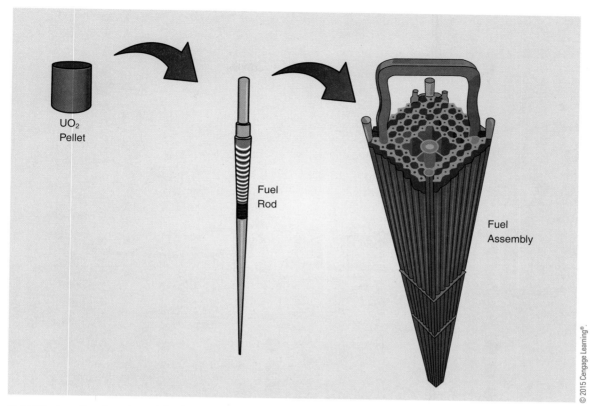

**FIGURE 7-7** Fuel rod and fuel assembly for a nuclear reactor.

passes through a heat exchanger to transfer the heat to another closed loop water system. This is then converted into steam that powers a turbine generator. The closed loop system of high-pressure water prevents any dangerous radiation from leaving the reactor vessel. The water itself helps cool the reactor core while also acting as a neutron absorber to control the fission reaction (Figure 7-8). Control rods or plates made of a neutron absorbing material like boron, can also be lowered into the reactor core to control the reaction rate. Boiling Water Reactors bathe the reactor core in low-pressure water that boils to form steam when it comes in contact with the fuel assemblies. This steam is then used directly to drive a turbine generator. BWR's require much more control of the fission chain reaction to prevent overheating. Also, the loss of the closed loop heat transfer system like the ones used in PWR's, increases the possibility of radioactive contamination getting out of the containment building. In the United States, 35 of the 104 nuclear reactors in operation are BWR's, with the remainder being PWR's. In other countries, different nuclear reactors are also used, including heavy water reactors, gas cooled reactors, and liquid metal reactors. Gas cooled reactors use gases like carbon dioxide or helium to cool the reactor core, but have not been successful. New gas cooled reactors are under development such as the pebble bed reactor that uses tiny spheres of uranium bathed in helium gas coolant. Liquid metal reactors that use liquid metals like sodium or lead have also been explored, but are expensive to operate and difficult to maintain.

Because of the potential dangers associated with nuclear power, all nuclear power plants in the United States are regulated by the Nuclear

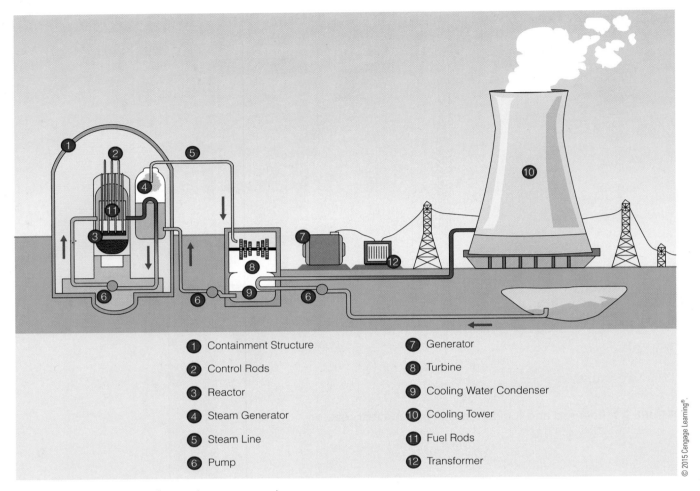

© 2015 Cengage Learning®

1. Containment Structure
2. Control Rods
3. Reactor
4. Steam Generator
5. Steam Line
6. Pump
7. Generator
8. Turbine
9. Cooling Water Condenser
10. Cooling Tower
11. Fuel Rods
12. Transformer

**FIGURE 7-8** Structure of a nuclear power plant.

### CAREER CONNECTIONS

## Nuclear Engineering

A nuclear engineer works with all aspects of using nuclear energy. Nuclear power plants require engineers to design, run, and maintain the plant in order to produce electricity. Four-year colleges and the U.S. Navy have majors in nuclear energy. Careers are also available in the application of nuclear medicine that uses radiation to cure diseases. Opportunities include training as a nuclear medicine technologist that takes between one to four years, or an advanced clinician that requires a Bachelor's degree and graduate degree in nuclear medicine. Nuclear engineering is also applied to food production using radiation to sterilize and preserve food. NASA also uses nuclear Engineers to help design the power systems for spacecraft. The U.S. Dept. of Labor estimates the need over the next ten years for nuclear engineers working in power plants could create many new employment opportunities for those who enjoy high-tech science.

Regulatory Commission (NRC). The purpose of the NRC is to oversee the use of nuclear material for energy, medicine, and industry and to assure the safety of the public and the environment. The last nuclear power plant to be built and approved by the NRC was in 1996. This plant, named the Watts Bar 1, operates in Tennessee. Currently, there is one other nuclear power plant under construction in Tennessee (Watts Bar 2) and is expected to begin operation 2013. This plant began construction in 1988! In 2012 the NRC approved the construction of 2 nuclear reactors near Augusta, Georgia. Known as the Vogtle Plant, these new units are expected to be operational by 2017 and are the first construction plans approved by the NRC in 30 years. Many nuclear power plants in the United States were built in the 1970s and 1980s. The strict regulations outlined by the NRC for building new nuclear power plants makes the process slow. Typically it takes three years for plans and licensing to be approved.

## ENVIRONMENTAL IMPACTS OF NUCLEAR ENERGY

The most popular argument for the use of nuclear power to generate electricity is that it does not produce greenhouse gases and atmospheric pollutants like those resulting from the burning of fossil fuels. Increasing the use of nuclear energy would decrease the emissions of atmospheric pollutants like carbon dioxide, sulfur oxides, nitrogen oxides, volatile organic compounds, and particulate matter; but at what cost? The biggest drawback of nuclear energy is the potential for damage to the environment by the release of radiation. The process used for uranium to yield heat, also produces harmful radiation that causes sickness and death to human beings, and long-term harm to ecosystems. Power plants in the United States use reactor containment vessels and containment buildings along with emergency cooling technologies to minimize the potential for the release of radioactive material. A typical containment building used in the United States is composed of a layered steel and concrete structure that is airtight, more than three feet thick, and can withstand the impact of a jet airliner. As of 2012, there have been three notable reactor accidents since the first nuclear power plant began operation in 1956. The Three Mile Island nuclear power plant located in Middletown, Pennsylvania experienced a nuclear accident in spring of 1979. A problem with the coolant system and human error resulted in the overheating of the fuel assemblies, and caused approximately half of them to melt and settle at the bottom of the reactor vessel. This accident almost created a meltdown that results when the extreme heat of the molten fuel causing it to melt through the floor of the reactor vessel and into the environment. This did not happen. The melted reactor core settled to the bottom of the reactor vessel and the reactor began to cool, preventing a total meltdown. Radioactive material was released in the containment building during the accident, but did not leak into the environment. This was the worst nuclear accident in the United States and did not result in any harm to humans or the environment. The partially damaged reactor was kept idle for ten years, so the melted fuel could be safely removed.

By far the worst nuclear accident to occur in the world occurred in 1986 in Chernobyl, located in the Ukraine. As a result of a test of the reactor system initiated by the plant's engineers, a large amount of steam was generated in the reactor core. This build-up of steam lead to a rapid buildup of pressure that caused the reactor to explode, sending molten radioactive material flying outward onto the roofs of surrounding buildings. This explosion created numerous fires, along with the introduction of massive amounts of radiation into the environment. Unlike nuclear power plants in the United States, the Chernobyl plant did not have a containment building built around the reactor to prevent the spread of radioactive material. Two workers were killed immediately by the explosion and 126 became exposed to dangerous levels of radiation. Eventually an 18-mile radius around the site was evacuated, causing the displacement of more than 200,000 people. The accident was so bad that increased levels of radiation were detected in parts of Northern Europe more than 3,000 miles away. It is estimated more than 4,000 people eventually died as a result of radiation exposure.

The Russian government eventually contained the radiation by erecting a massive concrete and steel structure called a sarcophagus to cover the site. Much has been learned from Chernobyl and Three Mile Island, and the NRC has used this information to improve the safety of nuclear power plants in the United States.

In March of 2011, a magnitude 9.0 earthquake off the coast of Japan generated a massive tsunami that washed over the Fukushima nuclear power plant. The surging wave of water cut the power to three reactors, causing them to overheat and this lead to three explosions inside the reactor's containment vessels. The explosions and resulting fires sent high levels of radiation into the surrounding environment. The reactor cores in three reactors then suffered partial meltdowns. Today, the reactors have cooled to safer levels and radiation has dropped significantly, although full knowledge of the damage is still to be determined.

Another concern over the use of nuclear power involves what to do with the spent fuel. Once the uranium fuel rods exhaust their fissionable material, they are then removed from the reactor. These fuel bundles are classified as spent fuel, but still remain highly radioactive. This radioactive waste is classified as High-Level Waste (HLW). Once removed from the reactor, the spent fuel is placed in either a storage pool called a spent fuel pool, or a dry containment vessel. The spent fuel pool cools the rods in at least 20 feet of water, and also absorbs the radiation. The spent fuel rods can stay in the pool for upwards of 15 years. Dry containment of the spent fuel uses concrete and steel vessels to house the rods that are cooled with circulating air. The nuclear material that makes up HLW can be dangerously radioactive for 1,000 years, so a stable, safe, long-term storage solution is a vital part of using nuclear power. An average nuclear power plant generates about 2,000 tons of spent fuel every year. Currently all HLW is stored on site at each nuclear power plant. In 2002, more than 51,000 tons of spent fuel was stored at the 65 nuclear power plants in the United States. In 1987, Congress instructed the Department of Energy to construct a central nuclear storage facility to be located in Yucca Mountain, Nevada; this was never built. In 2013, there are no formal plans for creating a permanent storage facility in the United States for safely housing spent fuel and HLW; therefore all spent fuel is currently stored on site at the nuclear power plant.

Other environmental concerns of nuclear power involve the mining and milling processes. The extraction of uranium ore using surface mining techniques is destructive to the landscape. Ecosystem disruption, erosion, runoff, and water pollution are the problems associated with surface uranium mining. The most common form of surface mining for uranium is the use of open-pit mining that is destructive to the landscape. The use of in situ leach mining also has its concerns. The potential for the injection solutions used to wash out uranium ore to contaminate groundwater and drinking water aquifers has been raised. Supporters of the process argue the injection wells are much deeper in the ground than the drinking water aquifers (more than 8,000 feet below the surface), and are often separated by layers of impermeable rock. Careful study of the underlying geology and groundwater flow in regions identified for uranium mining is an important first step in the process to prevent any groundwater contamination. Also the milling and processing of the uranium ore that separates the desired uranium oxide can have potential negative environmental effects. Remember that uranium ore contains only 3%–0.5% uranium oxide, so a large amount of waste rock called tailings is generated during the milling process. The tailings can contain radioactive materials like radium. Radium decays to produce radon gas that is also radioactive. Other potentially harmful metals can be present in tailings like thorium, selenium, molybdenum, and trace amounts of uranium. Today in the United States there are 26 sites that contain waste tailing for uranium processing. Each site contains more than three–million tons of waste. Twenty-four of these locations have been abandoned and are managed by the EPA. The waste piles are usually put in pits lined with an impermeable barrier to prevent the leaching of radioactive materials into groundwater. Also, the abandoned piles are often covered to prevent the introduction of water into the tailings and to lessen the potential for wind erosion of the harmful materials.

Although the long-term storage of radioactive spent fuel has been discussed as a potential environmental threat, there are other forms of radioactive waste produced by using nuclear power. Many of the mechanical parts themselves that need to be replaced as part of routine maintenance of the power plant become radioactive and need to be handled carefully as well. This material is stored in containment vessels on site at the plant. Eventually, when the entire power plant itself gets too old to safely operate it needs to be decommissioned. Currently there are three strategies approved by the NRC for nuclear power plant decommission. The first is called DECON; the immediate dismantling of the plant and safe storage of the radioactive parts of the plant. Typically a special structure is built on site for the long-term storage of the radioactive material that comes from dismantling the plant. The second method of decommissioning is known as SAFSTOR. This is known as "mothballing" and mean the plant is shut down and left "as is" for a period of at least 15 years in order to allow the radioactivity to subside. Eventually the plant will be dismantled safely. Currently there are 12 power plants in SAFSTOR mode in the United States. Six of these plants have been idle since the 1970s. The third method to decommissioning is called ENTOMB. This involves the construction of a permanent concrete and steel barrier to completely cover the site. Currently this method has not been used in the United States.

# REVIEW OF KEY CONCEPTS

1. Nuclear power uses the strong nuclear force that binds the nucleus of an atom together to generate power. There are two types of nuclear power: nuclear fission and nuclear fusion. Nuclear fission involves splitting apart the atom of a large element like Uranium 235 with a neutron. The result is the release of a great amount of energy, and when additional neutrons break apart more uranium atoms that cause a chain reaction. The result is a sustained amount of heat energy. Nuclear fusion occurs when two atoms of hydrogen are fused together to form a new element; helium and a great amount of energy. This type of nuclear reaction is difficult to sustain and is currently in development.

2. Uranium 235 is rare, and usually found in trace amounts in uranium ore that exists in the Earth's crust in the form of uranium oxide. Uranium is mined using traditional surface mining techniques, or in situ leach mining that uses water and chemicals to dissolve the uranium ore that is then pumped to the surface. Once the uranium oxide is extracted it needs to be processed and enriched. Enrichment separates the uranium 235 from uranium 238. Typically the enrichment process increases the percentage of uranium 235 from 0.7% to 2–5%. The enriched uranium is then made into pellets that are stacked in long cylindrical fuel rods, and bundled together in fuel assemblies used in nuclear power plants.

3. Nuclear power plants use nuclear fission to produce heat that creates steam that then powers an electrical turbine. The two main types of nuclear power plants are a Boiling Water Reactor (BWR) and a Pressurized Water Reactor (PWR). Both of these plants use uranium 235 as their fuel source.

4. A light water reactor operates with the use of regular water in the reactor ($H_2O$); water made up of 2 hydrogen atoms with an atomic mass of approximately 2 and an oxygen atom with a mass of 16. A heavy water reactor operates with the use of water containing 2 deuterium isotopes of hydrogen bonded to one oxygen molecule ($D_2O$).

5. The advantage of using nuclear power is the plants do not emit greenhouse gases or other atmospheric pollutants.

6. The risk to the environment from nuclear fission is the threat of radiation contamination leaking from the plant as a result of an accident known as a meltdown. Other environmental risks include the disposal of the spent nuclear fuel and the shutdown of the power plants themselves when they are no longer operable. Nuclear waste and spent fuel stay dangerously radioactive for a thousand years or more, so safe, long-term storage is necessary to protect human life and the environment. Currently all spent fuel is housed on-site at the power plant where it was used. The Nuclear Regulatory Commission (NRC), a U.S. government agency, is responsible for developing a long-term storage site to safely store our nation's nuclear waste. Uranium mining and processing can also cause potential pollution to land and water.

# CHAPTER REVIEW

## Short Answer

1. What is the difference between nuclear fusion and nuclear fission?

2. Which fuels are used for nuclear fission and nuclear fusion?

3. Describe the processes used to extract and enrich uranium.

4. How is nuclear fission used to produce electricity?

5. What is the difference between light water and heavy water?

6. Describe the differences between the operation of pressurized and boiling water reactors.

7. What are two positive effects on the environment of using nuclear power?

8. What are four negative effects that using nuclear power have on the environment?

9. Name two advantages and two disadvantages of using nuclear power.

## Energy Math

1. What percentage of the world's nuclear power plants is located in the United States?

2. If a large nuclear power plant contains 2 million uranium fuel pellets, how much coal would you need to burn to produce the same amount of energy?

## Multiple Choice

1. Which type of nuclear reaction is used to generate electricity?
   a. Nuclear fission
   b. Nuclear fusion
   c. Heavy water
   d. Enriched uranium

2. What is the primary fuel used in most nuclear power plants?
   a. Plutonium
   b. Uranium
   c. Heavy water
   d. Radium

3. What percentage of our electricity is generated by nuclear power?
   a. 5%
   b. 10%
   c. 20%
   d. 50%

4. What is the isotope of hydrogen that makes up heavy water?
   a. Hydrogen
   b. Tritium
   c. Deuterium
   d. Barium

5. Approximately how many nuclear power plants are in the United States?
   a. 25
   b. 65
   c. 100
   d. 165

## Matching

*Match the terms with the correct definitions*

a. nuclear fission
b. nuclear fusion

c. nucleon

d. heavy water

1. _____ The sub-atomic particles that make up an atom's nucleus, that include protons and neutrons.

2. _____ The process of using nucleons to split apart an atom.

3. _____ Water that contains 2 deuterium isotopes of hydrogen bonded to one oxygen molecule ($D_2O$)

4. _____ The process by which two atoms are fused together to form a new element and a great amount of energy.

# UNIT III

# Contemporary Renewable Energy Resources

## OVERVIEW

Although contemporary renewable energy resources provide only 9% of energy used in the United States today, the significance they will play in our energy future is going to be crucial in order to sustain our technological society. Concerns over dwindling reserves of fossil fuels and their negative environmental impacts are making the development and use of renewable energy resources more appealing. The term renewable means it can be replenished in a reasonable amount of time, and is therefore sustainable. In most cases a reasonable amount of time equates to the length of one human lifetime. Contemporary renewable energy resources usually include solar power, wind power, and hydropower. Great scientists and thinkers have often looked to nature in order to solve problems. If this technique is applied to energy use on earth, then tapping into the Earth's natural energy sources could help us supply clean, sustainable means to meet our power needs.

## CHAPTER 8

# Solar Power

## KEY CONCEPTS

*After reading this chapter, you should be able to:*

1. Explain three methods of passive solar energy.

2. Define active solar power and give one example of an active solar power system.

3. Identify the three types of concentrated solar power.

4. Explain the photoelectric effect.

5. Describe photovoltaic solar power.

6. Identify the two types of photovoltaic systems.

7. Explain why the use of solar power may be limited in some areas.

## TERMS TO KNOW

passive solar energy
active solar energy

concentrated solar power

photovoltaics

# INTRODUCTION

Energy from the sun is the main source of almost all the energy used by living and non-living systems on the earth. Use of fossil fuels like coal, oil, and natural gas, that provide 81% of our energy, are simply stored forms of solar energy captured by photosynthesis millions of year ago. Harnessing the incredible amount of energy produced by the sun seems like the next logical step for our modern society.

## PASSIVE SOLAR ENERGY

**Passive solar energy** is a means of harnessing the natural light and heat energy produced by the sun with no other input of energy (Figure 8-1). In the Northern Hemisphere, south-facing windows allow sunlight to penetrate into a structure. Even in the cold months of winter, this light and heat can reduce the amount of electricity needed to light the interior spaces and the amount of fuel used to heat the space. Walls and floors within the building take in this solar energy and radiate to heat the interior of the structure. Typically the walls and floors that face the southern exposure windows should be a dark color to maximize the heat absorbing potential of the material. The simple use of south facing windows to capture the light and heat energy of the sun directly is also known as *direct gain solar heating*. Greenhouses have used passive solar heat for centuries to heat their

**passive solar energy**—a means of harnessing the natural light and heat energy produced by the sun, with no other input of energy

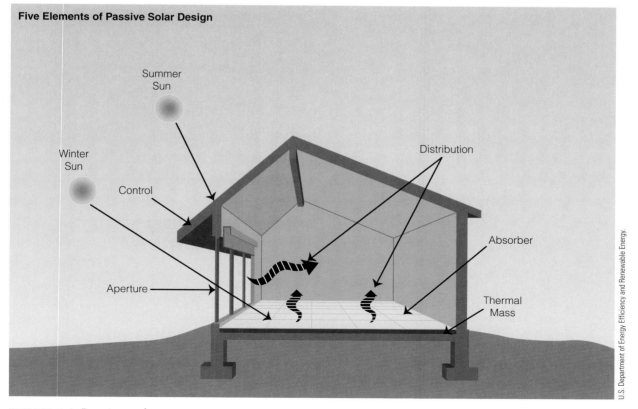

Five Elements of Passive Solar Design

Summer Sun

Winter Sun

Control

Aperture

Distribution

Absorber

Thermal Mass

U.S. Department of Energy Efficiency and Renewable Energy.

**FIGURE 8-1** Passive solar energy.

interiors. Passive solar is also used in barns and other agriculture buildings as a supplemental source of heat and light.

An advanced method of using passive solar energy is known as a Trombe wall. Named for a French architect who employed the idea, this passive solar structure uses a 4 to16–inch thick (10–40 cm), dark–colored masonry wall that is set off about 2 inches (5 cm) from a south facing window. The dark colored masonry absorbs the solar energy during the day and radiates it out into the structure at night (Figure 8-2). This type of passive system is known as an indirect gain because it absorbs the sun's energy and radiates it a later time. Some Trombe walls use vents set along the top and bottom of the wall to allow the flow of air to absorb the heat and distribute it into the room. To minimize the possibility of overheating the space during the warmer months, the windows can be shaded, or the overhang of the roof's eaves can be aligned to block the sun when it reaches a higher angle.

Another type of passive solar energy is known as isolated gain. This method uses a specific south facing room called a sunroom that is much like a greenhouse. This room can use heat absorbing materials in the floors and walls to capture the sun's heat energy that can then be distributed to the rest of the house via convection currents. The goal of passive solar is not to heat the entire structure but reduce the amount of energy needed by conventional heating systems to heat the home. This can save a lot of energy in the long term.

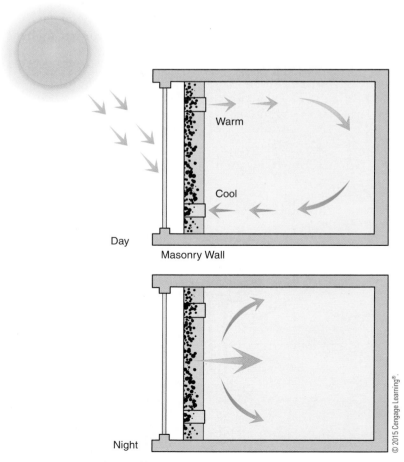

Day

Masonry Wall

Warm

Cool

Night

© 2015 Cengage Learning®

**FIGURE 8-2** Trombe wall.

# ACTIVE SOLAR ENERGY

**Active solar energy** is the use of a collection device that absorbs the sun's heat energy, then circulates the heat by a fluid or gas, such as air or water that is driven by fans or pumps (Figure 8-3). Typically, active solar devices are used to heat water. Most active solar systems use a flat collector plate made up of a heat absorbing material that is usually some type of metal. The plate contains small pipes or channels that allow heat transfer fluids to pass through and absorb the heat. The plate is painted a dark color to increase its heat absorbing capacity. The collector is then covered in transparent glass or plastic to prevent heat loss and the back and sides of the collector are well insulated. There are a variety of different heat transfer fluids that can be used for active solar heating. They include water, antifreeze solutions, oils, or refrigerant gases. Each type of fluid has its advantages and disadvantages and is chosen based on the specific system being used. There are two types of active solar systems; direct circulation and indirect circulation. Direct circulation involves the circulation of water or air to be heated directly into the solar collector where it absorbs the heat. The fluid is then moved out of the collector via a fan or circulating pump that delivers the heat. These systems are only useful in areas that do not experience freezing temperatures, because the fluid could be cooled or frozen during colder months making the heater ineffective. Indirect circulation uses a closed system that transfers the heat to a storage tank. This type of system prevents the unwanted cooling of the heat-transfer fluid that can occur in cooler climates. The use of an automatic shut-off valve to cut the flow of fluid to the collector occurs when a specific temperature is reached. Indirect circulation can also use a gravity drainage system that removes the fluid from the collector to prevent it from unwanted cooling.

A newer, efficient type of solar collector that can be used for active heating is called an evacuated tube solar collector (Figure 8-4). This technology uses dark, heat absorbing copper or aluminum pipes with attached fins that are housed within a glass vacuum tube. Because the tube is within a vacuum,

**active solar energy**—the use of a collection device that absorbs the sun's heat energy that is then circulated by a fluid or air driven by fans or pumps

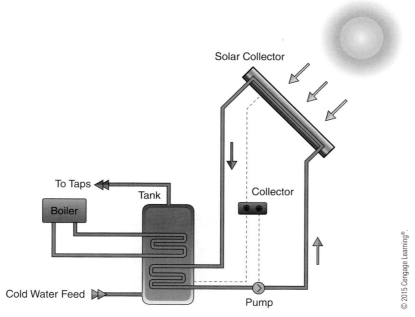

© 2015 Cengage Learning®.

**FIGURE 8-3** Active solar energy.

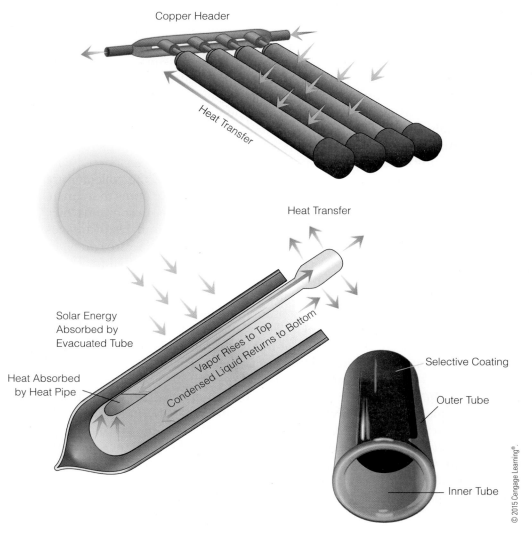

**FIGURE 8-4** Evacuated tube solar collector.

it cannot exchange heat with the environment, making it extremely efficient. Even on the coldest winter days, the temperature of the heat transfer fluid can reach more than 200°F (93°C)! Once again, like passive solar systems, active solar does not have to cause the heat transfer fluid to be superheated to be effective. Small increases in the temperature of water mean less energy used to bring the temperature of the water up to the desired level. For example if your water heater is pulling water in at 56°F (13°C), that needs to heated to 120°F (49°C), then your conventional electric or gas-fired water heater must put in enough heat energy to raise the temperature by 64°F (18°C). However, if you use an active solar-powered water heater able to preheat your water to 70°F (21°C), then you've decreased the amount of energy required to raise the water temperature to 120° by 14 degrees. This equates to a substantial savings in energy over time. Along with using active solar energy to heat air and water, there are also active systems that can be used for cooling as well. These are known as absorption heat pumps that use fluids heated by active solar collectors to operate a heat pump. A heat pump is a device that uses the compression and expansion of gases to either remove heat from a system in order to cool it or add heat to a system to heat it.

# CONCENTRATED SOLAR POWER

Another form of solar energy is known as **concentrated solar power**, also known as high temperature solar. Concentrated solar power uses mirrors or lenses to focus the energy of the sun in order to generate high temperatures. This heat is then typically used to boil water or other fluids to produce high-pressure steam that turns an electric turbine. This type of solar power is also referred to as solar thermal power. Solar power towers are one method of concentrated solar used to produce electricity. In this design, hundreds or even thousands of mirrors are used to focus the sun's rays on a central tower called a receiver. The mirrors are known as heliostats that are designed to focus and reflect the sun's energy. The receiver is located on top of a tower at the center of the field of heliostats (Figure 8-5A and B). Inside the receiver is a heat absorbing fluid heated by the focused power of the sun to extremely high heat. The fluids used in power towers are usually water or molten salt. In a water fluid system, the focused energy heats the water directly to steam then powers an electric turbine. This technology was applied to a project in Southern California called Solar One in 1982.

Solar One's power tower employed 1,818 heliostats to generate temperatures of around 900°F (482°C). The power output of this plant was 10 megawatts and produced power for four hours per day in winter and 8 hours per day in summer. To keep the water heated during the nighttime hours an oil heat transfer system used a storage tank filled with rocks and sand to act as a heat sink to preserve some of its heat. The use of molten salts as the operating fluid for power towers lead to development of the Solar Two project. The Solar One power tower facility converted to a molten salt system that used a mixture of 60% sodium nitrate and 40% potassium nitrate. The heliostats focus the energy to the receiver tower that superheats the molten salts to temperatures as high as 1,000°F (537°C). The superheated salt mixture then transfers its heat via a closed loop heat transfer system to heat water into steam that powers an electric turbine. The salt mixture is then stored in a cold storage tank that keeps it at approximately 550°F (287°C), so the plant can also produce power at night. Currently there are two solar power tower plants being built in California that together are expected to generate almost 600 megawatts of electricity, about the same output of one coal-fired power plant.

Another method of concentrated solar power is called the parabolic trough (Figure 8-6A and B). This system uses a series of parabolic mirrors to focus the sun's energy on a heat absorbing pipe filled with a heat transfer fluid; usually oil. The network of pipes runs through hundreds of mirrors that heat the oil to temperatures ranging from 212°F to 750°F (100°C–399°C). The heat is used to convert water into steam to power an electric turbine. The network of parabolic mirrors uses a one-dimensional, east-west automated tracking system that follows the sun to allow for the maximum amount of solar power to be absorbed. The troughs are aligned on a north-south axis and then follow the sun as it tracks from east to west. Most parabolic trough systems generate up to 1,000 megawatts of electricity for 10 to 12 hours a day. To extend the operating time of these solar plants to a full 24 hours, a hybrid system, known as an Integrated Combine Solar Cycle, uses a natural gas-fired electric turbine to be used when the troughs are not hot enough to produce power.

**concentrated solar power**— the use of mirrors or lenses to focus the energy of the sun in order to generate high temperatures

**(A)**

**Solar Power Tower**

Receiver

Steam Condenser

Feedwater
Reheater

Electricity

Generator

Turbine

Steam Drum

Heliostats

**(B)**

U.S. Department of Energy.

© 2015 Cengage Learning®.

**FIGURE 8-5** (A) A solar power tower at Solar Two Power Plant.
(B) Concentrated solar power.

A third method of concentrated solar is called the Solar Dish Engine
(Figure 8-7). This system uses one large parabolic mirrored dish to concentrate the sun's energy to a receiver mounted at the dish's focal point.
The unit employs a two-dimensional tacking system that follows the sun
through the sky to maximize the solar energy collected. The receiver is
filled with a heat absorbing fluid such as oil, water, liquefied sodium, or
hydrogen and helium gas that can be heated to more than 1,000°F (537°C).

**(A)**

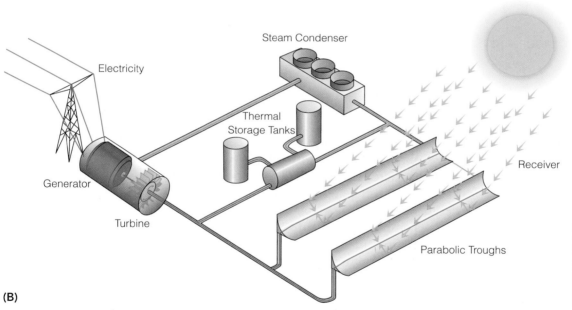

**(B)**

**FIGURE 8-6** (A) Function of a parabolic trough. (B) Parabolic trough.

The heated fluid can then be used to produce mechanical power to turn an electric turbine. The use of a Stirling engine as part of the Solar Dish Engine receiver is considered to be one of the most effective ways of transferring the sun's energy into mechanical power. A Stirling engine uses the heating and cooling and the expansion and contraction of a gas in a closed system to power a piston. The mechanical energy of the moving piston is then used to power an electric turbine. Solar Dish Engines can only generate small amounts of electricity; typically in between 5 and 50 kilowatts. Because of their relative low power output, they are the most useful for small energy applications or multiple units can be linked together in an array.

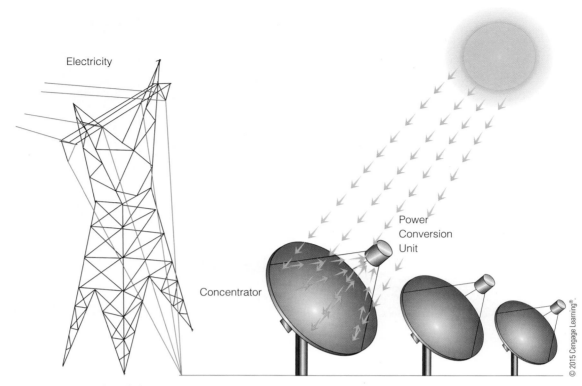

**FIGURE 8-7** Solar dish engine.

## PHOTOVOLTAICS

**photovoltaics**—the use of a semiconductor like silicon to convert the sun's energy directly into electricity

**Photovoltaics** is the use of a semiconductor like silicon to convert the sun's energy directly into electricity. A semiconductor is a substance that allows an intermediate flow of electrons, somewhere between that of a good conductor like copper and a good insulator like plastic. The process that sunlight can be used to produce electricity is known as the photoelectric effect. The photoelectric effect describes how energy in the form of photons emitted from light sources such as the sun, causes electrons to be dislodged from metal elements. The photons strike the electrons in a conductor and begin to flow. The electrons can be used to create an electrical current. Solar cells typically consist of two layers of a silicon semiconductor sandwiched together (Figure 8-8). One layer is known as the n-type layer, or negative layer. This is made by mixing phosphorus atoms with silicon in a process known as doping. The other layer is a p-type layer, or positive layer, that is a mixture of silicon and boron. The two layers are connected to form a circuit. When the n-type layer is pointed toward a light source, photons strike it, causing electrons to flow from the n-type layer, through an electrical circuit, to the p-type layer.

Typically one solar cell can produce 1 watt of power when exposed to a strong light source like the sun. The output of electricity of the solar cell is directly related to the intensity of sunlight; the greater the intensity, the greater the electric output. This is an important fact because it reveals solar cells can produce electricity even with diffuse light, such as that on a cloudy day. Individual cells are linked together to form solar modules that are encased in a weather-proof material. The modules are connected together in a specific arrangement to produce the desired electrical output.

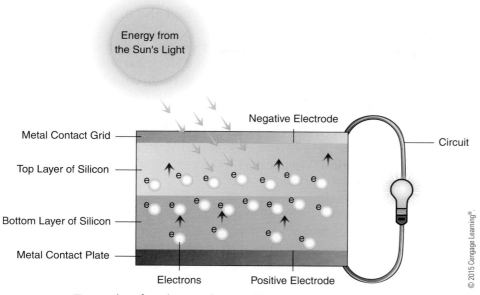

**FIGURE 8-8** Example of a photovoltaic cell.

Most solar cells are 10%–15% efficient, meaning only 10%–15% of the solar energy received by the cell is converted to electricity. Although photovoltaic (PV) cells are typically made from silicon, new thin film PV cells made from compounds like cadmium telluride (CdTe) which are flexible and can be applied to many different surfaces are now coming into use. There are also solar PV inks being developed that potentially could be used to coat objects, making it easier to apply photovoltaic capabilities to any surface is exposed to sunlight.

Currently there are two types of PV solar modules; Flat Plate Collectors and Concentrated PV. Flat plate collectors use a series of PV modules mounted on a flat surface that is pointed at the optimum angle to the sun (Figure 8-9). There are tracking systems for flat plate PV systems that follow

**FIGURE 8-9** Flat plate solar cells in use on a farm.

the sun throughout the day, therefore maximizing the solar angle. For optimal performance and energy output, a perpendicular angle between the PV cells and the sun is desired.

Concentrated PV systems use a lens to focus the sun's energy on the solar cells, thereby increasing its intensity and electrical output. This type of PV system requires lesser amounts of solar cells than conventional flat plate systems to produce the same amount of electricity. Concentrated PV is also more cost effective because of the smaller amount of solar cells required. The drawback to using a Concentrated PV system is they only work in direct sunlight, not diffuse light like the flat plate PV. This limits their applications to areas that have relatively clear skies throughout the year like the American Southwest.

There are two common ways PV systems can be used to provide electricity; the stand-alone and the grid-connect systems. Stand-alone systems produce electricity when the sun is shining and in order to gain access to electricity at night, the energy must be stored during the day for use later. This is usually done by using a bank of batteries. The grid-connect system ties the solar array to your electrical utilities grid system. During the day when the solar cells are producing power, any extra electricity that is not used goes out to the electrical grid, and your array acts like a mini power station. Your electric meter reverses, and you are credited for the amount of energy you produce that goes into the grid. At night, when the PV system is not producing electricity, you power your house from the grid. This process is known as net metering, and can result in a low utility bill, no utility bill, or the power company paying you for the excess power you produced depending on how much excess power you have contributed to the grid. The advantage of the grid-connect system is you do not have to purchase and maintain a battery bank to store your energy. Both the stand-alone and grid connect systems produce direct current or DC that must be converted to alternating current, or AC. This is done with a device called an inverter. The unit of measure used for the amount of solar output received from the sun at the earth's surface, also known as insolation (incoming solar radiation), is the kilowatt per square meter ($kW/m^2$). The value of insolation received at the earth's surface at a specific location is dependent on a few factors. First is the latitude location of the area. Typically areas near the tropics (between 23.5 degrees N and S latitude) receive the greatest amount of solar radiation throughout the year. In the middle latitudes (~40 degrees N and S latitude) the greatest amount of solar energy is received during the summer months. This does not mean that you cannot gain enough sunlight to justify using solar power in the higher latitudes. Even in the winter months there is enough insolation to supplement your electrical needs. Another factor that affects insolation, and therefore the output of solar cells, is the weather. Cloudy days will obviously lower the output of PV arrays. The American Southwest has the greatest potential for the use of solar power because of its southern location and general lack of cloud cover (Figure 8-10). The use of all types of solar power has great potential for being a renewable energy resource. Currently less than 1% of the energy used in the United States is supplied by solar power.

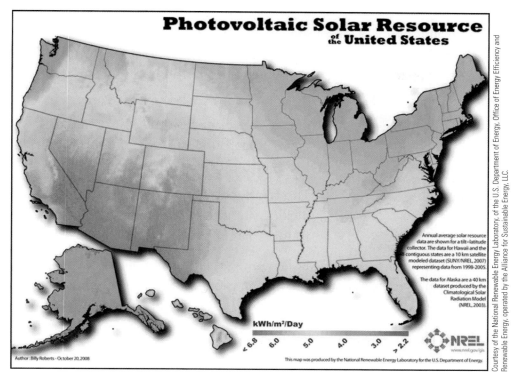

**FIGURE 8-10** Photovoltaic solar resource of the United States.

## CAREER CONNECTIONS

### Solar Installer/Technician

One of the fastest growing careers in renewable energy is a Solar Cell Technician. These technicians install photovoltaic arrays for a variety of uses both residential and commercial. These jobs require people to have a background in electronics that can be gained from attending a two year technical college. Also the ability to work on roofs and high surfaces is required. Specialized training in the installation of solar systems is required as a well. The technician needs to be able to work with a variety of tools and must like working outdoors. This career involves both the installation and maintenance of photovoltaic systems that includes both the mechanical and electrical aspects of photovoltaic technology.

## REVIEW OF KEY CONCEPTS

1. Passive solar power includes the use of windows and building materials to capture the heat energy of the sun within a structure. There are three main ways to use passive solar: direct gain, the Trombe wall, and isolated gain.

2. Active solar energy uses solar collectors and a circulating fluid to heat water or air. There are two types of active solar systems: direct and indirect circulation.

3. High temperature solar focuses the sun's energy with mirror-like devices called heliostats, to concentrate solar energy in order to generate extremely high temperatures. This heat can then be used to create steam that drives an electric turbine. The three types of high temperature solar include the power tower, parabolic trough, or dish engine.

4. The photoelectric effect describes how energy in the form of photons emitted from light sources such as the sun, causes electrons to be dislodged from metal elements. The photons strike the electrons in a conductor and begin to flow. The electrons can then be used to create an electrical current.

5. Photovoltaics exploit the photoelectric effect to capture photons emitted from the sun to produce an electric current.

6. The two types of photovoltaic systems include flat plate collectors and concentrated solar.

7. Solar power is limited to areas that receive ample amounts of sunlight and can only be used during daylight hours.

## CHAPTER REVIEW

### Short Answer

1. What are the three methods of passive solar energy?

2. Define active solar power and give one example of an active solar power system.

3. What are the three types of Concentrated Solar Power?

4. How do photovoltaic cells produce energy?

6. What are the two types of photovoltaic systems?

7. What are some of the limitations for using solar power in some areas?

### Energy Math

1. If an active solar water heater preheats 1 gallon of 56°F water to 70°F, how much energy is saved in BTU's by using solar power if it takes 8.5 BTU's to raise the temperature of 1 gallon of water by 1°F?

2. One square meter of Texas receives 5 kilowatts of solar energy per hour during summer. How much electricity would be produced by 3, one-square meter solar panels per hour on a sunny day assuming the efficiency of the panels are 10%?

### Multiple Choice

1. Solar energy gained through windows and absorbed by building materials is known as:
   a. active solar
   b. passive solar
   c. concentrated solar power
   d. photovoltaics

2. The use of solar collectors to heat a circulating fluid in order to heat water is called:
   a. active solar
   b. passive solar
   c. concentrated solar power
   d. photovoltaics

**3.** This involves the use of a heliostat:
  a. active solar
  b. passive solar
  c. concentrated solar power
  d. photovoltaics

**4.** Converting the sun's energy into electricity is called:
  a. active solar
  b. passive solar
  c. concentrated solar power
  d. photovoltaics

## Matching

*Match the terms with the correct definitions*

a. passive solar energy
b. active solar energy

c. concentrated solar power

d. photovoltaics

**1.** ____ The use of a collection device that absorbs the sun's heat energy that is then circulated by a fluid, like water or air and is driven by fans or pumps.

**2.** ____ The use of mirrors or lenses to focus the energy of the sun in order to generate high temperatures.

**3.** ____ The use of sunlight to produce an electric current.

**4.** ____ A means of harnessing the natural light and heat energy produced by the sun with no other input of energy.

## CHAPTER 9

# Hydroelectric Power

## KEY CONCEPTS

*After reading this chapter, you should be able to:*

1. Identify the two types of large-scale hydroelectric power systems and briefly explain how they operate.

2. Define the term penstock.

3. Identify what percentage of electricity in the United States is produced by hydropower.

4. Explain how a Pelton Wheel is used to generate power.

5. Describe how a pumped storage system operates.

6. Identify two positive aspects of using hydropower.

7. Identify four potential negative aspects of using hydropower.

8. Describe how agriculture can benefit from hydropower.

## TERMS TO KNOW

hydroelectric power
Venturi effect

pumped storage
thermal pollution

anoxic
hydrologist

# INTRODUCTION

Harnessing the power of flowing water has been exploited by humans for thousands of years. The remnants of old mills that once used the kinetic energy of moving water scatter the landscape and stand as old monuments to the long dependence that civilization has had to hydropower. Today, hydropower is mostly used for generating electricity and is the most widely used form of renewable energy. On both the large- and small-scale, the use of hydropower is going to continue to act as an important source of renewable energy.

## LARGE-SCALE HYDROELECTRIC

**Hydroelectric power** uses the force of flowing water to drive an electric turbine. The force that drives the water is gravity that causes water to flow from areas of high elevation to low, thus providing it with the power to move a turbine. Ultimately, it is the power of the sun that causes water to flow along the surface of the Earth. Solar energy drives the hydrologic cycle that causes water to evaporate into the atmosphere and be transported over higher elevations where it returns to the surface as precipitation. This water then flows back to the oceans and its kinetic energy can then be harnessed to supply power. The use of flowing water as a power source has been exploited by humans for thousands of years (see Chapter 1), but it wasn't until the 1880s that water was used to produce electricity. The first large-scale hydroelectric power plant was built in Wisconsin in 1882, and only seven years later there were more than 200 hydroelectric power plants in operation.

Large-scale hydroelectric power plants use two methods to harness the power of water. The first is called an impoundment system. An impoundment system uses a large dam to stop the flow of a river to create a large storage lake of water known as a reservoir. The water from the reservoir is then allowed to flow through an inlet and down into a large pipe or channel called a penstock. The penstock directs the water to a turbine that begins to spin by the force of the flowing water and generates electricity. The water then leaves the turbine near the base of the dam and is returned to the river. This area is known as the tailrace. The flow of the water can be controlled at the head of the penstock using valves to regulate the amount of electricity generated by the water. Impoundment dams are usually large structures constructed from concrete and are designed to last from 50 to 100 years (Figure 9-1).

There are approximately 2,400 hydroelectric dams in the United States that produce 8% of our electricity. The largest hydroelectric plant in the United States is the Grand Coulee Dam, on the Columbia River in Washington State. This impoundment dam is 550 feet high and has created a reservoir that encompasses 82,000 acres and stretches more than 150 miles and named Franklin D. Roosevelt Lake. The dam was completed

**hydroelectric power**—the use of flowing water to drive an electric turbine

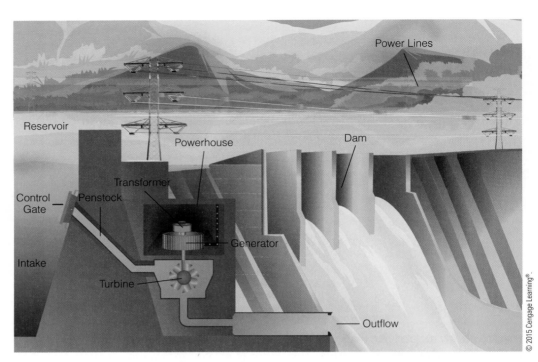

**FIGURE 9-1** The structure of a hydroelectric plant.

in 1941 and today produces more than 6,800 megawatts of power, making it the tenth largest hydroelectric dam in the world. The largest hydroelectric power plant in the world is the Three Gorges Dam in China. This hydroelectric dam is more than 590 feet high and produces more than 18,000 megawatts of electricity.

Another type of large-scale hydroelectric power is called the Diversion System, also known as Run of the River hydroelectric (Figure 9-2). This design does not require a dam to harness the power of water to produce

**FIGURE 9-2** Example of a diversion dam.

electricity. Instead a diversion system uses a canal to divert a portion of the river into a penstock that leads to an electric turbine. This type of hydroelectric system allows the river to flow uninterrupted. An example of a large-scale diversion system is the Niagara Power Project in New York State. This hydroelectric facility diverts part of the Niagara River to drive 13 electrical turbines that produce 2,400 megawatts of electricity.

## SMALL-SCALE HYDROELECTRIC

Not all hydroelectric power plants need to be large-scale diversions or impoundments. The same techniques applied to a large Run of the River project can be used on a much smaller scale to generate electricity as well. Diversions of smaller rivers or creeks may be enough to supply up to 100 kilowatts of power to be used by a home, farm, or business. Like larger projects, the flowing water must be diverted into a penstock that leads to a turbine generator. To get an idea of the potential power output of a flowing water resource that runs through a particular property, the following information is required. First determine the head of the water source. The head refers to the vertical distance that the water drops over a specific length. Next, determine the average flow rate of the water. This is how fast the water is moving past a particular point per unit area, and is typically measured in gallons per second or cubic feet per second. To determine the potential power output, multiply the head value by the flow value, and divide the answer by 10. The result would be the estimated power output in watts, assuming a 52% efficiency of the hydroelectric system. The use of a Pelton Wheel type turbine is often a good choice for small-scale hydroelectric. A Pelton Wheel diverts the water from the penstock into smaller diameter jet nozzles that increase the velocity of the water that forces a turbine to spin (Figure 9-3). This is called the **Venturi effect**. The Venturi effect can be demonstrated when you put your thumb over the end of a garden hose, to increase the velocity of water exiting the hose. Permits are required before a small-scale system is constructed to allow the diversion of water, that is often regulated, regardless of size. Regulations prevent the diversion or use of flowing water in the United States without first obtaining a permit.

**Venturi effect**—the increase in the velocity of water as it flows through a constricted area

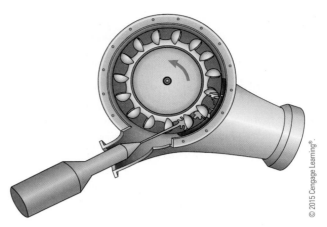

**FIGURE 9-3** Pelton Wheel.

## PUMPED STORAGE

**pumped storage**—the use of excess electricity to pump water from a lower elevation reservoir into a higher elevation reservoir. When electricity is needed, the water can flow back to the lower reservoir and generate electricity

Another method of hydroelectric power is known as **pumped storage**. Pumped storage uses excess electricity generated during off peak hours to pump water to a higher elevation through a penstock where it is stored in a reservoir (Figure 9-4). Then, during peak hours of electricity consumption, the water is released from the reservoir and flows back down through a penstock powering a turbine. The water is stored in a lower reservoir. In this system the turbine doubles as both a pump and generator depending on the direction of the flow of water. Pumped storage acts just like a battery, storing the potential energy of water at higher elevations. Typically a difference in elevation between 150–400 feet is needed between the upper and lower reservoir. Pumped storage can be used to store electricity from any energy source. The efficiency of a pumped storage system is around 85%, meaning that 15% of the stored power is lost to the friction of pumping and the evaporation of the water stored in the upper reservoir.

A large pumped storage facility is in operation as part of the Niagara Power Project in New York State. Here, water is pumped, usually at night, from the Niagara River to 1,900 acre reservoir above the Lewiston Pump-Generator Station. When extra power is required, the flow reverses and 12 turbines are powered. Pumped storage is an attractive low tech solution

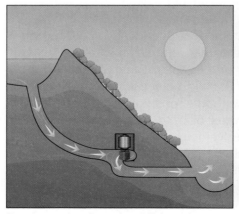

Day time—water flows downhill through turbines, producing electricity.

Night time—water pumped uphill to reservoir for tomorrow's use.

**FIGURE 9-4** Pumped storage.

for storing the power produced by variable sources of energy like solar and wind. Although most pumped storage facilities usually use two surface reservoirs, new systems using underground storage are being developed. In these systems, water is pumped up from aquifers through a well where the water table is greater than 200 feet deep. At times of excess power, the water from the aquifer is pumped up to a surface reservoir where it is stored. When the power is required, the water flows back down the well into the aquifer, generating electricity. Other plans for underground pumped storage include the use of abandoned mines and underground caverns both natural and excavated. Abandoned surface open-pit mines as well can be converted into pumped storage facilities. The use of ocean water for pumped storage is also under development. In this case, water is pumped out of the ocean to a higher elevation reservoir that can then be used later to produce electricity.

## ENVIRONMENTAL IMPACTS OF HYDROPOWER

The use of hydropower is considered a renewable, clean source of energy because it does not produce atmospheric pollutants like conventional fossil fuels. The hydrologic cycling of water on Earth continually drives the movement of water, making it an inexhaustible resource. Today 62% of our renewable electricity generation is delivered by hydropower, making it the largest renewable energy resource in America. The potential to increase the amount of hydroelectric dams in the United States is also possible. Currently there are approximately 80,000 dams of various sizes in the United States. Only about 2,400 are used to produce electricity; the other are for flood control but create the potential for using these existing dams as future power sources. Even though hydropower is considered to be a clean source of electricity, it still has potential negative effects on the environment. Probably the biggest drawback of hydroelectric power is the construction of an impoundment dam and its resulting reservoir. Large-scale dams create large reservoirs that flood terrestrial ecosystems. This can displace people and wildlife and also removes land from possible agricultural productivity. When the reservoir is first created and begins to flood the landscape, toxic metals that may be present in the soil and vegetation can leach into the water, causing harm to the food chain. The James Bay Project, a large-scale hydroelectric project in Canada, created controversy as a result of this environmental impact when it was first proposed (Figure 9-5). This massive project will eventually flood more than 60,000 square miles of coniferous forest and tundra ecosystems as a result of creating more than 200 dams. The reservoirs not only displaced the indigenous Cree Indians that live and hunt in the area to be flooded, but tens of thousands of caribou and other wildlife will also be affected as their habitats are taken over by the reservoirs created by the project's many dams. The flooding of the large tracts of forest and tundra is feared to cause the release of toxic mercury into the water. Coniferous trees in the areas to be flooded are believed to have mercury locked within their wood that will be released in the water when the flooded forests begin to decompose and thereby contaminating the food chain. The mercury was absorbed by the trees as a result of the massive amounts of coal burned by both Canada and the United States over the past one hundred years and deposited on the trees from the atmosphere. The James Bay hydroelectric project became so controversial that in 1992 New

**FIGURE 9-5** Hydro electric dam—Hydro-Quebec.

York State and Vermont both cancelled their agreement to purchase electricity from Hydro-Quebec, the energy company that runs the James Bay project. Today, these decisions have been re-evaluated because both states look to increase their use of renewable sources of energy.

Another potential environmental effect of reservoirs involves changing the water temperature of the river system. This is known as **thermal pollution**; a lowering or increase of the temperature of a body of water causing adverse effects. Thermal pollution is especially a problem associated with dams in the American Southwest. Large, deep reservoirs cause the temperature of the bottom third of the water column to be cooler than normal because only the upper portions of the reservoirs absorb heat from the sun. This cooler water is then drawn into the penstock to be used for turbine operation and released in the tailrace at a lower temperature than the river. Studies conducted on the Colorado and Green Rivers in the United States show temperatures ranging between 15°F–24°F (–9°C to –4°C) colder than normal summer temperatures in the tailraces below the dams. This correlates with the decline in both fish species and macroinvertebrates in those rivers.

Ways to minimize the thermal pollution of water caused by dams and reservoirs may include the use of multi-level intake systems that draw water from different levels within the reservoir in order to match the natural downstream water temperatures (Figure 9-6). This type of strategy has been successful when employed at the Flaming Gorge Dam on the Green River along the Utah–Wyoming border. Average summer river temperatures were increased by 13°F (–10°C) as a result.

Other negative effects of hydroelectric dams include the impoundment dam itself. Dams can block the migration of fish up and down stream. This problem has been addressed by employing a device called a fish ladder, or fishway to allow the fish to travel around the dam. A fish ladder is a series of steps near the dam where water flows, allowing fish to migrate up or down stream and bypass the dam (Figure 9-7).

Another environmental impact involves the annual silt load a river transports downstream. These sediments provide the aquatic ecosystem

**thermal pollution**—the lowering or increase of the temperature of a body of water, causing adverse effects

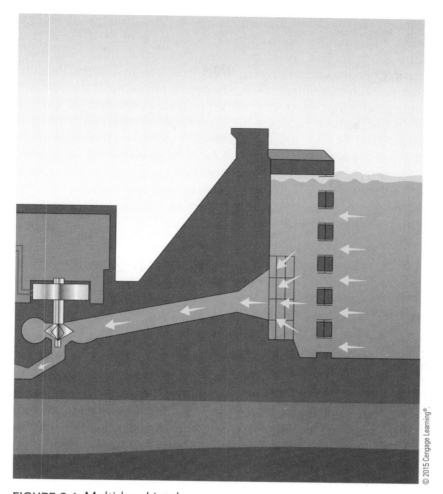

**FIGURE 9-6** Multi-level intake.

**FIGURE 9-7** Fish ladder.

anoxic—lacking oxygen

with important nutrients that drives aquatic food chains. The construction of an impoundment dam halts the transport of this sediment. This also causes the siltation, or build-up of sediments behind the dam, that must be periodically dredged to prevent the reservoir from filling with sediment. Another possible side effect of the siltation of reservoirs is the potential for methane production. Methane can be generated by anaerobic bacteria living in **anoxic** sediments at the bottom of a deep reservoir. This methane can be released into the atmosphere where it acts as a powerful greenhouse gas. Studies done in Brazil, where 90% of its electricity is produced from hydroelectric dams, show large amounts of methane emitted from reservoirs. Some researchers believe there may be a way to collect this type of methane to use as a fuel source and prevent it from going into the atmosphere where it contributes to global warming.

## HYDROPOWER AND AGRICULTURE

By far the biggest use of water resources in the world is for agriculture that accounts for 70% of all fresh water use. Much of this water is used for irrigation, with smaller amounts for livestock production. Because of the heavy reliance of agriculture on water resources, the use of hydropower coupled with irrigation may provide opportunities for farmers to increase their energy efficiency in the future. The irrigation process is energy intensive, requiring both electricity and fossil fuels to drive pumps that power irrigation systems. The use of pumped storage systems in conjunction with solar power might provide an efficient storage system of energy to be used with irrigation systems. Also, the creation of a reservoir at a higher elevation than the cropland to be irrigated can be used as a source of irrigation water delivered at high pressure without the need for a powered pump. The power of the flowing water may also drive a specially designed hydro-turbine that can produce electricity from the flowing irrigation water, and in turn, power a center pivot irrigation system. This type of integrated hydropower irrigation can only be used in areas with specific topography. Farms or agricultural facilities that lie near flowing water could also potentially benefit from employing small-scale, run of the river or diversion hydroelectric systems (Figure 9-8). Depending on the specific site, a small-scale hydroelectric system could supplement or provide all the electrical needs for an agricultural operation.

hydrologist—someone who studies all aspects of water

## CAREER CONNECTIONS
### Hydrologist

A **hydrologist** is someone who works on all areas of water use. Aspects of their job include maintaining water quality and supplies for irrigation, industry, recreation, and drinking. They study the effects of population growth, agriculture, pollution, and droughts on water supplies both above and below ground. Hydrologists can work in the public and private sector. A background in science and math in high school is a good start, along with a love of being outdoors. Hydrologists attend a four-year college and major in natural resource management and hydrology.

Intake Gates

Power Canal

Penstock

Powerhouse

Control Panel

Turbine

Draft Tube

Generator

Tail Race

© 2015 Cengage Learning®.

**FIGURE 9-8** A mini hydropower system used for irrigation.

## REVIEW OF KEY CONCEPTS

1. Hydropower uses the force of moving water to create power. Water flows from areas of high elevation to low by the force of gravity. The most common form of hydropower uses impoundment or diversion dams to channel flowing water through electric turbines.

2. A large pipe or tunnel through which water flows through a dam is called a penstock.

3. Approximately 8% of the electricity in the United States is supplied by hydropower.

4. A Pelton Wheel is a small-scale hydropower turbine turned by a ring of jets that shoot water out at high pressure.

5. Other methods of hydropower include pumped storage that uses excess electrical power to pump water up to higher

elevation storage ponds. Later, when more power is needed, the water is released and flows down turning a turbine that produces electricity.

6. The positive aspects of hydropower are that it is renewable and does not produce carbon dioxide.

7. The negative aspects of hydropower include the flooding of ecosystems when creating a reservoir, the prevention of sediment flowing downstream in a river, the leaching of toxic metals into a reservoir, thermal pollution, and the interruption of fish migration up river.

8. Agriculture can benefit from hydropower by using a combined irrigation-hydroelectric system or employing small-scale hydropower.

## CHAPTER REVIEW

### Short Answer

1. What are the two types of large-scale hydroelectric power systems and briefly describe how they operate?

2. What is a penstock?

3. Approximately what percentage of electricity in the United States is produced by hydropower?

4. How does a Pelton Wheel operate?

5. Describe how a pumped storage system operates.

6. What are two advantages of using hydropower?

7. Describe four negative effects hydroelectric power can have on the environment.

8. Describe two examples of how agriculture can use hydroelectric power.

### Energy Math

1. Determine the potential power output in watts of a small-scale hydroelectric system, if the head is 60 feet and the flow rate of the water is 100 gallons per second using the following formula: head value × flow value /10.

### Multiple Choice

1. The use of large dams and reservoirs to generate electricity is known as:
   a. pumped storage
   b. in-stream hydropower
   c. impoundment hydropower
   d. diversion hydropower

2. Which type of renewable energy resource can cause thermal pollution?
   a. Impoundment hydropower
   b. Pumped storage
   c. In-stream hydropower
   d. Diversion hydropower

3. What approximate percentage of electricity in the United States is generated by hydropower?
   a. 8%
   b. 21%
   c. 38%
   d. 50%

4. The small-scale hydropower system that uses the Venturi effect is:
   a. impoundment dam
   b. diversion dam
   c. Pelton Wheel
   d. pumped storage

## Matching

*Match the terms with the correct definitions*

a. hydroelectric power
b. pumped storage

c. thermal pollution
d. Venturi effect

e. anoxic

1. _____ The increase in the velocity of water as it travels through a constricted space.

2. _____ The use of excess electricity to pump water from a lower elevation reservoir into a higher elevation reservoir.

3. _____ Lacking oxygen.

4. _____ The lowering or increase of the temperature of a body of water, causing adverse effects.

5. _____ The use of the force of flowing water to drive an electric turbine.

## CHAPTER 10

# Hydrokinetic Power

## KEY CONCEPTS

*After reading this chapter, you should be able to:*

1. Define the term hydrokinetic power.

2. Explain the process of in-stream power and describe one advantage of using it.

3. Identify two methods of using tides to generate electricity.

4. Identify methods of using waves to generate electricity.

5. Describe the Ocean Thermal Energy Conversion process.

## TERMS TO KNOW

hydrokinetic power

in-stream hydropower

tidal power

wave power

Ocean Thermal Energy Conversion (OTEC)

thermocline

# INTRODUCTION

The traditional use of water to provide energy usually takes the form of a dam controlling the flow of a river to drive an electric turbine; however, new technologies are being developed to harness not only the kinetic energy of flowing water, but also of waves as well. This is known as hydrokinetic power, and has great potential for tapping into the energy stored within the world's oceans and rivers, while also having a reduced impact on the environment.

## IN-STREAM HYDROPOWER

**Hydrokinetic power** is the use of tidal energy, currents, wave energy, and differences in the temperature of water to supply power. **In-stream hydropower** is a form of hydrokinetic power that uses underwater turbines with rotor blades that spin as a result of the flow of the water (Figure 10-1). Similar to a wind turbine, the in-stream hydropower transfers the rotary motion of the blades to a generator housed within the unit that produces electricity.

This type of submerged hydropower has many advantages over conventional hydropower because it does not require large impoundment dams or diversion channels. Instead the units are secured to the bottom of a river or tidal basin where the blades are turned by the movement of the water. Another advantage is the turbines are not visible and do not block the flow of the body of water that can impede the migration of fish. In-stream hydropower is currently being tested in locations both in the United States and Canada. One location is in the East River that runs along the east side of Manhattan in New York. Six 15-foot diameter rotor blade units were installed in the East River near Roosevelt Island, producing about 7.7 kilowatts of power per hour (Figure 10-2). The average speed of the rotors is about 35 revolutions per minute. The next phase of the project is to install 30 river turbines in the river to supply one megawatt of power. Another in-stream hydropower project is also underway on the St. Lawrence River in Canada. Here two 15-foot diameter rotary turbines will be submerged in the St. Lawrence to produce between 60–80 kilowatts of electricity. If the testing phase is successful, then more in-stream turbines will be installed bringing the capacity up to 15 megawatts.

Another application of in-stream hydropower involves the use of a submerged turbine in the tailrace of an existing impoundment dam. The tailrace is the high velocity outlet water at the base of a hydroelectric dam. Installing a submerged in-stream turbine can capture this energy and transfer it to electric grid making the hydroelectric power plant more efficient. This technology is being used in the Mississippi River near the outlet of a dam in Hastings, Minnesota. This in-stream turbine produces an average of 35 kilowatts of power. The potential to use this new form of hydropower is great considering it has a low impact on the environment and uses the free energy of flowing water. Studies conducted on the impacts of in-stream turbines on aquatic organisms reveal an extremely low negative potential.

**hydrokinetic power**—the use of tidal energy, currents, wave energy, and differences in the temperature of water to supply power

**in-stream hydropower**—a form of hydrokinetic power that uses underwater turbines that have rotor blades that spin as a result of the flow of water

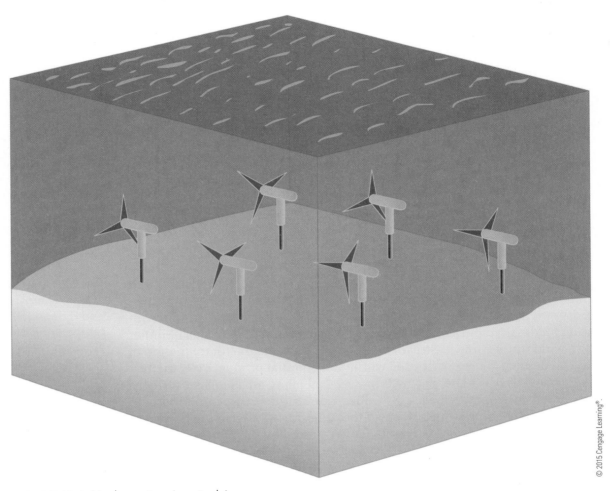

**FIGURE 10-1** Underwater river turbines.

**FIGURE 10-2** Installation of underwater turbines in the East River in New York.

The slow moving nature of the turbines gives fish and other aquatic organisms time to avoid the structures. Unlike impoundment dams, the turbines do not impede the natural flow of the river.

## TIDAL POWER

**Tidal power** involves the use of the rise and fall of tides to generate electricity. Tides on the Earth are caused by its rotation and the gravitational attraction of both the sun and the moon. This force causes the ocean to form a tidal bulge; a prominence of water that forms when gravity tugs on the earth. As the Earth spins, the tidal bulge causes the level of the ocean to rise and fall. Typically there are two high tides and low tides each day. This is known as a diurnal tide. The difference in sea level elevation between the high and low tide is known as the tidal range. Tidal ranges vary according to the local coastline topography. The largest tidal range in the world occurs in the Bay of Fundy in Nova Scotia, where the tidal range is more than 50 feet! In order to use tidal power, a range of 10 to 20 feet is needed, making only specific areas along the coasts able to harness this power. There are currently two systems that can be used to convert tidal energy into electricity. The first is known as the tidal barrage. A tidal barrage is a low-lying dam that runs across a tidal basin. Within the dam are sluice gates that allow water to flow through when the tide is coming in. When the tide begins to reverse, the sluice gates are closed, forcing the water to flow out through electrical turbines to produce electricity. Currently only two barrages are in operation, one in France that generates 240 megawatts of electricity, and another in Nova Scotia that produces 20 megawatts. The tidal barrage has potential negative effects to the marine environment. The dam-like structure of the barrage that controls the flow of tidal basins can also interrupt the movement of aquatic organisms living there. This is a concern because many marine species use estuaries for reproduction and feeding grounds.

> **tidal power**—the use of the rise and fall of tides to generate electricity

The other type of tidal system is known as the tidal fence. A tidal fence is a structure built across a tidal basin that uses a vertical blade turbine to capture the movement of water flowing through the fence when the tides are going in or out (Figure 10-3). The benefit of this system is that it does not completely block the flow of the water and allows the electrical generating equipment to be housed above the water line. Although no tidal fences are currently in operation, plans for constructing them are underway in both England and the Philippines. The application of in-stream hydropower systems is also being considered for use in the oceans to capture tidal energy.

## WAVE POWER

**Wave power** is the use of the rise and fall of ocean waves to generate electricity. Waves are produced in the oceans by the frictional force of wind moving along the ocean surface. Strong storms and sustained planetary scale winds create waves that travel great distances through the ocean. When a wave comes close to shore, the change in the depth of water causes its crest to rise upward before crashing on shore. The energy of the rising crest of a wave can be harnessed to produce electricity. Several methods of converting wave energy into electricity are being developed. One is known as Oscillating Water Column (OWC) wave technology. This type of wave power uses the up and down force of waves to force air or water through

> **wave power**—the use of the rise and fall of ocean waves to generate electricity

FIGURE 10-3 Example of a tidal fence power system.

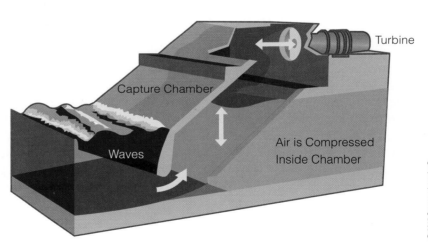

FIGURE 10-4 Example of a wave chamber turbine.

a tapered intake that drives a turbine (Figure 10-4). The continued force of wave action pushes and pulls water or air through a turbine that turns in the same direction regardless of the direction of flow of the fluid. This type of unidirectional turbine is known as a swells turbine. Angled wave chambers built along coastlines capture the force of waves that push air up through the chamber causing the turbine to turn and create electricity. This technology has been successfully tested in Scotland where an OWC system has been proven to generate 100 kilowatts of electricity.

Another type of wave power is known as float system wave energy. This works using a fixed or floating device that rises and falls from wave action. This vertical motion is then transferred into rotary motion. One such device using this principle is called the Salter Duck, invented in the 1970s. The "duck" is anchored to the bottom of the sea and bobs up and down with the swells causing a pendulum inside the duck to swing back and forth powering a generator. Another float system is known as a Pelamis "Wave Snake". This device is a long series of connected tubes that float on the surface of the ocean (Figure 10-5). Wave action causes the individual tubes to rise and fall, making the device look like a moving snake. Each section of the "snake" is hinged together by hydraulic pistons that force hydraulic fluid through a turbine. The continual movement of the device powers turbines that generate electricity. This concept is being tested off the coast of Portugal and is capable of producing 2,250 kilowatts of power.

In 2003, a unique wave energy system was tested off the coast of Denmark. The Wave Dragon, as it's called, is a floating platform that directs waves to break up and over a ramp that collects water in a reservoir. The water then flows down through several turbines and back into the ocean and generating electricity. A large-scale Wave Dragon is slated for testing off the coast of Wales. This particular unit is designed to generate seven megawatts of power by using 18 turbines. The biggest challenge to the use of OWC technology is the ability for the devices to withstand the strong forces generated by ocean storm systems. This certainly must be taken into consideration when designing and deploying these types of energy facilities.

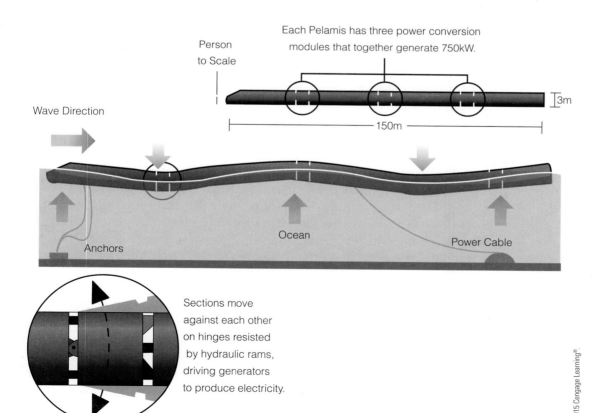

© 2015 Cengage Learning®.

**FIGURE 10-5** Illustration of a wave snake power system.

## OCEAN THERMAL ENERGY CONVERSION

**Ocean Thermal Energy Conversion (OTEC)**—the use of the differences in the temperature of water in the ocean to generate electricity

**thermocline**—the rapid change in the temperature of water with depth

The world's oceans absorb vast amounts of solar energy every day, especially in the tropics. **Ocean Thermal Energy Conversion (OTEC)** takes this stored heat and turns it into electricity by using temperature differences within the ocean. In tropical areas, between 23.5 degrees north and south latitude, the surface temperature of the ocean ranges between 64°F and 90°F (18°C–42°C). The average depth of the ocean is 2.5 miles and the temperature drops rapidly with depth. The location where the temperature of the ocean declines rapidly is known as the **thermocline**. The location of the thermocline varies seasonally and by location, but on average it occurs between 300 and 600 feet below the surface. Below this depth the temperature drops to a cold 40°F (4°C). This rapid change in temperature in depth is known as a thermal gradient that can be employed to generate power. Currently, there are three systems that can use the thermal gradient of the world's oceans to produce electricity.

An Open Cycle OTEC System uses seawater water drawn into an evaporator with a partial vacuum. The atmospheric pressure within the evaporator is about 1–3% that of normal atmospheric pressure. This causes water to vaporize at much lower temperatures. The warm surface water of the ocean is rapidly evaporated in a process known as flash evaporation that causes water inside to increase its pressure, to turn an electric turbine. The water is then condensed by drawing up cold water from below the thermocline, returning it to a liquid. The process is then repeated to create a Rakine Cycle heat engine that is commonly used in coal-fired electric plants. The difference lies in the lower heat required to boil the water while it is at a greatly reduced pressure.

The Closed Cycle OTEC is another type of Ocean Thermal Energy Conversion System. This system replaces water as the working fluid with other substances like ammonia or refrigerant gases that vaporize at lower temperatures (Figure 10-6). The fluid is contained in a loop where a heat exchanger takes the heat from the warm surface water and causes the

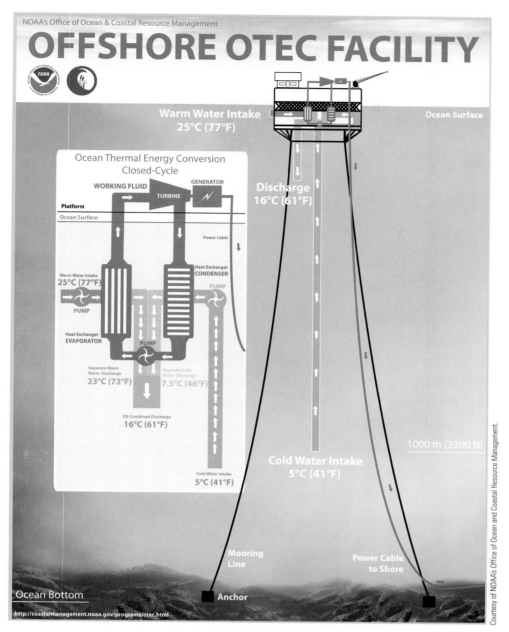

**FIGURE 10-6** A closed-cycle ocean thermal energy conversion system.

working fluid to vaporize. The vapor turns a turbine to produce power, then another heat exchanger draws up cold water from below the thermocline to cool a condenser, and the cycle is repeated.

A third option is being explored called the Hybrid Cycle OTEC. This system not only produces electricity but also creates fresh water to be used for agriculture or drinking water. The Hybrid Cycle draws in warm seawater to be flash-evaporated like the Open Cycle, but instead of powering a turbine directly, the heat is transferred by a heat exchanger to a closed loop system using a working fluid like ammonia that vaporizes and powers a turbine. The seawater is then condensed by the cold deep water back into a liquid. The process of flash-evaporation of the seawater also works as a desalinization method, and creates freshwater that can be pumped onto land. This hybrid system has the potential to generate both electricity and drinking or irrigation water at the same time. Besides the technical and engineering challenges posed by building and maintaining a large, sea-based energy system, concerns about the use of toxic working fluids like ammonia getting into the environment have been raised. This is especially relevant to the hybrid system that could potentially generate drinking and irrigation water. Also, the potential damage the structures could sustain during violent storms is a potential drawback.

## REVIEW OF KEY CONCEPTS

1. Hydrokinetic power is the use of flowing water in rivers or the ocean to produce power.

2. The use of submersible turbines, also known as in-stream hydropower, to transfer the motion of water in rivers or the tidal flow of the ocean into electricity is possible. One of the advantages of in-stream hydropower is it doesn't require impoundment dams or diversion channels.

3. Capturing the flow of tides using damns or tidal fences that harness the flow of water to turn electric turbines is also an option.

4. Wave action can be used to convert the motion of water into electricity by using devices that transfer the rolling motion of the ocean into electrical energy called oscillating water column (OWC) technology.

5. Ocean Thermal Energy Conversion is a method that uses the large temperature differences between the surface and the deep ocean to produce electricity.

## CHAPTER REVIEW

### Short Answer

1. What is hydrokinetic power?

2. How does in-stream power produce electricity, and describe one advantage of using it?

3. What are two methods of using tides to generate electricity?

4. How does the Ocean Thermal Energy Conversion process produce electricity?

5. Describe three examples of hydrokinetic power.

### Energy Math

1. What is the thermal gradient for the tropical ocean if the surface water is 25°C, and the temperature of the water at 300 meters is 8°C?

## Multiple Choice

1. The use of large underwater "wind" turbines is a form of:
   a. wave power
   b. Ocean Thermal Energy Conversion
   c. tidal power
   d. in-stream power

2. The "duck" and the "wave snake" are examples of:
   a. oscillating wave column technology
   b. Ocean Thermal Energy Conversion
   c. tidal power
   d. hydrokinetic power

3. What is the average depth of the thermocline within the oceans?
   a. 10 feet
   b. 50 feet
   c. 300 feet
   d. 1,000 feet

## Matching

*Match the terms with the correct definitions*

a. hydrokinetic power
b. in-stream hydropower
c. tidal power

d. wave power
e. Ocean Thermal Energy Conversion

f. thermocline

1. _____ The use of the rise and fall of ocean waves to generate electricity.

2. _____ The rapid change in the temperature of water with depth.

3. _____ A form of hydrokinetic power that uses underwater turbines with rotor blades that spin as a result of the flow of the water.

4. _____ The use of temperature differences within the ocean to produce electricity.

5. _____ The use of tidal energy, currents, wave energy, and differences in the temperature of water to supply power.

6. _____ The use of the rise and fall of tides to generate electricity.

# CHAPTER 11

# Wind Power

## KEY CONCEPTS

*After reading this chapter, you should be able to:*

1. Describe the process of how wind forms on the Earth.

2. Identify the specific areas where wind velocities are greatest and the average minimum wind velocity required for a wind turbine.

3. Describe the three main parts of a horizontal wind turbine.

4. Define the terms, cut-in speed, rated speed, and cut-out speed.

5. Identify three negative effects the use of wind power can have on the environment.

6. Explain how production agriculture can benefit from wind power.

## TERMS TO KNOW

wind
frictional force

wind turbine
nacelle

wind farm

# INTRODUCTION

The images of the old windmills on farms scattered across America have become an icon of rural agriculture. Today, wind power is the second fastest growing energy resource, behind natural gas, used to produce electricity. The use of wind turbines is no longer limited to pumping water for irrigation on America's farms. Since 1999, wind power has gone from generating less than 4,488 megawatts of electricity to almost 94,000 megawatts. In the United States there are more than 35,000 wind turbines in operation in 38 states, and the use of this renewable resource is only going to continue to grow into the future. The use of wind power on America's farms will not only provide farmers with opportunity to produce their own power, but agricultural land may also be able to be used as both cropland and as sites for wind farms.

## WIND RESOURCES

Humans have long exploited the force of the wind, most prominently as a means of propulsion for ships. The use of sails to power boats goes back more than a thousand years and illustrates the power potential of wind (see Chapter 1). **Wind** is formed on the Earth as a result of the unequal heating of the Earth's surface, and is therefore a byproduct of solar energy. Because the Earth is spherically shaped, the intensity of sunlight absorbed at the surface is greater per unit area near the equator, and much less near the poles. This unequal heating results in differences in atmospheric pressure. This pressure difference leads to the formation of wind as air moves from areas of high pressure to areas of low pressure (Figure 11-1). The definition of wind is the horizontal movement of air, also known as a pressure gradient

**wind**—the horizontal movement of air from areas of high atmospheric pressure to low

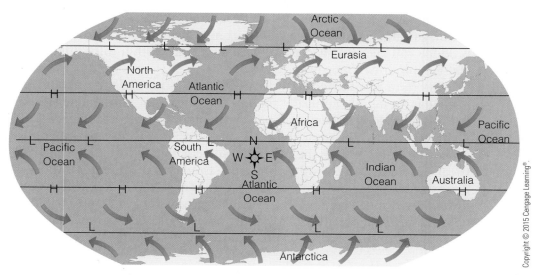

**FIGURE 11-1** Map of planetary winds.

**frictional force**—an influence that reduces the velocity of something

force. The force of the wind is dependent on the pressure differences that exist in an area. This is measured as a pressure gradient that is simply the change in air pressure over a specific distance. The greater the pressure gradient, the greater the force of the wind. Wind is more commonly measured by determining its velocity, in either miles per hour (mph) or kilometers per hour (km/hr). The force of the wind varies both seasonally and geographically. Generally speaking, wind velocity increases with altitude, and is also greater over water. This is the result of the lack of frictional forces that reduce wind speed. A **frictional force** is something that resists the motion of an object, in this case, wind. Surface topography, trees, and buildings all act as frictional forces that reduce the velocity of the wind. Because of this, the greatest wind velocities on Earth tend to be at higher elevations, or near large bodies of water. In the United States, the greatest potential for wind power exists along the coastlines, in certain high elevation areas, and the north central plains.

## WIND TURBINES

**wind turbine**—a device that converts the force of the wind into rotational motion that is used to turn an electric generator

The most effective way to harness the power of wind is to convert the pressure gradient force of the atmosphere into electricity. This involves the use of a **wind turbine**. A wind turbine is a device that converts the force of the wind into rotational motion used to turn an electric generator (Figure 11-2). Modern wind turbines consist of three main components, the rotor, nacelle, and the tower. The rotor is composed of blades similar in shape to those on a propeller that are attached to a hub. Rotors usually have either two or three blades that are aerodynamically designed to convert the horizontal movement of the air into rotational motion. The rotor blades are shaped to create pressure differences in the air moving past them causing lift that is transferred into rotation. The individual blades on some rotors can be controlled to change their pitch or angle in order to control their speed. Today most rotors are three-bladed, and are constructed from a variety of materials including metal, wood, fiberglass, carbon fiber, and composite plastics. Generally, the greater the diameter of the rotor, the higher the potential electrical output. A wind turbine having a 250-foot diameter rotor can deliver 1.5 megawatts, while a 340-foot diameter rotor generates 3.6 megawatts. Smaller scale wind turbines have diameters of 150-feet to produce 750 kilowatts while large-scale turbines with a rotor diameter of 400-feet produce 5 megawatts of power.

**nacelle**—the aerodynamically shaped enclosure of a wind turbine housing the gearbox, generator, braking system, and power conditioning systems

The rotor is attached to a driveshaft that is part of the **nacelle**. The nacelle is an aerodynamically shaped enclosure that houses the gearbox, generator, braking system, and power conditioning systems (Figure 11-3). As the wind turns the rotor, the drive shaft's rotation is transferred through a gearbox that increases the rotational velocity that in turn drives the electrical generator. Most wind turbines use fixed speed generators, although new variable speed generators may play a larger role in wind turbines in the future. Fixed speed generators are usually alternating current (AC) that are able to deliver electricity directly to the power grid. Variable speed generators are mostly direct current and require a power inverter to convert to AC current.

The wind speed required to produce electricity is called the cut-in speed. Most wind turbines have cut-in speeds around 7 miles per hour. This illustrates the basic requirements for wind energy. The average wind speed

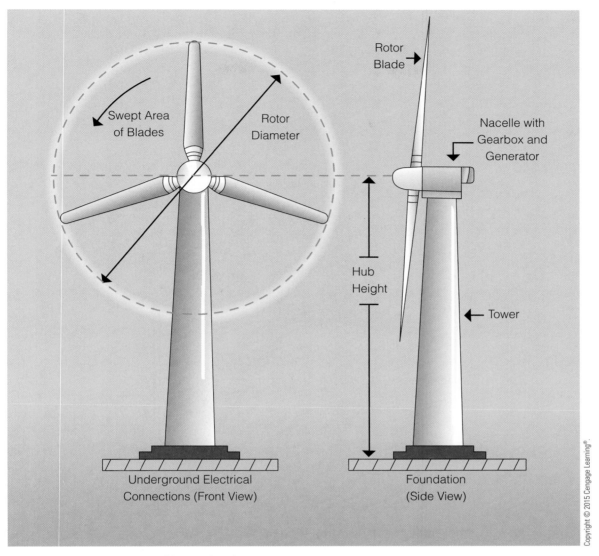

**FIGURE 11-2** Illustration of a wind turbine.

needs to be above 7 miles per hour for an area to be considered for installation of a wind turbine. After the cut-in speed is reached and the velocity of the wind increases, so does the electrical output of the turbine until the rated speed is met. The rated speed is the velocity of the wind required for the designed electrical output. A 3.6 megawatt rated turbine will produce 3.6 megawatts of power at 31 mph. All wind turbines also have a cut-out speed; the velocity of the wind reached in order to slow down the turbine. The cut-out speed is designed to prevent excess winds from damaging the structure. The cut-out speed is typically around 50–60 mph. When this speed is reached there are a few methods used to slow down the rotor. This can be done by a braking system, changing the pitch of the blades, or by a process known as furling that moves the nacelle so the rotors are not pointed directly in to the wind. Figure 11-4 graphically shows cut-in, rated, and cut-out speeds.

The nacelle is mounted on a tower that holds the wind turbine off the ground. Towers are usually tubular steel cylinders that sit atop a concrete base. The distance from the base of the tower to the nacelle is known as the hub height. Most wind turbines have hub heights roughly equal to

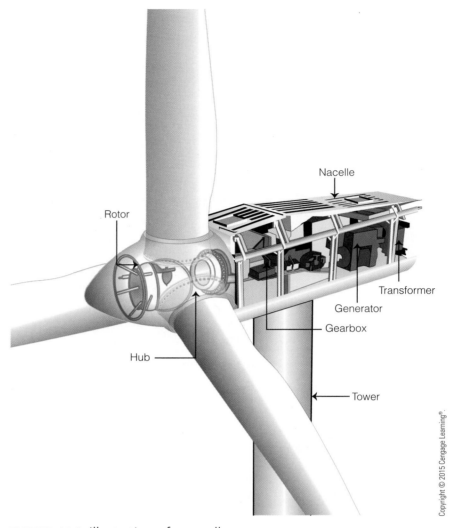

Rotor

Nacelle

Transformer

Generator

Gearbox

Hub

Tower

Copyright © 2015 Cengage Learning®.

**FIGURE 11-3** Illustration of a nacelle.

**wind farm**—a group of wind turbines in a concentrated area that are linked together to produce large amounts of electricity

their rotor diameter. The taller the turbine the greater the wind velocity, so taller wind turbines are able to produce more power. There is about a 0.4% increase in wind turbine power output for every 1% increase in hub height. Currently, about 3% of the total electricity used in the U. S is generated by wind, providing 140 billion kilowatts in 2012. Wind supplied 62% of all the renewable electricity produced in the U.S. in 2011. The top producing wind power states are Texas, Iowa, California, Oklahoma, and Illinois. Together individual wind turbines installed together in an area are known as a **wind farm**. As of 2013, all wind farms in the U.S. are land based. Coastal wind farms have greater potential for wind power because of the greater velocity winds associated offshore. The first offshore wind farm in the United States is being constructed off of Cape Cod, Massachusetts. Known as the Cape Wind Project, this offshore wind farm will sit approximately 5–7 miles south of Cape Cod in the Nantucket Sound (Figure 11-5). When installation is complete, the wind farm will have 130, 3.6 megawatt wind turbines spaced about 0.5 miles apart from one another. The wind farm will produce an average output of 170 megawatts; about half of the electricity used by Cape Cod.

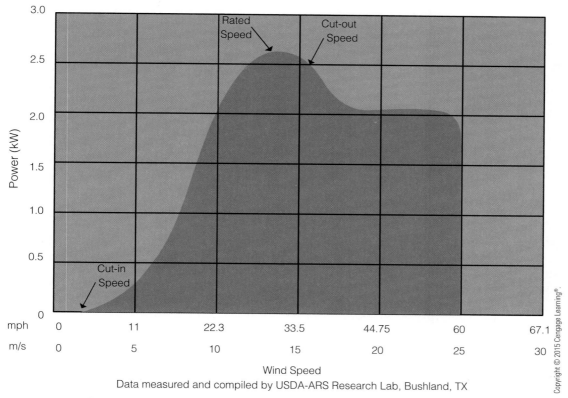

Data measured and compiled by USDA-ARS Research Lab, Bushland, TX

**FIGURE 11-4** Cut-in, rated, and cut-out speeds.

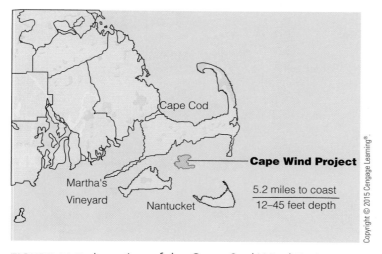

**FIGURE 11-5** Location of the Cape Cod Wind Project wind offshore wind farm.

If all the wind power potential was used in the United States, wind could provide as much as 50% of our electricity. This estimate was obtained after a survey of all the average wind speeds in areas around the country where determined. Wind power density classifications were created to identify regions that could be used for wind power. The classification uses the average wind speeds recorded at 33 feet and 156 feet above the ground. A wind power density classification of 3 has an average wind speed of 11.5 miles per hour at 33 feet. Wind power classes of 3 or better are required for most wind turbines (Figure 11-6).

## United States - Wind Resource Map

This map shows the annual average wind power estimates at a height of 50 meters. It is a combination of high resolution and low resolution datasets produced by NREL and other organizations. The data was screened to eliminate areas unlikely to be developed onshore due to land use or environmental issues. In many states, the wind resource on this map is visually enhanced to better show the distribution on ridge crests and other features.

U.S. Department of Energy
National Renewable Energy Laboratory

06-MAY-2009 1.1.9

### Wind Power Classification

| Wind Power Class | Resource Potential | Wind Power Density at 50 m W/m² | Wind Speed[a] at 50 m m/s | Wind Speed[a] at 50 m mph |
|---|---|---|---|---|
| 3 | Fair | 300 - 400 | 6.4 - 7.0 | 14.3 - 15.7 |
| 4 | Good | 400 - 500 | 7.0 - 7.5 | 15.7 - 16.8 |
| 5 | Excellent | 500 - 600 | 7.5 - 8.0 | 16.8 - 17.9 |
| 6 | Outstanding | 600 - 800 | 8.0 - 8.8 | 17.9 - 19.7 |
| 7 | Superb | 800 - 1600 | 8.8 - 11.1 | 19.7 - 24.8 |

[a] Wind speeds are based on a Weibull k value of 2.0

**FIGURE 11-6** Wind power classifications in the United States.

One of the problems associated with wind power is that it's inconsistent. The wind needs to be blowing all the time in order for the turbine to produce power and winds are variable, even in so called "windy" places. To address this problem, methods have been proposed to store electricity produced by wind in order to make it available when it's needed. Two methods under consideration include the pumped storage of water (see Chapter 9) and compressed air energy storage. Compressed air storage uses compressors that are powered during times of excess power to inject air at pressures more than 700 psi into storage areas. Storage areas can include tanks or abandoned salt mines. Later, during times of peak electrical demand, the air is released and is used to power an electric turbine. Because air heats up when it is compressed, the loss of this kinetic energy reduces the efficiency of the system. When the air is released, it must be slightly reheated upon expansion in order to restore its lost energy. The preheating can be done using natural gas. California plans to build a compressed air energy storage facility designed to store enough energy to produce 300 megawatts of power for up to ten hours.  This plant will use approximately 1 kilowatt of natural gas energy to produce 3 kilowatts of compressed air electricity. Most wind turbines are the horizontal axis type, there are also vertical axis turbines used to generate electricity. The two most popular vertical designs are the two bladed Darrieus wind turbine and the scooped Savonius turbine. The Darrieus turbines are also known as "egg beaters" because of the arrangement of their blades (Figure 11-7). Both types of vertical turbines are not as efficient to operate as horizontal shaft turbines.

## ENVIRONMENTAL IMPACTS OF WIND POWER

Although wind power is a renewable source of energy that does not directly pollute the atmosphere like conventional fossil fuels, there are still a few drawbacks that impact the environment in negative ways. One of the main problems associated with the use of wind turbines is the possibility of bird and bat deaths when the animals fly into the rotating turbines. These flying creatures have difficulty sensing the spinning rotors and can strike them during flight. When sighting a potential wind farm, studies on the types of birds and bat species that move through the area, and their flying heights,

### CAREER CONNECTIONS

## Wind Turbine Technician

A Wind Turbine Technician is a person who installs and maintains wind turbines. This specialized field requires training in both electronics and mechanics. Wind turbine technicians can be schooled and licensed at specialty technical colleges that prepare students for jobs in the wind industry. Wind turbines require constant preventive maintenance and safety inspections. As large-scale wind farms containing hundreds of large turbines are constructed across the United States, these technicians are going to be in greater demand. Along with the technical requirements required for this career, the ability to climb long ladders and having no fear of heights is a must.

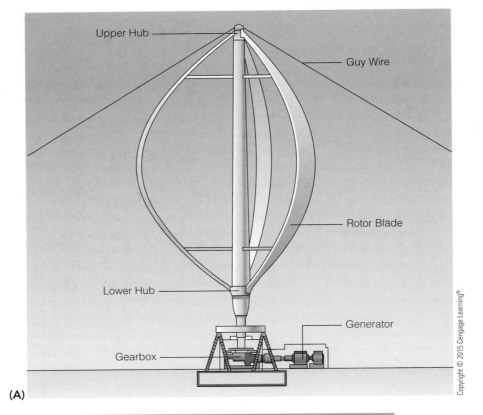

(A)

(B)

**FIGURE 11-7** (A) Darrieus vertical wind turbine. (B) Darrieus vertical wind turbine.

help to minimize the possibilities of these animals striking the rotors. Also, tracking the areas of migratory bird species and identifying the migration routes passing through might reduce any negative consequences on wildlife as a result of wind power. Another technique to minimize the impact on flying species is to identify the times of the year when the creatures tend to move through the area on their annual migration routes and shut the turbines down accordingly. Estimates of annual down times to allow migration of birds are usually around 15 days per year. Other negative aspects of wind power include the noise made by the turbines. Typically this is a low volume hum. The sound of the turbine is greatly diminished after a distance of 1,600 feet. Some complaints have also been raised about the visual impact the turbines have on the landscape. Many possible wind farm locations are in scenic areas such as the mountains or the coast and may cause a visual disruption to the landscape. This was one of the problems that arose when America's first offshore wind farm located off of Cape Cod, was proposed. Studies were conducted on the visual impact of the 130 wind turbines as viewed from the shore. The results showed the wind turbines would appear about one inch high along the horizon on a clear day. Many proponents of wind energy actually like the way the turbines look on the landscape, as their rotors gently turn with the wind. One of the benefits of wind turbines is the relatively low land area impact they have. Because the towers are so tall, they have a small land-based footprint. This allows the land below the wind farm to be used for other purposes, such as agriculture and solar power.

## WIND POWER AND AGRICULTURE

Agriculture has had a long-standing relationship with wind power stretching back into the late 1800s (Figure 11-8). The use of windmills to operate water pumps was widely successful until the 1930s when they were replaced by electric and fossil-fuel powered motors.

**FIGURE 11-8** Old agricultural wind mill.

Today, there are many opportunities for production agriculture to benefit from wind power. Farmers can use small or large-scale wind turbines to produce a portion or even all of their electricity needs if their farm is located in an area with sufficient winds. Even small-scale wind turbines can be used to power water pumps or electric fences. Having wind turbines tied to the electrical grid enables farmers to use net metering. Net metering pays a producer for the amount of excess power their wind turbine puts into the grid. This causes their electric meter to run backward, crediting their energy bill. At times of low wind, the meter turns forward again as power is consumed from the grid. Their utility bill is either canceled out, greatly reduced, or the electric utility company will actually owe the farmer for the excess power they have produced. Because large-scale wind turbines require so little land area, many farmers are leasing their land to power companies to construct wind turbines. The farmer can continue to plant crops below the turbines while collecting annual lease payments from the energy company (Figure 11-9).

Farmers may decide to become power producers themselves by financing a wind farm and selling the power to the local utility company. Although this has large upfront costs associated with it, some farmers are joining together and forming wind power cooperatives to help defray the costs.

**FIGURE 11-9** Wind turbines in farm fields.

© BESTWEB/Shutterstock.com

## REVIEW OF KEY CONCEPTS

1. Wind forms on the Earth as a result in the differences in air pressure. Air flows from areas of high pressure to low pressure.

2. The best places for wind power include higher elevations, flat areas, or coastal regions. Typically the average wind speed needs to be greater than 7 miles per hour to generate electricity.

3. Wind turbines are electrical generators that capture the horizontal movement of air by using an aerodynamic rotor. The three main parts of a wind turbine include the tower, nacelle, and rotor.

4. The cut-in speed is the speed of the wind at when electricity begins to be generated by a wind turbine. The rated speed is the speed of the wind when a wind turbine produces the greatest amount of electricity. The cut-out speed is the speed of the wind when a wind turbine begins to slow itself down.

5. Three disadvantages for using wind power include the noise they generate, the possible threat they pose to birds and bats, and their impact on the visual landscape.

6. Agriculture can benefit from wind power because cropland can be used for both growing crops and as a site for a wind farm. Also, wind turbines can produce electricity for farms directly.

## CHAPTER REVIEW

### Short Answer

1. How does wind form on the Earth?

2. Where are the specific areas that wind velocities are greatest, and what is the average minimum wind velocity required for a wind turbine?

3. What are the three main parts of a horizontal wind turbine?

4. Define the terms, cut-in seed, rated speed, and cut-out speed.

5. What are three negative effects that the use of wind power can have on the environment?

6. How can production agriculture industry benefit from wind power?

### Energy Math

1. Determine the total area swept by a wind turbine's rotor if its diameter is 400 feet?

2. What is the cut-in speed in miles per hour for a commercial wind turbine that has a cut-in rating of 4 meters per second?

### Multiple Choice

1. The speed at which a wind turbine begins to produce electricity is known as:
   a. cut-in speed
   b. rated speed
   c. cut-out speed
   d. wind speed

2. What is the average wind speed required for most wind turbines?
   a. 3 mph
   b. 7 mph
   c. 20 mph
   d. 30 mph

**3.** Which area is best suited for wind turbines?
a. Lower elevations
b. Cities
c. Plateaus
d. Coastal regions

**4.** Approximately how much of the electricity in the United States is produced by wind?
a. 2%
b. 5%
c. 25%
d. 50%

## Matching

*Match the terms with the correct definitions*

a. wind
b. frictional force

c. wind farm
d. nacelle

e. wind turbine

**1.** \_\_\_\_ A force that slows the velocity of something.

**2.** \_\_\_\_ The aerodynamic housing of a wind turbine that protects the gears and generator.

**3.** \_\_\_\_ The horizontal movement of air as a result of differences in atmospheric pressure.

**4** \_\_\_\_ A device that converts the force of the wind into rotational motion used to turn an electric generator.

**5.** \_\_\_\_ A concentrated grouping of wind turbines that are linked together to produce electricity.

## CHAPTER 12

# Geothermal

## KEY CONCEPTS

*After reading this chapter, you should be able to:*

1. Describe the four ways by which the Earth's interior is heated.

2. Define the term geothermal gradient.

3. Identify the four methods of hydrothermal power.

4. Explain the process of how a geothermal heat pump produces heat in winter and cools a house in summer.

5. Describe how agriculture can directly use geothermal energy.

6. Identify three possible negative effects the use of geothermal power can have on the environment.

7. Explain the effect of chlorofluorocarbon gases on the ozone layer and the actions taken to address the issues.

## TERMS TO KNOW

geothermal gradient
hydrothermal power

heat pump
direct use geothermal

ozone layer

# INTRODUCTION

The Earth's interior is hotter than the surface of the sun, and there are many ways by which this energy can be harnessed. Known as geothermal energy, this vast reservoir of heat lies not too far from the surface. Although not applicable in all areas, the use of the geothermal power locked deep within the Earth can supply a nearly inexhaustible supply of energy to be used directly as a heat source, as a means to produce electricity, and to supply sources of both heating and cooling for indoor structures.

## EARTH'S INTERIOR HEAT

The term, geothermal literally means "Earth's heat". The Earth's interior is extremely hot. The approximate temperature of the Earth's core is almost 12,000°F (6,600°C). This temperature is actually hotter than the surface of the Sun. The extreme heat of the Earth's interior is believed to be generated by a number of sources including radioactive decay, accretion, tidal friction, and gravitational compression. The decay of radioactive elements like uranium 238, potassium 40, and thorium 232 are responsible for most of the heat contained within the earth. As these elements decay into more stable elements, a large amount of heat is given off. The early stages of the Earth's formation also generated a lot of heat through the process of accretion. Accretion is the growth of a body, such as a planet, by the addition of mass through gravitational attraction. Asteroids, meteorites, and comet impacts during the Earth's formation converted large amounts of kinetic energy into heat. The early Earth was also much less dense than it is today and as accretion continued to add more mass to the Earth, its gravity began to increase. This caused the planet to pull its mass toward the Earth's center, thereby compressing itself and generating heat. Finally, the constant tidal stress put on the Earth from both the sun and the moon is believed to generate heat from friction as the planet is flexed by their combined gravitational force. The rate at which the temperature changes with depth in the Earth is called the **geothermal gradient** (Figure 12-1). Areas near tectonic plate boundaries can have high geothermal gradients. This is because magma is fairly close to the surface. The geothermal gradient in these geographic zones average about 11°F (20°C) per 100 feet. Therefore with an average surface rock temperature of 56°F (113°C), you would only have to drill approximately 1,400 feet to tap into water that is at the boiling point. Most places on the Earth have much lower geothermal gradients, somewhere between 0.6°F and 4.4°F (1°C–8°C) per 100 feet that require much deeper drilling. In either case, it is the high temperatures beneath the Earth's surface that geothermal power tries to employ.

geothermal gradient—the rate the temperature changes with depth in the Earth

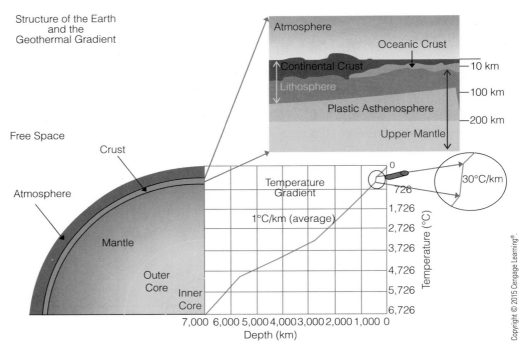

FIGURE 12-1 Diagram of geothermal gradient.

## HYDROTHERMAL POWER

Copyright © 2015 Cengage Learning®.

**hydrothermal power**—the use of superheated water from the ground to produce electricity

Places on the Earth located near tectonic plate boundaries have magma close to the surface (Figure 12-2). As groundwater percolates through the ground, it can encounter these superheated areas and vaporize. This steam then blasts out of the Earth in areas known as geysers and hydrothermal vents. Hydrothermal literally means "water hot". **Hydrothermal power** is a method of tapping into superheated underground water that is then used to power electrical turbines. There are three types of hydrothermal power (Figure 12-3). The first is known as dry steam geothermal that uses steam directly produced from geysers to drive electric turbines. Pipes capture the steam being ejected from the ground and is used to turn a generator. Currently there are 20 dry steam hydrothermal power plants in the U.S. that produce an average of 70 megawatts each. These are located mostly in California.

Another method of hydrothermal power is called lash steam. Flash steam power plants use wells drilled deep into hydrothermal reservoirs where the temperature of the water is between 350°F–575°F (177°C–302°C). This high pressure, superheated water is then expanded rapidly at the surface in a chamber where it "flashes" into steam, and then used to power an electric turbine. There are 30 of these power plants operating in the United States, mostly in California. The final type of hydrothermal power is called the binary system that uses lower temperature water between 190°F–350°F (88°C–177°C). This hydrothermal fluid is pumped through a heat exchanger that causes a hydrocarbon fluid held within a closed loop to vaporize. The high-pressure vapor then turns an electric turbine. The hydrocarbon fluid typically has a low boiling point like isobutene. There are about 20 of these types of hydrothermal plants operating in the United States.

Some areas have sufficient heat reasonably close to the surface but lack enough groundwater to be used for a hydrothermal system. This type

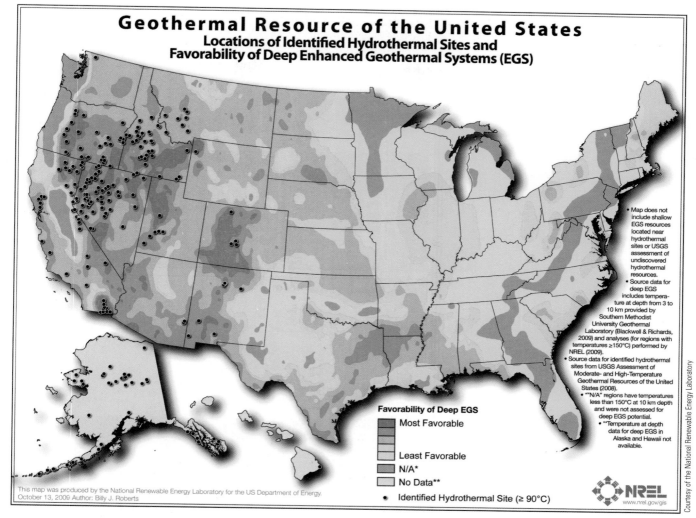

FIGURE 12-2 Map showing the geothermal resources in the United States.

of geothermal resource is known as hot dry rock or HDR. For this to be employed, a well can be drilled into the heated rocks. The rocks are then fractured to open up pore spaces, and another well is drilled into the area. Finally, water is pumped into and flows through the fractured rock. The heated water is then returned to the surface by the second well that is hot enough to use. Although this system is promising in some areas, it is not yet cost effective. Fig 12-4 shows a map of areas with current or planned geothermal power operations.

## GEOTHERMAL HEAT PUMPS

Geothermal heat pumps use the constant temperature of the ground near the surface of the Earth as a method of heating or cooling depending on the time of year. The average temperature of the ground to a depth of a few hundred feet ranges between 45°F–75°F (7°C–24°C) depending on the latitude location. In the mid-latitudes the temperature at three feet is typically 56°F (13°C). During summer, this is much cooler than the average surface temperature, and in winter much warmer. This heat difference can then be used by a **heat pump**.

**heat pump**—a device that is used to move heat, typically from a cooler region to a warmer region using mechanical energy

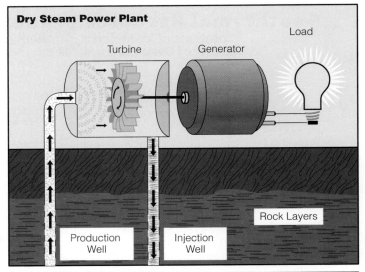

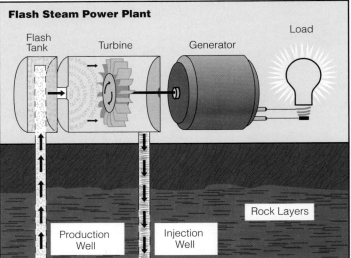

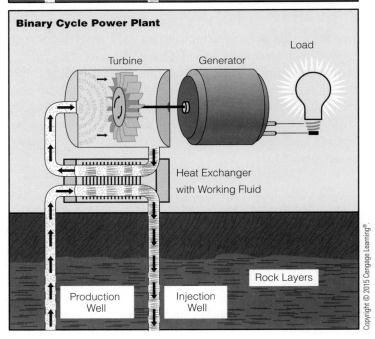

**FIGURE 12-3** Diagrams of (A) dry steam, (B) flash, and (C) binary geothermal power plants.

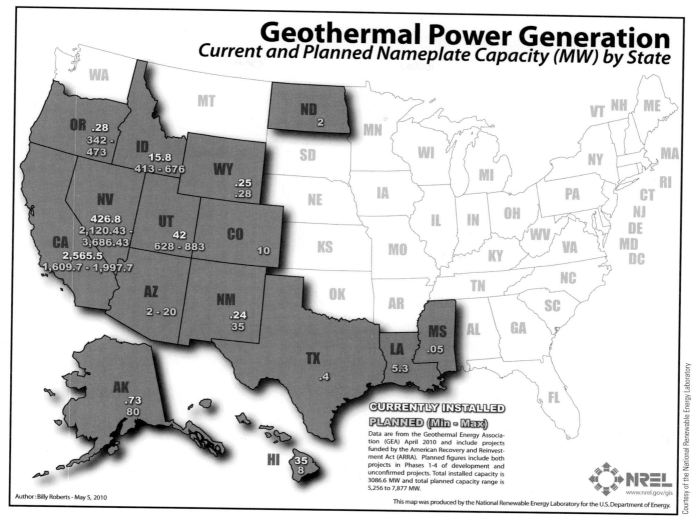

**FIGURE 12-4** Map of geothermal power plants in the United States.

Heat pumps use a working fluid with a low boiling point and a compressor. Commonly used working fluids in heat pumps include hydrochlorofluorocarbon 22, also known as HCFC–22. This compound is useful in heat pumps because it can be vaporized at low temperatures. During winter, when outside temperatures are below freezing, a geothermal heat pump transfers the heat from the ground, through a heat exchanger, to the working fluid within a closed loop. The fluid's temperature is raised to the point where it vaporizes. The vapor is then compressed to increase its temperature that is hot enough to be passed through another heat exchanger that transfers the heat to a blower that produces hot air. The fluid is then condensed and the cycle repeats itself (Figure 12-5). This type of heating system is extremely efficient compared to traditional fossil fuel or electric heat. Typically, geothermal heat pumps use between 25%–50% less electricity than conventional fossil fuel–based heating systems. Some geothermal heat pumps use the system to heat water as well as air. The heated water can be used as a source of hot water. Another benefit to these systems is their use as a source of cooling during warmer months. In this case the process is reversed and the heat from the atmosphere is transferred into the ground.

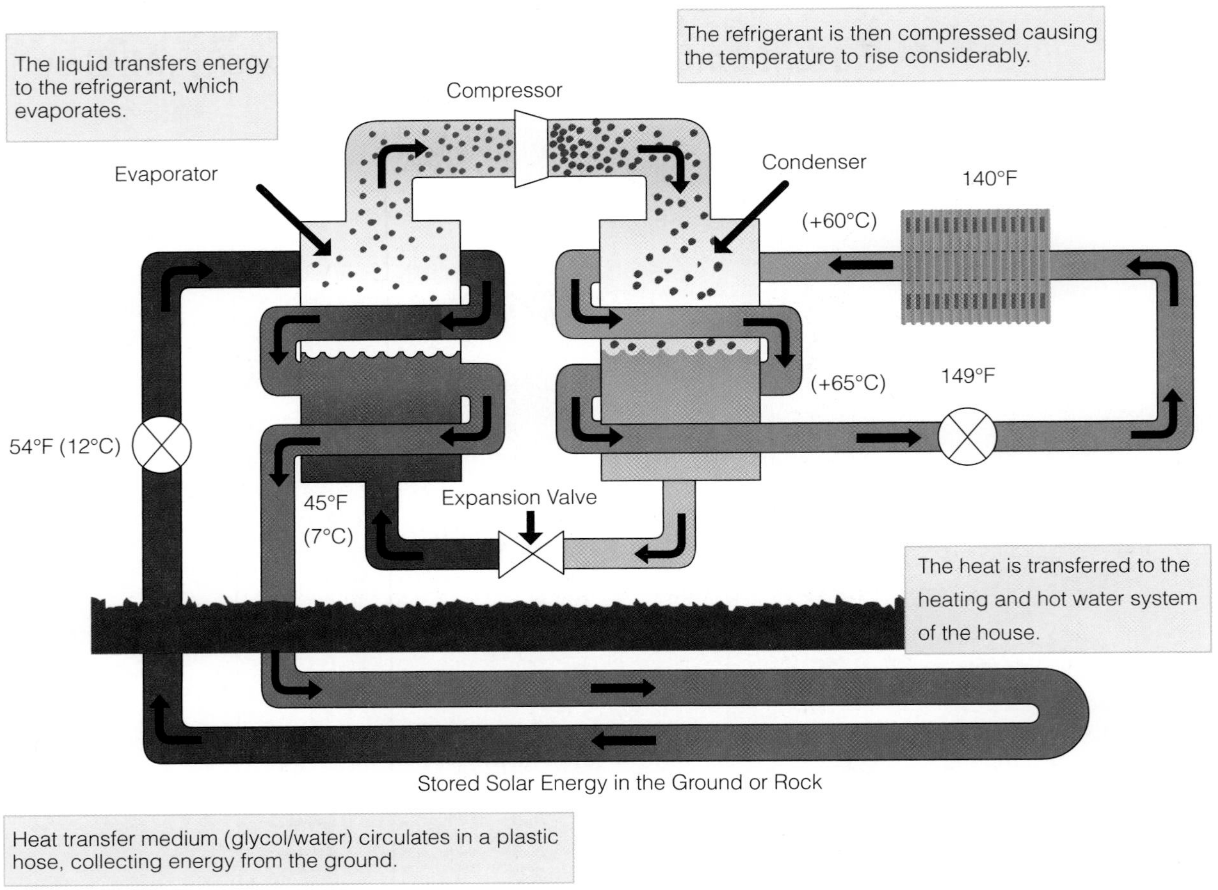

The liquid transfers energy to the refrigerant, which evaporates.

The refrigerant is then compressed causing the temperature to rise considerably.

Compressor

Evaporator

Condenser

140°F

(+60°C)

(+65°C)    149°F

54°F (12°C)

45°F
(7°C)

Expansion Valve

The heat is transferred to the heating and hot water system of the house.

Stored Solar Energy in the Ground or Rock

Heat transfer medium (glycol/water) circulates in a plastic hose, collecting energy from the ground.

**FIGURE 12-5** Diagram of a geothermal heat pump heating system.

There are four types of heat transfer systems used with geothermal heat pumps to exchange heat to or from the ground (Figure 12-6). These are known as ground loop systems. A horizontal ground loop system uses loops of pipes buried in trenches about five feet deep. The pipes are filled with an antifreeze solution that circulates through the ground. This heat is then exchanged with the heat pump. The horizontal system is the least expensive because it does not require deep excavation, although it requires a fairly large area to lay out the loop of piping. The vertical system uses loops of piping buried deep into the ground to depths between 100–400 feet. The fluid within the pipes transfers the heat within the ground to the heat pump. Instead of using the ground as a source of heat, large bodies of water can also be used. This closed loop method is called a pond/lake system. In this type of system, the loop of pipes goes into a lake or pond and exchanges heat with the water.

A large-scale example of a pond/lake system is the Cornell University Lake Source Cooling Project. This system uses the cold bottom water of Cayuga Lake in upstate New York to cool the buildings of the Cornell University Campus (Figure 12-7). Water is drawn from the lake at a depth of 250 feet where the temperature is 39°F (4°C). This cold water is passed through a heat exchanger where it cools water in a closed-loop system pumped through buildings on campus. The use of this system resulted in

**Geothermal Energy for the Home**

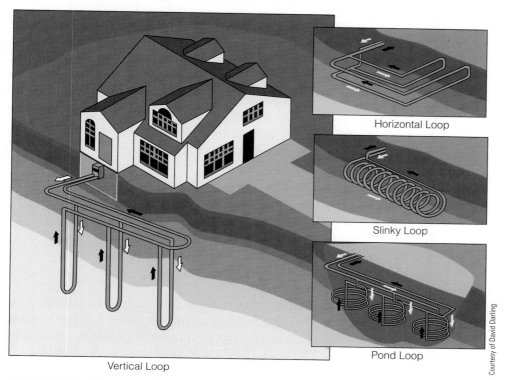

FIGURE 12-6 Diagram of geothermal ground loop systems.

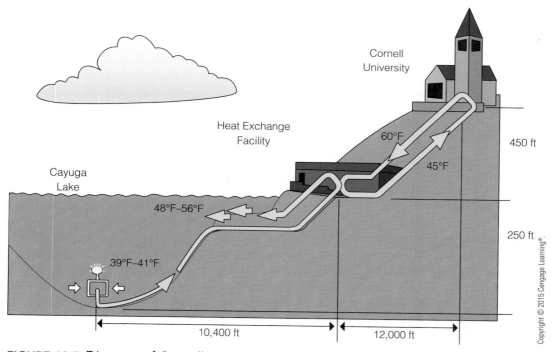

FIGURE 12-7 Diagram of Cornell University's lake water cooling system.

significant environmental benefits and energy savings for the university. The use of the cold lake water to cool buildings instead of using conventional air conditioners reduced the electricity use associated with cooling the campus by 86% and an overall reduction in electrical use by 10%. Also, the need for ozone destroying refrigerant gases is eliminated.

A similar larger scale lake water cooling project was constructed to provide cooling for the city of Toronto, Canada. In 2009, Toronto's Deep Water Lake Cooling System was completed, and now provides cooling to 60 downtown buildings. The system reduced the electricity demand for air conditioning by 90%, saving the city 85 million kilowatt hours of electricity each year. The system draws in cold water from Lake Ontario at 39°F (4°C) from a depth of 272 feet. The water is passed through a heat exchanger that cools a closed-loop system to cools buildings.

Another heat transfer method is called the open loop system. Unlike the three closed-loop systems that use a circulating anti–freeze solution to transfer the heat, the open loop system pumps cold water from a deep well directly into the heat exchanger within the heat pump. The water is then discharged back into the ground to another well. All four systems have advantages and disadvantages and selecting the best one is dependent on the specific site characteristics where the geothermal heat pump is going to be used. Other than the heat stored within the ground, the use of a geothermal heat pump requires electricity. Many systems are now being installed in conjunction with photovoltaic systems that help to supply the electricity, making them efficient.

## AGRICULTURAL USE OF GEOTHERMAL ENERGY

**direct use geothermal**—a method of geothermal energy which utilizes hydrothermal fluids directly from the ground as a source of heat

There are many agricultural practices that benefit from the use of geothermal energy. The most widespread form of geothermal energy used for agriculture is called **direct use geothermal**. This method uses hydrothermal fluids directly from the ground as a source of heat. Unlike hydrothermal electric plants that use the hydrothermal fluids to power a turbine, direct use systems employ hot water as a heat source. This geothermal energy can be used to heat greenhouses, aquaculture facilities, and agricultural processing (Figure 12-8). Currently there are almost 70 greenhouses and aquaculture operations in the United States that use this type of geothermal. In these systems, hot water between 70°F–300°F (21°C–149°C) is pumped through the buildings to heat indoor spaces or water tanks. Direct use geothermal is used to dry grains, grow mushrooms, and dehydrate onions and

FIGURE 12-8 A geothermal greenhouse, Hveragerdi, South Iceland.

garlic. Typically, wells are drilled to depths where the heated water exists. This varies geographically, but is usually more than 3,000 feet. The hot water is then pumped to the surface where it transfers its energy through a heat exchanger that supplies the heat.

## ENVIRONMENTAL IMPACTS OF GEOTHERMAL ENERGY

Most geothermal systems have no direct negative effects on the environment. This is especially true for hydrothermal power plants and direct use hydrothermal. The only real concern about the use of hydrothermal energy is the release of the spent hot water. Typically this water is discharged into cooling ponds that allows the temperature of the water to adjust to ambient conditions. Precaution must be taken to prevent the potential of the thermal pollution of surface waters as a result of the use of hydrothermal fluids. Also, some hydrothermal fluids contain dissolved gases like methane, carbon dioxide, ammonia, and hydrogen sulfide that can potentially pollute the atmosphere. Dissolved minerals, especially sodium chloride may also exist in hydrothermal fluid. These are usually removed during the vaporization process, but must also be disposed of in some way. The biggest environmental disadvantage to the use of geothermal heat pump technology is its need for a working heat transfer fluid in the form of refrigerant gases. The most commonly used refrigerant gases for almost one hundred years are known as chlorofluorcarbons (CFCs). One in particular, CFC–113 known as Freon was widely used for refrigerators and air conditioners. CFC–113 was eventually linked to the destruction of the **ozone layer**. The ozone layer is a region of the upper atmosphere known as the stratosphere where a thin layer of ozone exists (Figure 12-9). Stratospheric ozone that protects the planet by filtering ultraviolet radiation should not to be confused with ground-level ozone (smog) that is detrimental to health.

**ozone layer**—a layer in the stratosphere of an unstable form of oxygen ($O_3$) that filters out ultraviolet radiation

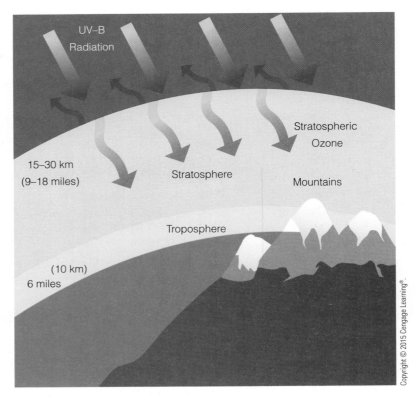

**FIGURE 12-9** Ozone layer in the atmosphere.

Ozone is an unstable form of oxygen ($O_3$) that acts as a filter for dangerous ultraviolet light emitted from the sun. Ozone absorbs solar UV radiation and prevents it from reaching the Earth's surface where it could harm living things. Exposure to high energy UV radiation causes damage to living cells. In humans it leads to the development of skin cancer. In the late 1950s, a British meteorologist, Gordon Dobson, began to measure ozone levels in the stratosphere. Twenty years later, data on stratospheric ozone concentrations revealed the depletion of the ozone layer as a result of CFC gases. In 1996, ozone levels had been reduced by almost 60% causing as much as 150% more UV radiation striking the Earth's surface. CFCs were an excellent refrigerant gas, especially if they leaked because they are non-toxic to humans. This was an advantage over other early refrigerants like ammonia that is highly toxic. Although CFCs pose no direct harm to life, once they escape into the atmosphere, they migrate to the stratosphere where they are bombarded by UV radiation. This breaks apart the CFC molecule and frees up chlorine atoms that were once part of the CFC (Figure 12-10). The chlorine is reactive and bonds with oxygen within the ozone layer. This oxygen now becomes part of a chlorine monoxide molecule and reduces the amount of oxygen available to make ozone.

The result is a depletion of the ozone layer. Scientists recognized this problem and in 1987 an international treaty was signed in Montreal, Canada by all participating countries in the United Nations to phase out ozone destroying CFCs. This landmark international agreement is known as the Montreal Protocol. As a result, Freon (CFC–113) can no longer be used as a refrigerant. To replace Freon, scientists developed HCFC refrigerants

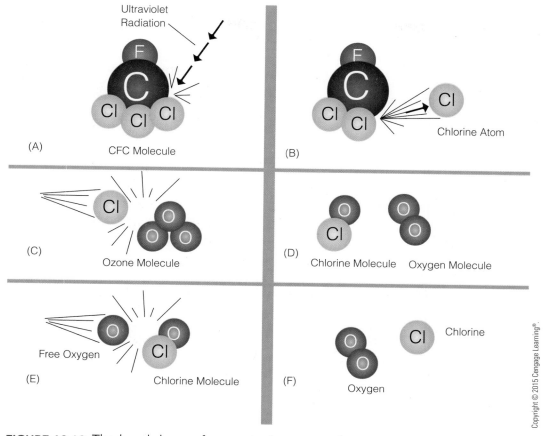

**FIGURE 12-10** The breakdown of ozone in the stratosphere.

like HCFC–22 that is a hydrofluorochlorocarbon. These gases are not as destructive as CFCs and will eventually be phased out and replaced by 2015. Because of actions initiated by the Montreal Protocol ozone levels within the stratosphere are now beginning to regenerate. This is good news for the environment, and also important because it shows how international agreements can be successful in combating worldwide environmental problems. Because geothermal heat pumps still use HCFC–22, there is still the potential for the destruction of ozone if they leak. As of January 2010, any new heat pumps will be required to use a safer refrigerant gas. The most likely candidate will be a Hydrofluorcarbon (HFC) gas that contains no chlorine. Another drawback to CFCs and HFCs is that they are greenhouse gases that can contribute to global warming.

## REVIEW OF KEY CONCEPTS

1. The Earth is heated by radioactive decay, accretion, tidal friction, and gravitational compression.

2. The geothermal gradient is the rate the temperature of the Earth changes with depth.

3. The four methods of hydrothermal energy include dry steam, flash steam, binary, and hot dry rock.

4. A geothermal heat pump is a device that uses mechanical energy to move heat stored in the ground to the inside a building to warm it. In times of hot weather, the heat pump works in reverse to transfer heat from inside a house to the ground.

5. Agriculture can use geothermal energy directly to heat greenhouses, aquaculture facilities, and other farm buildings.

6. Some negative aspects the use of geothermal can have on the environment include the possible release of gases like carbon dioxide, methane, ammonia, and hydrogen sulfide into the atmosphere. Also minerals that exist within hydrothermal fluids like sodium chloride must also be disposed of.

7. The use of refrigerant gases in heat pumps once posed a threat to the ozone layer. The ozone layer is a thin area of ozone gas ($O_3$) that exists in the stratosphere that filters harmful ultra-violet radiation given off by the sun. Chlorofluorocarbon gases (CFCs) commonly used as refrigerants were responsible for destroying stratospheric ozone. Once this was recognized, nations agreed to phase these gases out and replace them with ones that do not destroy ozone.

## CHAPTER REVIEW

### Short Answer

1. What are four ways the Earth's interior is heated?

2. Define the term geothermal gradient.

3. What are the four methods of hydrothermal power?

4. How does a geothermal heat pump produce heat in winter and cool a house in summer?

5. How can agriculture use direct use geothermal energy?

6. What are three negative effects that the use of geothermal power could have on the environment?

7. Explain why greenhouses and aquaculture facilities utilizing geothermal energy are mostly located in California.

### Energy Math

1. At what depth would you drill to reach the boiling point of water (212°F), if the temperature of the rocks at the surface is 56°F, and the geothermal gradient is 2.5°F per foot?

## Multiple Choice

1. Which of the following is not a source of the Earth's internal heat?
   a. Radioactive decay
   b. The sun
   c. Tidal stress
   d. Accretion

2. Which energy source can produce hot air in the winter and cool air during the summer?
   a. A geothermal heat pump
   b. Evacuated solar collectors
   c. Ocean thermal energy conversion
   d. A Tromb wall

3. What layer of the atmosphere is the ozone layer located?
   a. Troposphere
   b. Stratosphere
   c. Mesosphere
   d. Thermosphere

4. Which gas is primarily responsible for the destruction of ozone?
   a. Carbon dioxide
   b. Methane
   c. Hydrogen sulfide
   d. CFCs

5. Which type of geothermal system is used to heat greenhouses and aquaculture facilities?
   a. Dry steam
   b. Flash steam
   c. Binary system
   d. Direct use

## Matching

*Match the terms with the correct definitions*

a. geothermal gradient
b. hydrothermal power
c. heat pump
d. direct use geothermal
e. ozone layer

1. _____ A method of tapping into water that is superheated underground and is then used to power electrical turbines.

2. _____ A region of the upper atmosphere known as the stratosphere, where a thin layer of $O_3$ exists.

3. _____ A method utilizing hydrothermal fluids directly from the ground as a source of heat.

4. _____ The rate at which the temperature changes with depth in the Earth.

5. _____ A device, which is used to move heat, typically from a cooler region to a warmer region using mechanical energy.

## CHAPTER 13

# Secondary Sources of Energy

## KEY CONCEPTS

*After reading this chapter, you should be able to:*

1. Explain the difference between a primary and secondary energy source.

2. Define the term electricity.

3. Describe the three basic parts of an electric circuit.

4. Identify the units commonly used to measure electricity.

5. Explain how an electric current is produced by a generator.

6. Describe the difference between AC and DC current and the advantage of using AC over DC.

7. Explain how a battery produces electricity.

8. Describe the main components of the electrical grid.

9. Identify four processes used to produce hydrogen gas.

10. Describe three ways hydrogen can be used to produce energy.

11. Discuss three methods used to store and transport hydrogen.

12. Explain two benefits of using hydrogen as a fuel.

## TERMS TO KNOW

primary energy source
secondary source of
  energy

electricity
electric current
alternating current (AC)

direct current (DC)
electrolysis
fuel cell

# INTRODUCTION

Most of the forms of energy discussed in this text have one thing in common, they are used to produce electricity. Electricity is the cleanest and easiest form of energy to transport. It is also easy to manipulate and change into other forms of energy to generate light, heat, and power. Benjamin Franklin discovered lightning was a natural form of electricity and Thomas Edison harnessed it to produce light. In our modern world, electricity powers most of what we do. If you have ever experienced a power failure you suddenly realize how much we depend on electricity for our daily lives. However, natural forms of electricity are nearly impossible to harness. Both electricity and hydrogen are considered secondary sources of energy because they cannot be used directly to generate power. Hydrogen is mostly locked up within the water molecule and needs to be released in order for it to be used. Once available, hydrogen provides a clean, inexhaustible source of power that can produce both electricity and heat.

# ELECTRICITY

The main forms of non-renewable and renewable energy resources discussed in the preceding chapters are classified as **primary energy sources**. A primary energy source is a naturally occurring form of energy used directly to perform work. Typically we use these resources to provide heat such as using natural gas to heat water in your home, or burning gasoline in an automobile engine to move a car.

A **secondary source of energy** takes the energy generated from a primary source and transports it to where it can be used to perform work. The most common form of secondary energy used today is **electricity**. Electricity is the movement of electrons through a conductor. A conductor is a substance that allows its electrons to move easily from one atom to another (Figure 13-1). Good conductors are metals like copper or aluminum because they readily allow their electrons to move to a neighboring atom. The opposite of a conductor is an insulator. An insulator is a substance that does not allow its electrons to flow easily like wood, plastic, or dry air. Much of the things we use in our modern world are powered by electricity and the majority of primary sources of energy are used to generate electricity.

**primary energy source**—a naturally occurring form of energy used directly to perform work

**secondary energy source**—a form of energy used to transport the energy generated by a primary source

**electricity**—the movement of electrons through a conductor

## Production of Electricity

Electricity can be produced in four basic ways: statically, chemically, magnetically, and by using photovoltaics. Static electricity results when an electric charge builds up as two insulators are rubbed together. One insulator accumulates electrons and becomes negatively charged while the other becomes positively charged as the loss of electrons creates an imbalance

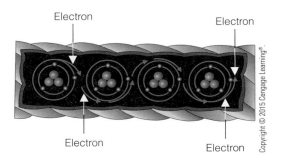

Electron                    Electron

Electron              Electron

**FIGURE 13-1** An illustration of a conductor of electricity.

of protons and electrons of an atom. The higher the charge difference is between the two, the greater the electric potential. Eventually when the two insulators touch, the electrons travel back to their source to equal out the charge and an electric current flows. This is what causes you to get a shock when you walk across a carpet in your socks and touch a doorknob. Your socks build up electrons along your body as they rub against the carpet. When you touch the doorknob, the electrons flow and give you a shock. The same principle is responsible for the production of lightning in the atmosphere (Figure 13-2). Opposite electric charges build up within clouds as wind rubs ice particles together. This causes positive electric charges to accumulate along the top part of the cloud, while negative charges build up near the base. Eventually a static discharge is released, forming lightning. The rapid heating of the air within the cloud by the sudden release of electricity causes the air to expand, creating thunder.

Electricity can also be generated by chemical reactions. This is how a battery works. The chemicals within a battery exchange electrons when they react with one another and begin the flow of electricity through a circuit. One of the most commonly used batteries that power many of our portable electronic devices are known as alkaline batteries. These use the exchange of electrons between magnesium dioxide and zinc powder within

**FIGURE 13-2** Production of lightning.

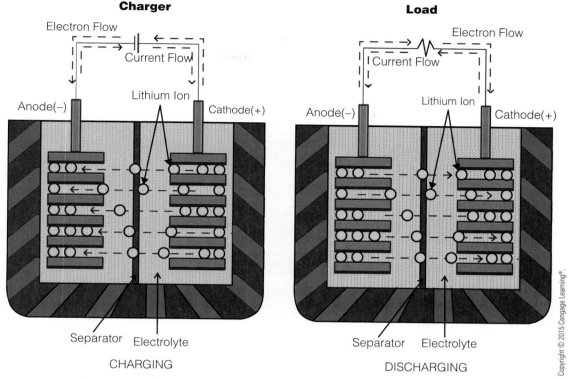

**FIGURE 13-3** Diagram of a lithium ion battery.

a potassium hydroxide electrolyte. These batteries are disposable, meaning once they release their charge, they must then be recycled. A common rechargeable battery is known as the lithium–ion battery (Figure 13-3). These use a lithium oxide compound cathode, electrolyte, and a graphite anode. Once the circuit is connected, lithium ions flow from the cathode through the electrolyte to the anode. When the battery is recharging, the electrical current reverses the flow of lithium ions back to the cathode, and the battery is recharged. Because the flow of electrons during the reaction within a battery is in one direction, all batteries generate only DC current. See Chapter 15 for more information concerning batteries.

Photovoltaic cells can also produce electricity using the photoelectric effect (Figure 13-4). Photons released from the sun strike a semiconductor like silicon that causes electrons to begin to flow. This is the way solar panels produce electricity (see Chapter 8).

Another form of generating electricity is by using magnetism. Magnetism is an attractive or repulsive force generated by the spinning of electrons as they move around the nucleus of an atom. Certain elements, like iron, generate magnetic fields with strong attractive and repulsive forces. When a magnetic field is passed over a good conductor, it can cause the electrons within the conductor to flow, therefore generating an **electric current**.

**electric current**—the movement of electrons through an electric circuit

## Electric Current

An electric circuit is a closed loop through which electrons flow to do work. The simplest electric circuit consists of a power source, a conducting wire, and a load (Figure 13-5). The power source begins the flow of electrons that travel through the conducting wire to arrive at the load that converts the electrical energy into work.

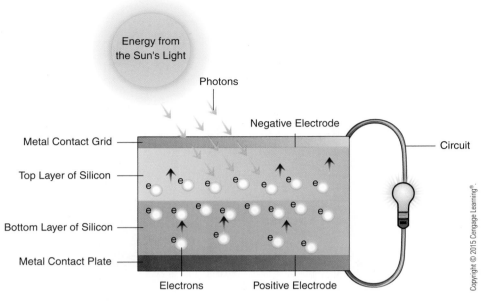

**FIGURE 13-4** Diagram of a photovoltaic cell.

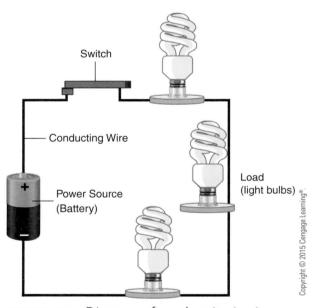

**FIGURE 13-5** Diagram of an electric circuit.

Electric currents are measured by two main units, the amp and the volt. An amp is a measure of how many electrons flow through a conducting wire while the volt is the force at which the electrons are moving. If you multiply the amount of amps by the amount of volts you can determine the wattage of the electric current. Watts are a common unit of measure for determining the power of an electric current that represents the rate the electrons are converted into another form of energy. Electricity is commonly recorded using the watt hour. A watt hour is the amount of wattage used by a load over the time period of an hour. For example if you use a 60 watt light bulb for two hours, then you have used 120 watt hours of electricity. Because we use so much electricity every day, the kilowatt hour is used to measure electrical consumption. One kilowatt hour is equal to one thousand watt hours. Another term associated with electricity is the ohm. An ohm is the amount of resistance the electrons experience within a circuit. When the resistance is high, the electrons convert their energy into heat and light. This is how an incandescent light bulb works. The wire in the light bulb has a high resistance; therefore it glows as the electrons collide, generating heat and light. Electricity is commonly produced by using the interaction of a magnetic field with electrons in conducting wires within a generator. A generator is a device that uses coils of conducting wire held around a rotating magnet. As the magnet spins, the electrons within the wire to flow as an electric current. There are two types of electric currents that can be produced by generators, **alternating current** or **AC**, and **direct current** or **DC**.

Alternating current (AC) alternates the flow of electrons between positive and negative poles many times a second. The rate at which the flow reverses for an AC current is known as its frequency that is measured in hertz. Common household electric currents run at 60 hertz, meaning the flow of electrons complete one complete cycle reversal 60 times each second. Most electrical appliances use AC current because it is more efficient with long distance transportation of electricity. AC current arrives in most homes at 120 volts which is what is required for the operation of most appliances.

Direct current (DC) generates an electric current that flows in one direction; negative to positive. DC generators can supply only one level of voltage that drops as you increase the distance of the circuit. AC generators can generate extremely high voltages that can be reduced by a transformer. This enables the electric current to travel far distances and then the voltage can be stepped down to a useable level at the load.

**alternating current (AC)**—an electric current where the flow of electrons reverses between positive and negative poles many times a second

**direct current (DC)**—an electric current that flows in one direction, from the negative to the positive

## Generation and Distribution of Electricity

Electricity used by homes, businesses, and industry is generated by a network of generating stations. The majority of power stations are classified as thermal generating plants that use heat to create steam that turns a turbine generator. In the United States during 2012, more than 87% of electricity generated used thermal power plants; 38% was made by burning coal, 30% by natural gas, and 19% by nuclear power (Figure 13-6). The efficiency

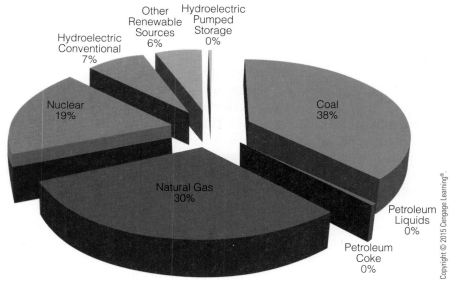

Net Electricity Generation by Energy Source 2012

**FIGURE 13-6** Net electricity generation by energy source, 2012.

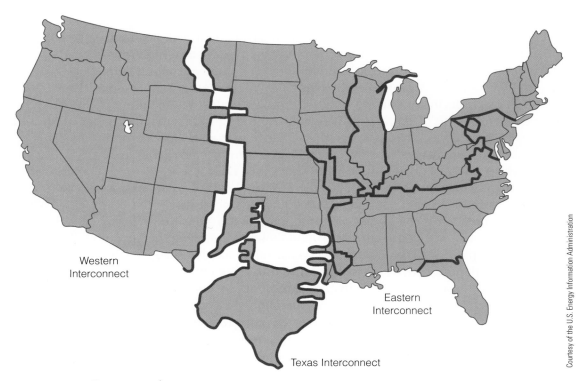

**FIGURE 13-7** Power grid.

of these types of thermal generating plants is typically between 35%–45%. This means 3%–45% of the fuel's energy is used to produce electricity, while the remaining 55%–65% is lost in the process to heat. New combined cycle gas turbine generators are able to reach higher efficiencies of around 60%. Kinetic energy generating plants produce a little more than 12% of our electricity and include hydroelectric and wind turbines.

The total amount of electricity used in 2011 in America was more than 3.8 billion kilowatt hours. All of this electricity is fed into a large network of transmission lines known as the electrical power grid (Figure 13-7). The power grid transports electricity throughout the country using high voltage cables at voltages between 115 and 765 thousand volts. This high voltage allows for the long distance transport of electricity. Substations and transformers then step down the voltage to levels used in homes, businesses, and industries. The grid is divided into three different regions that deliver power where it is needed. These regions include the Eastern Interconnection, Western Interconnection, and Texas Interconnection. Both the eastern and western grids connect with Canada, and the Texas and western have limited connection with Mexico. Peak demand is the time of the day when the most amount of electricity is needed. This usually occurs during the middle of the day. Management of the grid involves using the most efficient of generating stations that can supply the current load demand. These are usually nuclear or hydroelectric power plants. As more electricity is needed, other of generating stations are brought online to produce the required power.

The price of electricity is dependent on many factors including the cost of the fuel of the power plant, power plant maintenance, and transmission grid maintenance. In 2012 the average retail price of electricity in the United States was 9.65 cents per kilowatt hour. In the United States, the grid system is a centralized system of power distribution, meaning large power plants produce the electricity that is then fed out to end-users. Because renewable

## CAREER CONNECTIONS

### Journey Lineman

A Journey Lineman works on the main electrical transmission lines that distribute power. This job requires a love of the outdoors and of high places. Journey Lineman installs, maintains, and repairs high voltage electrical lines in all kinds of weather. Their job requires travel to remote places were electric lines need repair. Some jobs even require the use of a helicopter! The work can be physically demanding because it usually involves climbing utility poles and towers with tools to access the electrical grid. Jobs in this field exist in all parts of the country and require a high school diploma and specialized training by utility companies. Apprenticeships are also common to learn this valuable skill.

forms of energy are becoming more widespread, a decentralized power system may be developed in the future as individual users set up their own electrical power sources.

## HYDROGEN

Another form of secondary energy is hydrogen. Hydrogen is the most abundant and simplest element in the universe. The sun is composed of mostly hydrogen, and through the process of nuclear fusion produces energy that sustains the Earth's ecosystems. On Earth, hydrogen is an important part of many compounds including hydrocarbons like methane ($CH_4$), carbohydrates such as glucose sugar ($C_6H_{12}O_6$), and of course water ($H_2O$). In its pure form hydrogen is a diatomic gas ($H_2$). Diatomic means it is made up of two atoms. In this form, hydrogen is a combustible gas and when burned, produces heat and water vapor as a byproduct. Because of this, hydrogen is the cleanest burning of fuels. Today, hydrogen gas is used mostly for industrial processes such as fertilizer production and rocket fuel. Rockets use liquid hydrogen mixed with oxygen in a combustion chamber to produce a large amount of thrust to launch vehicles into space.

### Hydrogen Production

Most all hydrogen used for these purposes is extracted from methane gas in a process known as steam methane reforming. Methane is exposed to high-pressure steam that separates the hydrogen atoms from the carbon atoms. This energy intensive process produces carbon dioxide and requires the use of a non-renewable resource. Although steam methane reforming is the least expensive way to produce hydrogen today, there are other ways to isolate hydrogen. **Electrolysis** is a method of hydrogen production that uses an electric current that is passed through water to separate the water molecule ($H_2O$). When electricity is passed through water, hydrogen bubbles form at the negative terminal, also known as the cathode, while oxygen bubbles collect at the anode or positive terminal (Figure 13-8). The device that performs electrolysis is called an electrolyzer. The lowest optimum voltage of electricity required to produce hydrogen is 1.47 volts at around

**electrolysis**—the use of an electric current to separate water into hydrogen and oxygen gas

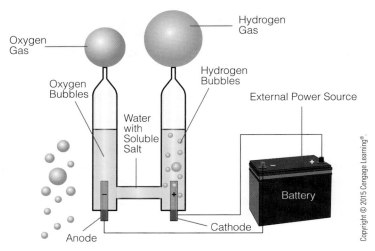

**FIGURE 13-8** Electrolysis.

68°F (20°C). This process becomes more efficient when the temperature of the water within the electrolyzer is increased. New electrolyzer technologies are being developed that use high temperature water during electrolysis to increase the efficiency of the process. Electrolysis is the easiest and cleanest method of producing hydrogen gas. The use of photovoltaic cells or other renewable resources to provide the electricity for electrolysis is an appealing idea, although currently it is still more expensive than extracting hydrogen from natural gas.

There are methods of using bacteria and algae that can produce hydrogen. These photobiologic methods are currently being researched. Sunlight is the energy source for organisms to produce hydrogen. One new development in the search for creating an efficient way to produce hydrogen is called the artificial leaf. The artificial leaf is being developed by Harvard scientist, Daniel Nocera. The leaf is actually a silicon based wafer that when placed in water produces hydrogen gas from sunlight.

Another method of hydrogen production is the gasification process. Hydrocarbons like coal or biomass can be heated in a low-oxygen, high-pressure steam environment to produce hydrogen gas. This process is energy intensive and produces carbon dioxide as a byproduct.

## Storage and Transport

Once the hydrogen is produced it must be stored and then transported. The storage of hydrogen is one of the challenges that widespread use of hydrogen faces. Hydrogen can be stored in three main ways. The first is as a pressurized gas within a tank. Current hydrogen storage tanks hold hydrogen at pressures between 3,600–10,000 pounds per square inch. The long distance transport of pressurized hydrogen through pipelines, like those used for natural gas, may become a means of distributing hydrogen in the future. In this case the hydrogen would be transported at pressures nearing 1,000 psi. Hydrogen can also be stored in a liquid form within a tank. For this method to be used however, the gas must be cooled to below −424°F (−253°C). This is known as cryogenic storage. Finally, the last method of hydrogen storage is within materials that absorb or bond with the hydrogen. Special tanks lined with or containing hydrogen absorbing compounds known as hydrides, can hold the hydrogen in a solid form, then when they are heated the hydrogen is released. This last method is being considered for hydrogen storage tanks used in motor vehicles because of the space requirements and potential dangers associated with storing pressurized hydrogen gas.

## Hydrogen as an Energy Carrier

Hydrogen can be used as an energy carrier in three basic ways. The first is in the form of a combustible gas. Hydrogen is flammable in the presence of oxygen and can be burned just like natural gas. The benefit of using hydrogen as a gaseous fuel is that the only byproduct is water vapor. Therefore the burning of hydrogen does not contribute to the production of greenhouse gases. A hydrogen internal combustion engine is currently under development that runs on hydrogen gas alone, or as a mixture of hydrogen and natural gas. This type of engine is basically the same as a gasoline internal combustion engine with only a few modifications. Burning hydrogen in these prototype engines resulted in the reduction of harmful emissions of nitrogen oxides and carbon dioxide, along with increased fuel efficiency. Another way hydrogen can be used is with a **fuel cell** to produce electricity (Figure 13-9).

A fuel cell creates the exact opposite reaction of electrolysis. The Polymer Electrolyte Membrane or PEM fuel cell, also known as a Proton Exchange Membrane fuel cell uses a catalyst to combine hydrogen gas with atmospheric oxygen to produce an electric current. Fuel cells act much like batteries, producing electricity as a result of a chemical reaction. The PEM fuel cell is an efficient way of producing an electric current with a 60% average efficiency. The other 40% is usually lost as heat. Each fuel cell can generate about 1 watt of power. Typically fuel cells are combined into what are known as fuel cell stacks. The PEM fuel cell is being developed for use in electric vehicles because of its reliability and efficiency. Gasoline engines in most

**fuel cell**—a device that uses a special membrane that combines hydrogen and oxygen gas to produce an electric current and water vapor as a byproduct

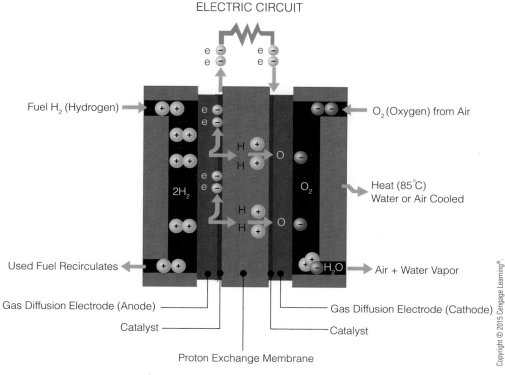

**FIGURE 13-9** Diagram of a hydrogen fuel cell.

cars have efficiencies below 20%. Fuel cells have been in use since the 1960s by NASA and have a great potential as becoming a clean source of electricity.

The final method hydrogen can be used as a fuel source is as an additive to gasoline or diesel. The injection of hydrogen gas into gasoline internal combustion or diesel engines is known as hydrogen boosting. The use of a device called a hydrogen gas reformer that takes hydrogen from the fuel itself, then injects it with the fuel, has greatly increased the efficiency and reduced harmful emissions of traditional engines. Another method of hydrogen boosting uses a device that creates hydrogen and oxygen from the electrolysis of distilled water. These gases are injected in the fuel and demonstrated increased efficiency and lowered harmful emissions. Whether you use hydrogen as a combustible fuel, or in a fuel cell to produce electricity, the fact that the only byproduct produced is water makes this fuel an attractive alternative energy resource. However, the challenges of production and delivery must be improved in order to make hydrogen the clean fuel of the future.

## REVIEW OF KEY CONCEPTS

1. A primary energy source is a naturally occurring form of energy used directly to perform work. Typically we use these resources to provide heat such as using natural gas to heat water in your home, or burning gasoline in an automobile engine to move a car. A secondary source of energy takes the energy generated from a primary source and transports it to where it can be used to perform work.

2. Electricity is the movement of electrons through a conductor.

3. An electric circuit consists of a power source, a conducting wire, and a load.

4. The common units used to measure electricity include amps, volts, watts, ohms, watts, and watt hours.

5. A generator produces electricity by moving a magnet over conducting wires.

6. Alternating current (AC) alternates the flow of electrons between positive and negative poles many times a second. Direct Current (DC) generates an electric current that flows in only one direction, from the negative to the positive. The advantage of using AC is it can be transported over long distances.

7. Batteries generate electricity by using chemical reactions that exchange electrons when they react with one another and that begins the flow of electricity through a circuit.

8. The power grid consists of generating stations, transmission lines, substations, and transformers.

9. Hydrogen can be produced by using steam methane reforming, electrolysis, fossil-fuel gasification, and biologically.

10. Hydrogen can be used as an energy source by burning it directly, to generate electricity in a fuel cell, or as an additive to gasoline or diesel engines.

11. Three methods used to store hydrogen include pressurizing it as a gas, cooling it to a liquid, or bonding it as a metal hydride.

12. Two advantages of using hydrogen include that it only produces water vapor as a byproduct, and it is abundant on the Earth.

## CHAPTER REVIEW

### Short Answer

1. What is the difference between a primary and secondary energy source?

2. Define the term, electricity.

3. What are the three basic parts of an electric circuit?

4. Which units are commonly used to measure electricity?

5. How is an electric current produced by a generator?

6. What is the difference between AC and DC current, and what is the advantage of using AC over DC?

7. How does a battery produce electricity?

8. What are the main components of the electrical grid?

9. Explain the four processes used to produce hydrogen gas.

10. What are three ways that hydrogen can be used for energy?

11. What are three methods used to store and transport hydrogen?

12. Explain two benefits of using hydrogen as a fuel.

13. What are some ways that agriculture uses electricity?

### Energy Math

1. How many watts would be required to run an appliance that draws 120 volts, at 15 amps, and how much would it cost to run for two hours if your electricity costs 10 cents per kilowatt hour?

2. Determine the amount of kilowatt hours of electricity that was produced by burning coal in the United States during 2011 assuming that coal produced 38% of our total use.

3. How many hydrogen atoms would be produced if you used electrolysis on 1,400 water molecules?

### Multiple Choice

1. Which of the following is a primary energy source?
   a. Hydrogen
   b. Coal
   c. Electricity
   d. A fuel cell

2. Which of the following is a secondary energy source?
   a. Coal
   b. Electricity
   c. Oil
   d. Natural gas

3. How many watts are in a kilowatt?
   a. 1
   b. 10
   c. 100
   d. 1,000

4. Which unit of electricity represents how many electrons are moving past a particular point?
   a. Volt
   b. Amp
   c. Watt
   d. Hertz

5. What type of electric current flows in only one direction?
   a. AC
   b. Hz
   c. DC
   d. kW

6. How many watts are required to run an appliance that draws 120 volts at 15 amps?
   a. 8 watts
   b. 105 watts
   c. 135 watts
   d. 1800 watts

**7.** Which method produces the most hydrogen in the United States?
a. Electrolysis
b. Electrolyzing
c. Photobiotic
d. Steam reforming

**8.** What is the byproduct of burning hydrogen gas?
a. Carbon dioxide
b. Methane
c. Water vapor
d. Nitrogen oxides

**9.** What device uses hydrogen and oxygen to produce electricity?
a. A fuel cell
b. A battery
c. A reformer
d. A hydrogen internal combustion engine

## Matching

*Match the terms with the correct definitions*

a. primary energy
b. secondary source of energy
c. electricity

d. electric current
e. alternating current
f. direct current

g. electrolysis
h. fuel cell

**1.** _____ The movement of electron through a conductor.

**2.** _____ A device that combines hydrogen and oxygen gas to produce an electric current and water.

**3.** _____ An electric current where the flow of electrons reverses between positive and negative poles many times a second.

**4.** _____ A method of hydrogen production that uses an electric current that is passed through water to separate the water molecule ($H_2O$).

**5.** _____ A naturally occurring form of energy.

**6.** _____ An electric current that flows in one direction, from the negative to the positive.

**7.** _____ Taking the energy generated from a primary source and transporting it to where it can be used to perform work.

**8.** _____ The flow of electrons through a conductor.

# UNIT IV

# The Science of Energy and Work

## TOPICS TO BE PRESENTED IN THIS UNIT INCLUDE:

- Thermodynamics
- The chemistry of energy
- Work and machines
- Heat engines
- Energy flow through living systems

## OVERVIEW

Energy is a fundamental aspect of the universe. Understanding its basic principles and how energy drives the living and non-living systems of our planet is vital in order to improve the way it can be used by our technological society. The basic principles of thermodynamics describe the different forms of energy and how they are transformed. Using this knowledge to create machines and engines that perform work can make the use of energy by society more efficient.

## CHAPTER 14

# The Laws of Thermodynamics

## KEY CONCEPTS

*After reading this chapter, you should be able to:*

1. Define the terms energy, work, and power.
2. Explain the first law of thermodynamics.
3. Describe the difference between potential and kinetic energy.
4. Identify the eight forms of energy.
5. Explain the second law of thermodynamics.
6. Identify the seven different forms of electromagnetic radiation.
7. Explain the five ways in which electromagnetic energy can interact with matter.
8. Describe the relationship between wavelength and energy of electromagnetic waves.
9. Describe how conduction transfers energy.
10. Explain the process of convection and the formation of a convection cell.
11. Explain radiation and the relationship to electromagnetic waves.

## TERMS TO KNOW

energy
work
power
kinetic energy
potential energy
gravitational energy
mechanical energy
electrical energy

electromagnetic energy
Absolute Zero
chemical energy
thermal energy
nuclear energy
sound energy
thermodynamics
first law of thermodynamics

second law of thermodynamics
entropy
conduction
thermal conductivity
R-value
convection

## INTRODUCTION

In order to develop new sources of power and improve the efficiency of our modern energy resources, a full understanding of the fundamental laws of energy is necessary. The rules by which energy is governed in the universe are known as the laws of thermodynamics. These universal principles define all aspects of energy production, transfer, and use in both living and non-living systems.

## ENERGY

The definition of **energy** itself can be somewhat complicated. Energy used in everyday life can mean a variety of things, but in science the definition of energy is simply the ability to perform work or cause change. A simple illustration of energy would be the movement of your pencil from one location on your desk to another. In order to move the pencil, you must use energy. In this case the energy comes from within your body. The **work** that was done moved the pencil from one location to another. Because energy is related to work, an understanding of the scientific definition of work is also needed. Once again, like the term energy, work has a variety of meanings. In science, work is defined as a change in position caused by a force. Therefore you performed work on your pencil when you moved it to a different location. A force is simply the power and direction needed to move something. The force that caused the work came from the **power** contained within the muscles of your hand and arm. The term power refers to the rate at which work is being done. Once again, using the pencil analogy, if you slowly move your pencil from one side of your desk to another, then you have used a small amount of power to do so. However, if you quickly move your pencil, then you have used more power. The common unit of power used for energy is the watt.

In Chapter 13 you learned the watt is a unit of electrical power that can be derived by multiplying the amount of volts in an electric current by its amperage. In physics, the definition of the watt is more basic; it is equal to 1 joule per second. The joule is a common unit used to measure energy like the calorie. One calorie is the amount of heat energy required to raise 1 milliliter of water by 1 degree celsius. One calorie is equal to 4.1868 joules. The units used to describe energy are outlined in Table 14-1. One thing is clear when looking in depth at energy, work, and power. There has always been a constant flow and transfer of energy in the universe. The laws of thermodynamics are designed to understand the transfer of energy and its nature.

### Forms of Energy

Before you can understand thermodynamics, you must first understand the different forms energy can come in. Generally speaking, energy is divided into two basic types, kinetic and potential. **Kinetic energy** is energy being used to perform work. An internal combustion engine that is moving a car down the road powered by gasoline is an example of kinetic energy.

**energy**—the ability to perform work or cause change

**work**—a change in position caused by a force

**power**—the rate that work is being done

**kinetic energy**—energy that is being used to perform work

## TABLE 14-1 Common Units of Energy

| | | |
|---|---|---|
| 1 joule (J) | = | the force of 1 newton acting through 1 meter |
| 1 watt (W) | = | the power of 1 joule of energy per second |
| watts | = | amps × volts (1 watt = 1 amp flowing through 1 volt) |
| 1 kilowatt (kW) | = | 1,000 watts |
| 1 megawatt (MW) | = | 1 million watts |
| 1 gigawatt (GW) | = | 1 billion watts |
| 1 terrawatt (TW) | = | 1 trillion watts |
| 1 kilowatt hour (kWh) | = | 1,000 watt hours |
| 1 calorie (cal) | = | 4,184 joules |
| 1 kilocalorie (Kcal) | = | 1,000 calories (1 calorie in food is actually 1 Kcal) |
| 1 barrel | = | 42 gallons |
| 1 British Thermal Unit (Btu) | = | the amount of energy to raise 1 pound of water by 1 degree Fahrenheit |
| 1 Btu | = | 1,055 joules or 252 calories |
| 1 Quad | = | 1 quadrillion Btu ($1 \times 10^{15}$ Btu) |
| 1 therm | = | 100,000 Btu |
| 1,000 kWh | = | 1,000 watt hours or 3.41 million Btu |
| 1 horsepower (hp) | = | 745.7 watts |

**potential energy**—energy that has the ability to perform work, but is not being used

**Potential energy** is energy that is "stored" and has the ability to perform work, but is not being used. Examples of potential energy include water being held behind a dam that can be used to power an electrical turbine, or a snowboarder at the top of a hill ready to carve down the slope (Figure 14-1). Along with these two general types, energy is also classified as being one of eight different forms.

**gravitational energy**—the attractive force all objects in the universe have toward one another

**Gravitational energy** is the attractive force all objects in the universe have toward one another. The force of gravity is dependent on the mass of the object, so the greater the mass, the greater the gravitational force. Gravity is what causes objects to fall down on the earth (Figure 14-2). One of the most pressing questions being researched by modern physicists is, "What force actually causes gravity?" Believe it or not, no one is sure! We know much about the effects and strengths of gravity, but not what actually causes it. One of the most important aspects of gravity is that the force of gravity is dependent on the distance between the two objects. This basically means the closer you get to an object, the stronger the gravitational force. Gravitational energy is what drives our modern hydropower resources (see Chapters 9 and 10), by causing water to flow down from areas of higher elevation.

**mechanical energy**—the energy produced by physical movement

Another form of energy is **mechanical energy**. Mechanical energy is the energy produced by physical movement. The force of the wind moving a wind turbine or the force used to lift a pencil from your desk are both examples of mechanical energy (Figure 14-3). Many modern engines and power plants use mechanical energy to produce electricity.

**electrical energy**—a form of energy that is the result of the movement of electrons through a conductor

**Electrical energy** is a form of energy that is the result of the movement of electrons through a conductor and is a major energy source of our modern world (Figure 14-4).

Potential

Kinetic

(A)

(B)

**FIGURE 14-1** (A) Potential versus kinetic energy. (B) Water flowing over a dam is an example of kinetic energy.

**FIGURE 14-2** The force of gravity will cause fruit to fall from a tree.

**electromagnetic energy**—energy emitted from the oscillations of the electrons surrounding the nucleus of an atom

**absolute zero**—the theoretical temperature when all atomic motion stops. This occurs at −459.67°F (−273.15°C)

**chemical energy**—the potential energy stored within the bonds between atoms that make up compounds

**thermal energy**—the kinetic energy of motion of vibrating atoms, also known as heat energy

**nuclear energy**—energy that is associated with the forces that bind the nucleus of atoms together

Radiant energy, also known as visible light, is a kind of **electromagnetic energy**. Electromagnetic energy is the energy emitted from the oscillations of the electrons surrounding the nucleus of an atom. All atoms in the universe having a temperature above that of **Absolute Zero** emit electromagnetic energy. Absolute Zero is the theoretical temperature at which all atomic motion stops and no electromagnetic energy is given off by atoms. This occurs at −459.67°F (273.15°C). The electromagnetic energy given off by atoms is temperature dependent, meaning the higher the temperature, the greater the electromagnetic energy produced.

Another form of energy within the universe is **chemical energy**. Chemical energy is the potential energy stored within the bonds between atoms that make up compounds. The energy within food is a form of chemical energy, as is that of a battery, and fuels like natural gas, gasoline, and coal. When certain compounds are broken down whether by combustion or as a result of a chemical reaction, energy is released.

**Thermal energy**, also known as heat energy, is the kinetic energy of vibrating atoms. As the temperature of a substance is increased, its kinetic energy also increases (Figure 14-5). Substances that are hotter have atoms that are vibrating faster than cooler atoms. The use of thermal energy as a means to provide power is widely used in our society today to produce electricity, propel motor vehicles, prepare food, and to warm our homes.

**Nuclear energy** is a type of energy associated with the forces that bind the nucleus of atoms together. These forces are known as the strong and weak nuclear forces. The strong force holds the protons and neutrons

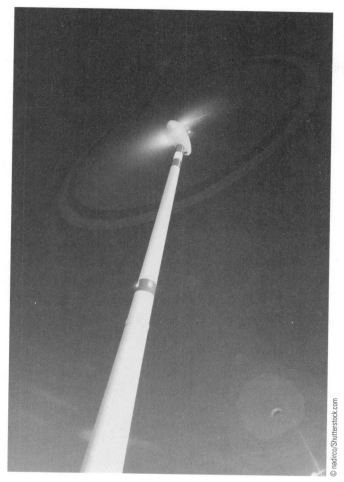

**FIGURE 14-3** Wind blowing a wind turbine is an example of mechanical energy.

**FIGURE 14-4** A light bulb is an example of electrical energy.

**FIGURE 14-5** A campfire is an example of thermal energy.

within the nucleus together while the weak force causes the radioactive decay of a nucleus. The sun is powered by fusion, or the combining of two atoms, and is thus a form of nuclear energy. Nuclear power plants that produce electricity by generating heat from nuclear fission, or the splitting of an atom, is also a form of nuclear energy.

**sound energy**—energy that travels in the form of a wave in a solid, liquid, or gas

The final form of energy is **sound energy**. Sound energy, like electromagnetic energy, travels in the form of a wave; however unlike electromagnetic waves that can travel through empty space, sound waves need matter in which to travel (Figure 14-6). Much like throwing a rock in a pond that generates waves outward from the impact, sound waves travel out from their energy source. Sound waves can move through solids, liquids, and gases. On average, sound travels at about 767 miles per hour through air.

## THERMODYNAMICS

**thermodynamics**—the study of the movement of heat and energy

**Thermodynamics** is the study of the movement of heat and energy within the universe. The term itself comes from the Greek words "therm", meaning heat, and "dunamis", meaning power.

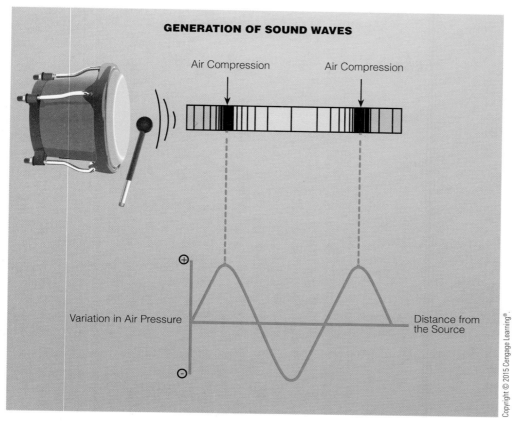

**FIGURE 14-6** Examples of sound waves.

## The First Law of Thermodynamics

The **first law of thermodynamics** states that energy cannot be created nor destroyed, it just changes from one form to another (Figure 14-7). For example, the sun is the source of more than 99% of all the energy used by living systems on earth, but energy is not created there. Recall from Chapter 7, the sun's energy comes from the process of nuclear fusion. This means the incredible amount of energy given off by the sun comes from the combining of two atoms of hydrogen. The result is the production of one atom of helium and energy. The energy is not "created" by the sun, but rather converted from the forces held within the original 2 hydrogen atoms. We may not realize it but the first law of thermodynamics happens around us every day. When you turn on the light in your room, the light bulb is not "creating" light energy. The energy has gone through many transformations from the power plant, through the electrical grid, and finally to the light bulb in your room.

**first law of thermodynamics—** the physical law that states energy cannot be created nor destroyed; it just changes from one form to another

## The Second Law of Thermodynamics

The **second law of thermodynamics** explains that with every transfer of energy, small amounts of heat are lost to the environment (Figure 14-8). This means when one form of energy is converted into another, the process cannot be 100% efficient. For example, all of the electrical energy that is powering the light bulb in your room is supposed to go to producing light energy; however much of this energy is lost as heat as it passes through the electrical wiring

**second law of thermodynamics—**the physical law explains that with every transfer of energy, small amounts of heat are lost to the environment

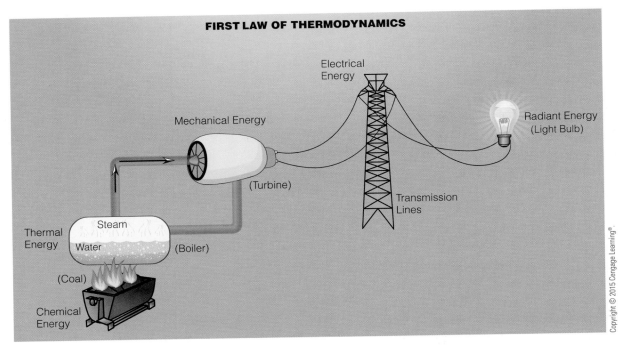

**FIGURE 14-7** The first law of thermodynamics.

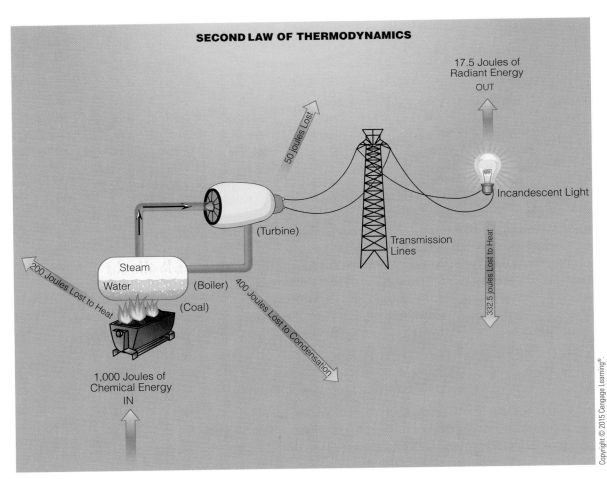

**FIGURE 14-8** The second law of thermodynamics.

and the light bulb itself. In fact, most incandescent light bulbs are only about 5% efficient. This means only 5% of the energy entering the light goes to lighting the room, with the remaining 95% being lost as heat. This is why incandescent light bulbs are being replaced by more efficient, compact fluorescent light bulbs. The energy that has been "lost" as heat is now "unusable" because it is spread out into the environment. The second law of thermodynamics explains this as **entropy**. Entropy is the increase of the disorder of a system, much like the way your room would look if you never cleaned it up. Over time, the disorder of a system is increased. The only way to decrease the disorder is to put energy back into the system. The second law is important when dealing with the efficiencies of energy systems. Reducing the amount of energy lost to entropy between transfer steps is an important way of making our energy resource more efficient. Many of our current energy producing processes are inefficient, like a steam turbine that is about 40% efficient or an internal combustion engine with efficiencies ranging between 15%–25%. The efficiency of a system can be calculated by dividing the energy produced by the total energy consumed, then multiplied by 100 to express it as a percentage. If an incandescent light bulb is supplied with 100 joules of electricity and produces only 5 joules of radiant energy, the efficiency of the light bulb is 5% (5/100 × 100). The low efficiency percentages of steam turbines and car engines show the importance of the second law of thermodynamics, and in the case of an incandescent light bulb, the efficiency of 5% is so low it makes the bulb more like a room heater instead of a light source!

**entropy**—the increase of the disorder of a system

## ELECTROMAGNETIC ENERGY

The transfer of energy via radiation involves electromagnetic energy that travels in the form of a wave through empty space, moving at a constant speed. The speed electromagnetic waves travel is known as the speed of light. This is equal to 186,000 miles per second (300,000 km/s). The speed of light is the fastest known speed achievable in the natural world. Because electromagnetic energy travels in the form of a wave, it comes in different forms based on their unique wavelengths. The highest energy, shortest wavelength form of electromagnetic energy is known as a gamma ray. The wavelengths of gamma radiation are smaller than 1 billionth of a centimeter. Gamma rays are emitted from extremely high-energy sources like the sun or a nuclear explosion. As you increase the wavelength of electromagnetic energy you move from higher energy forms, that include x-rays and ultraviolet, to lower energy forms like visible light, infrared, microwaves, and finally radio waves that have the longest wavelengths (Figure 14-9).

The earth receives energy from the sun in the form of electromagnetic waves mainly as ultraviolet and visible light. The different colors of the visible light spectrum are caused by the slight changes in the wavelength of their electromagnetic energy (Figure 14-10). Blue light has shorter wavelengths than red light for example.

The electromagnetic energy given off by the sun takes approximately 8 minutes to travel to the Earth at the speed of light. Plants have evolved to use this radiant energy through the process of photosynthesis, making this form of energy vital to life on earth. Electromagnetic energy interacts with matter in five basic ways (Figure 14-11). These include reflection, refraction, scattering, absorption, and transmission.

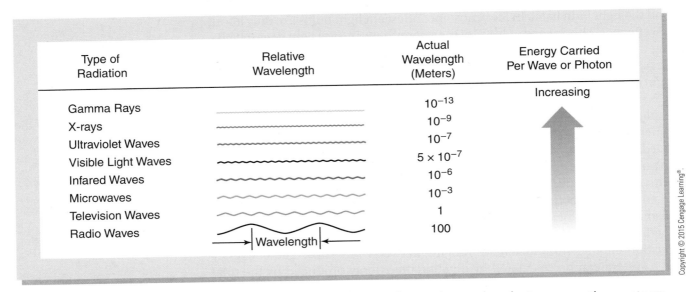

| Type of Radiation | Relative Wavelength | Actual Wavelength (Meters) | Energy Carried Per Wave or Photon |
|---|---|---|---|
| | | | Increasing |
| Gamma Rays | | $10^{-13}$ | |
| X-rays | | $10^{-9}$ | |
| Ultraviolet Waves | | $10^{-7}$ | |
| Visible Light Waves | | $5 \times 10^{-7}$ | |
| Infared Waves | | $10^{-6}$ | |
| Microwaves | | $10^{-3}$ | |
| Television Waves | | 1 | |
| Radio Waves | Wavelength | 100 | |

**FIGURE 14-9** A chart showing the different types of radiation that make up the electromagnetic spectrum, their wavelengths, and energy levels.

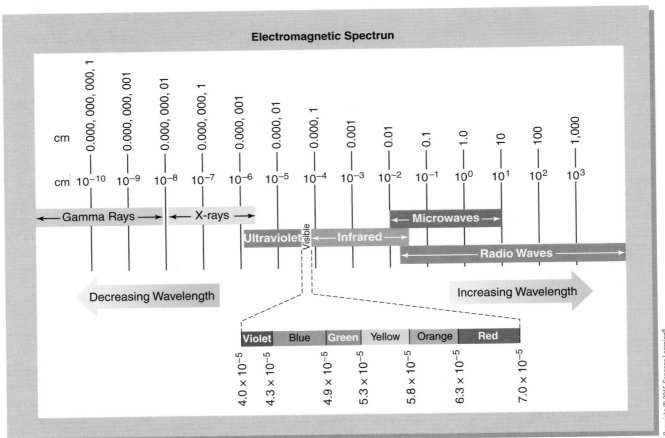

**FIGURE 14-10** The electromagnetic spectrum.

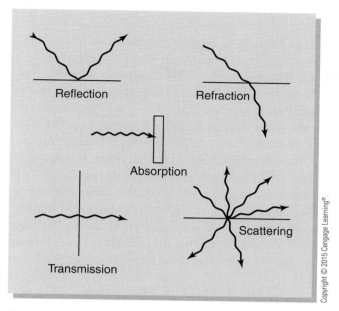

**FIGURE 14-11** Five ways in which electromagnetic energy interacts with matter.

Reflection occurs when a wave bounces off of an object. The reflection of your image in a mirror is the result of visible light waves bouncing off the reflective mirror material.

Refraction is the process by which a wave changes its direction of travel. When light enters water for example, the change from the gas to air causes the waves to redirect. This causes the "bending" of light. Placing a pencil in a half full glass of water illustrates refraction, because the pencil appears to bend at the interface between the air and the water.

Scattering occurs when an electromagnetic wave gets redirected in many directions. The scattering of blue wavelength light within the Earth's atmosphere is what causes the sky to appear blue during the day.

Absorption occurs when an electromagnetic wave is taken in by a substance, and causes its temperature to increase. Good absorbers of electromagnetic radiation are usually darker in color and have rough surfaces. It also tends to be that a good absorber of radiation is also a good emitter. This means if an object absorbs energy quickly, it will also give off that energy quickly, thus an object that heats rapidly will cool rapidly. Transmission is the final way electromagnetic energy can interact with matter. This is the process of a wave passing through a substance without interacting with it at all. An example of transmission is when light waves pass through glass, enabling you to see through it. The waves of light move through the molecules within the glass and do not interact with them at all.

## ENERGY TRANSFER

The flow of energy in, on, and around the earth drives all of its living or non-living systems. The transfer of energy within these systems occurs by conduction, convection, or radiation.

## Conduction

**conduction**—a method of energy transfer that involves the movement of energy by direct molecular contact

**Conduction** is a method of energy transfer that involves the movement of energy by direct molecular contact. The atoms that make up all matter are in a constant state of motion. This motion, or kinetic energy, is directly related to the temperature. When a substance is heated, the kinetic energy of the atoms within it increases. In a solid, the atoms begin to vibrate more rapidly as more thermal energy is applied. These vibrations cause atoms to bump into one another, thereby transferring the heat energy. The process is similar to using a bowling bowl to knock down bowling pins. When the bowling ball rolls down the lane and knocks down the first pin, hitting the pin next to it causing it to fall, and eventually the energy from the bowling ball is transferred to all the pins making them all fall—but only if you throw a strike of course! The same process transfers energy through matter. For example, if you hold an iron rod in your hand and place it in a fire; the heat energy from the flames will eventually travel through the rod to your hand (Figure 14-12).

The atoms located at the tip of the iron rod exposed to the flame begin to increase their kinetic energy quickly then transferred up the rod by the vibrations of the iron atoms. Generally, conduction is most efficient in solids and least efficient in gases. This is because the atoms are much closer together in a solid then they are in a gas, making the transfer of energy much easier.

**thermal conductivity**—the ability for a substance to conduct heat

The ability for a substance to conduct heat is known as its **thermal conductivity**. Thermal conductivity is somewhat related to electrical conductivity, because substances that are good electrical conductors are also good thermal conductors. The best thermal conductors are metals because the mobility of their electrons also helps them to transfer heat well. Non-metal elements, even in a solid form are not good thermal

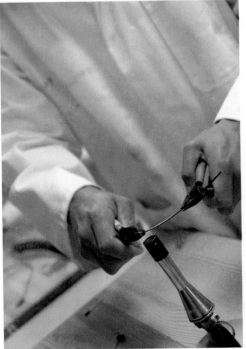

**FIGURE 14-12** Conduction of heat from a flame to a metal wire.

conductors because their atomic structures do not allow the efficient transfer of their kinetic energy. The worst conductors are gases because of the wide distance that exists between atoms. The best thermal conductor found naturally on the earth is a diamond. The close arrangement of carbon atoms within a diamond makes it an efficient conductor. Thermal conductivity is an important aspect of engineering. The loss of heat through building materials impacts the designs of many systems. A measure of the rate a material conducts heat is known as it U-value. Steel is a good thermal conductor with a U-value of about 1.2, while a 1 foot thick section of fiberglass insulation has a U-value of 0.03.

The ability for a substance to reduce heat loss and have a low thermal conductivity is known as thermal resistance. Thermal resistance, also known as an **R-value**, is an important part of building construction. Engineers and architects designing new buildings are concerned with a building material's R-value in order to prevent heat loss or gain through walls, windows, or roofs. The higher the R-value, the greater its thermal resistance; so a substance with a high R-value is a poor conductor. Typical fiberglass insulation has an R-value of about 4, while concrete has an R-value of around 0.8. The highest R-values are those associated with vacuum panels. These materials have spaces where a vacuum exists within them causing their R-value to range between 30 and 50. Table 14-2 lists various building materials and their R-values. Increasing the R-value of your house is a good first step to making your home more energy efficient. The relationship between good thermal conductors and good electrical conductors is also influenced by heat. When heat is applied to an electrical conductor, it decreases its electrical conductivity. This is because the electrons normally responsible for transferring the electrical energy are now in the process of transferring heat energy, thus decreasing its electrical conductivity. The opposite effect is also true, because electrons flow more readily when the conductor is cooled down. This is how superconductors work. Their temperatures are so low it allows for the efficient transfer of electrical energy.

**R-value**—the ability for a substance to reduce heat loss, also known as thermal resistance

## Convection

**Convection** is the third form of energy transfer on and around the earth. Convection is the transfer of energy as a result of differences

**TABLE 14-2  R-Values of Various Building Materials**

| Building Material | R-Value (ft²/°F/Btu/hour per inch thickness) |
| --- | --- |
| Fiberglass insulation | 3.14–4.3 |
| Brick | 0.2 |
| 12 inch concrete block | 0.1 |
| Plywood sheathing | 1.25 |
| Single paned glass | 0.9 |
| Double paned glass (1/2 inch air space) | 4.08 |
| Triple paned glass (1/2 inch air space) | 6.46 |
| 1 inch polystyrene foam board | 4–5 |
| Vacuum insulated panels | 30–50 |
| Metal insulating door | 7 |

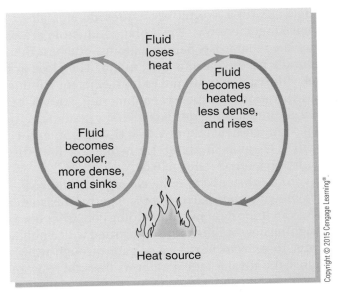

**FIGURE 14-13** The formation of a convection cell in a fluid as a result of differences in temperature and density.

**FIGURE 14-14** A lava lamp is really a convection lamp.

in temperature and density. Unlike conduction that takes place in all states of matter, convection can only occur in a fluid state that includes liquids or gases. When a fluid is heated its volume begins to increase as a result of thermal expansion. Thermal expansion is the increase in the volume of a substance caused by heating. As a substance's volume increases as a result of the increasing kinetic energy of its atoms, the increase in kinetic energy begins to push each atom farther apart (Figure 14-13). Because the heated substance has the same mass, but with a larger volume, its density decreases. The heated substance now has a lower density than its surrounding fluid and it begins to rise as a result because objects that are less dense float on objects that are greater in density. This causes the heated material to move upwards, away from the heat source, thus transferring the energy.

An example of convection can be illustrated by observing a "lava lamp". A lava lamp is a mixture of solid wax held within a glass container filled with a solution that has a density slightly lower than the wax (Figure 14-14). At room temperature the wax is a solid, and is greater in density than the solution so it sinks to the bottom of the container. At the base of the container is a light bulb. When the light bulb is turned on, it begins to heat the wax that starts to expand, decreasing its density. Eventually the wax heats to the point where its density becomes less than the density of the solution and it begins to rise to the top of the container. When the wax arrives at the top, a metal cone attached to the top of the glass container dissipates the heat from the wax (through conduction) to the air within the room. The wax begins to cool, its density increases again and sinks back down toward the light. So a lava lamp is actually a convection lamp. If you continue to watch your lava lamp you will notice the cycle of the wax rising and falling repeats itself over and over, forming a circular movement.

This is known as a convection cell, and is one of the most important energy transfer methods that drive many earth processes. In fact most all weather on earth occurs because of convection. This includes the formation of clouds, precipitation, and winds! Because the equator receives the greatest amount of incoming solar radiation throughout the year, the land and water absorb a lot of heat energy. This energy heats the air above it that becomes less dense and rises up into the atmosphere forming clouds and precipitation. Because the air at the equator is warm and less dense, it forms an area of low atmospheric pressure. As the air rises high into the atmosphere, it begins to cool. This causes its density to increase, and the air begins to sink back down toward the surface again. The air descends in the sub-tropics near 30 degrees north and south latitude, where it is now cooler and denser, forming an area of high atmospheric pressure at the surface. This creates two distinct pressure centers between the equator and the sub-tropics. Low pressure exists at the equator and high pressure exists at 30 degrees north and south latitude. Air then moves from the areas of high pressure to low, forming wind that returns the cooler air back to the equator where it is once again heated. So the atmosphere is acting like one big lava lamp near the tropics, forming a convection cell known as a Hadley Cell (Figure 14-15).

Convection not only affects the atmosphere and weather, but it is also an important energy transfer mechanism that drives plate tectonics. Plate tectonics is the theory that describes the Earth's crust divided into large

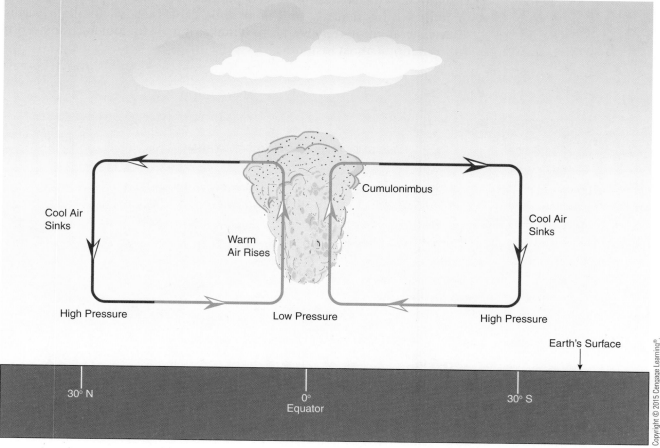

**FIGURE 14-15** Formation of a Hadley Cell.

sections or plates. These plates float and move relative to one another on a plastic-like rock material called the asthenosphere. In this case, extreme heat generated from the Earth's interior heats the asthenosphere causing it to become less dense and rise toward the crust. The superheated fluid rock breaks through the crust at divergent plate boundaries, where it forms new crust and dissipates its heat. The rock then cools and sinks back down again to be re-heated, forming a convection cell within the Earth. There are many of these convection cells underlying the Earth's crust that help transfer heat from the interior to the surface. It is this mechanism that leads to the formation of many geologic features and events.

## Radiation

The transfer of energy by radiation occurs via electromagnetic waves. Electromagnetic waves can travel through a vacuum and therefore do not need a medium for transfer like convection or conduction. Because of this quality, electromagnetic energy can move through empty space. Electromagnetic waves travel at the speed of light; the fastest known speed to occur in the universe (186,000 miles per second). More than 95% of the all the energy that drives the Earth's natural systems is transferred from the sun via radiation. Radiation is also an important way we transfer information in our modern technological world. Radios, cell phones, satellites, and computers all use different forms of radiation to transfer information by sending it through electromagnetic waves (Figure 14-16). For example cell phones and satellites use radio and microwave frequencies, and fiber optic cables use visible light waves. Because radiation travels so fast, it makes it a useful tool for transferring information almost instantly.

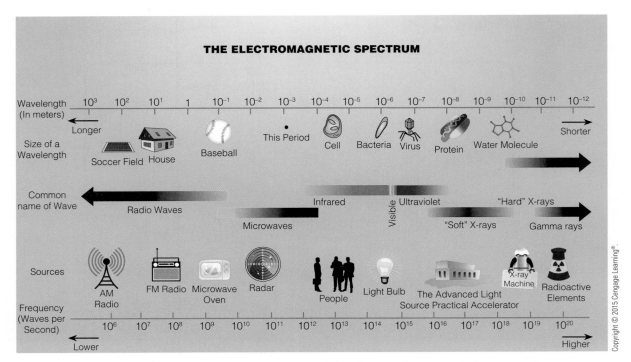

**FIGURE 14-16** Types of electromagnetic waves and the forms of energy they can radiate.

## REVIEW OF KEY CONCEPTS

1. Energy is the ability to do work, and work is defined as a change in position caused by a force. Power is the amount of work being done.

2. The flow of energy within the universe is governed by the fundamental laws of thermodynamics. The first law of thermodynamics states that energy cannot be created nor destroyed but just changes form.

3. Kinetic energy is energy being used to perform work. Potential energy is energy that is "stored" and has the potential to do work.

4. There are eight basic forms of energy in the universe. These include mechanical, chemical, thermal, gravitational, electrical, sound, nuclear, and electromagnetic.

5. The second law of thermodynamics states when energy transfers from one form to another, small amounts are lost as heat to the surrounding environment.

6. Electromagnetic energy is divided into seven different categories based on their wavelengths. These include gamma rays, x-rays, ultraviolet, visible light, infrared, microwave, and radio waves.

7. Electromagnetic energy interacts with matter in five ways: refraction, reflection, scattering, absorption, and transmission.

8. Electromagnetic energy travels in the form of a wave at a fixed speed known as the speed of light; 186,000 miles per second. Gamma waves are the shortest wavelength, highest energy form of electromagnetic energy, and radio waves are the longest wavelength, lowest energy form.

9. Conduction is the transfer of the kinetic energy of molecules by direct molecular contact.

10. Convection is the transfer of energy as a result of differences in temperature and density. Heat applied to a fluid (liquid or gas) causes it to become less dense and rise. The fluid then cools, becomes denser and sinks down to repeat the cycle again. This is known as a convection cell.

11. Radiation is the movement of energy by electromagnetic waves.

## CHAPTER REVIEW

### Short Answer

1. Define the terms energy, work, and power.

2. What is the first law of thermodynamics?

3. Describe the difference between potential and kinetic energy.

4. What are the eight forms of energy?

5. What is the second law of thermodynamics?

6. What are the seven basic forms of electromagnetic energy?

7. Explain the five ways electromagnetic energy can interact with matter.

8. Describe the relationship between wavelength and energy of electromagnetic waves.

9. How is heat transferred by conduction?

10. Describe how convection transfers energy.

11. How does convection lead to the formation of wind?

**12.** Identify the changes in the form of energy that occur from a coal-fired power plant to a light bulb in your house.

## Energy Math

**1.** How far does a sound wave travel every second?

**2.** How long would it take a radio wave to travel to Mars, which is about 141,610,000 miles from Earth?

## Multiple Choice

**1.** The ability to do work or cause change is:
a. power
b. watts
c. energy
d. force

**2.** The rate work is being done is known as:
a. power
b. watts
c. energy
d. force

**3.** The water stored behind a dam is an example of:
a. active energy
b. kinetic energy
c. potential energy
d. mechanical energy

**4.** An electric current powering a light bulb is an example of:
a. potential energy
b. mechanical energy
c. the first law of thermodynamics
d. the second law of thermodynamics

**5.** The loss of energy to the environment is an example of:
a. potential energy
b. mechanical energy
c. the first law of thermodynamics
d. the second law of thermodynamics

**6.** Eating a sandwich is an example of:
a. gravitational energy
b. mechanical energy
c. electromagnetic energy
d. thermal energy

**7.** This form of energy travels at the speed of light:
a. gravitational energy
b. mechanical energy
c. electromagnetic energy
d. thermal energy

**13.** Provide one example of energy transfer by radiation, conduction, and convection that occurs on the earth.

**3.** If the density of liquid water is 1.0 grams per cubic centimeter, determine if an object that weighs 160 grams and has a volume of 80 cubic centimeters would sink or float if placed in the water.

**8.** The redirection of an electromagnetic wave is called:
a. absorption
b. refraction
c. reflection
d. scattering

**9.** Which statement is correct about electromagnetic energy?
a. The greater the wavelength, the lesser the energy
b. The shorter the wavelength, the lesser the energy
c. Blue light is long wave radiation
d. Red light is short wave radiation

**10.** Which energy is related to the kinetic energy of an atom?
a. Gravitational energy
b. Mechanical energy
c. Electrical energy
d. Thermal energy

**11.** The handle of a metal frying pan heating up is an example of:
a. conduction
b. convection
c. radiation
d. potential energy

**12.** The mechanism that drives plate tectonics is partly caused by:
a. conduction
b. convection
c. radiation
d. potential energy

# Matching

*Match the terms with the correct definitions:*

a. thermodynamics
b. energy
c. work
d. power
e. first law of thermodynamics
f. kinetic energy
g. potential energy
h. gravitational energy

i. mechanical energy
j. electrical energy
k. electromagnetic energy
l. Absolute Zero
m. chemical energy
n. thermal energy
o. nuclear energy
p. sound energy

q. second law of thermodynamics
r. entropy
s. conduction
t. convection
u. thermal conductivity
v. R-value

1. _____ The flow of electrons through a conductor.

2. _____ When energy is transferred from one form to another, small amounts are lost as heat.

3. _____ The movement of waves through a solid, liquid, or gas.

4. _____ The change in position caused by a force.

5. _____ The attraction of one body in the universe to another.

6. _____ The study of the properties of energy.

7. _____ The theoretical temperature at which all atomic motion stops.

8. _____ The energy contained within the bonds between atoms.

9. _____ Stored energy that has the ability to do work.

10. _____ Energy that travels in the form of a wave trough empty space.

11. _____ The ability to do work or cause change.

12. _____ The energy contained within the nucleus of an atom.

13. _____ The increase in the disorder of a system.

14. _____ Energy that is doing work.

15. _____ Energy cannot be created nor destroyed, but just changes form.

16. _____ Energy produced by physical motion.

17. _____ The rate at which energy is doing work.

18. _____ The energy associated with the kinetic motion of atoms.

19. _____ The transfer of energy by direct molecular contact.

20. _____ The transfer of energy by changes in temperature and density.

21. _____ A measure of the thermal resistance of a substance.

22. _____ The rate at which a substance transfers heat energy via conduction.

## CHAPTER 15

# The Chemistry of Energy

## KEY CONCEPTS

*After reading this chapter, you should be able to:*

1. Define the term exothermic reaction and provide one example of an exothermic reaction.

2. Explain what is meant by the heat of combustion.

3. Define the term endothermic reaction and provide one example of an endothermic reaction.

4. Describe how energy is related to phase changes.

5. Identify which change in phase of water requires the greatest amount of heat.

6. Explain the oxidation/reduction reaction.

7. Identify the three main parts of a battery.

8. Describe how a simple voltaic cell produces an electric current.

## TERMS TO KNOW

| | | |
|---|---|---|
| thermochemistry | condensation | ion |
| exothermic reaction | electrochemistry | cation |
| combustion | oxidation | anion |
| endothermic reaction | reduction | anode |
| vaporization | electrolyte | cathode |

# INTRODUCTION

The chemical energy locked up in compounds plays an important role in not only supplying energy for living things on the earth, but for technological processes as well. Whether it is the energy you get from eating a sandwich during lunch, the ability to listen to your iPod, or driving a car down the road, all of these things rely on the energy stored within molecules. The study of these aspects of energy is known as thermochemistry and electrochemistry.

## THERMOCHEMISTRY

**Thermochemistry** is the study of the energy absorbed or given off during a chemical reaction. All compounds have potential energy stored within their molecular bonds. This potential energy can be converted into thermal energy as a result of a chemical reaction. A chemical reaction occurs when two or more substances are mixed together causing a change in their chemical structure. The study of thermochemistry looks at the way energy is related to chemical reactions. When two or more substances are mixed, the resulting chemical reaction can either give off heat or absorb heat.

An **exothermic reaction** occurs when heat is given off during a chemical reaction. Exothermic reactions take the potential energy stored within molecular bonds and convert them into thermal energy. Many of our energy resources use exothermic reactions. For example the burning of methane gas to supply heat is an example of an exothermic reaction. The chemical reaction for the combustion of one pound of methane gas is as follows: $CH_4 + 2O_2 = CO_2 + 2H_2O$ + 6,000 kilocalories of energy or 23,900 Btus. One Btu is approximately the same amount of heat energy produced by burning one match. The heat that is given off when a substance is burned within the presence of oxygen is called the heat of combustion (Figure 15-1).

**Combustion** is the process of producing heat and light by the rapid combining of oxygen with a substance. This is a common way we derive energy from many fossil fuels like coal, gasoline, and diesel fuel. The heat of combustion is not the only means to get energy from exothermic reactions however. Another type of common reaction that gives off heat energy occurs when iron is combined with oxygen, without combustion occurring. This is the way a simple hand warmer generates heat (Figure 15-2). Iron filings held within a porous pouch are enclosed within an airtight bag. Once the bag is opened, oxygen from the atmosphere combines with the iron in the bag to form iron oxide through the following reaction, $4Fe + 3O_2 = 2Fe_2O_3$ + 37 kilocalories of energy per 30 grams of iron. This is about 147 Btus or enough heat to keep your hand warm for a few hours.

Although most exothermic reactions involve the rearrangement of atoms or compounds, the exothermic reaction resulting from nuclear fission yields an incredible amount of heat. Recall from Chapter 7 the process of nuclear fission uses neutrons to split apart an atom of Uranium 235 to produce heat. In this case molecular bonds are not broken to generate heat, but rather the nucleus of the atom is broken. The exothermic reaction of

**thermochemistry**—the study of the energy absorbed or given off during a chemical reaction

**exothermic reaction**—a chemical reaction that produces heat

**combustion**—the process of producing heat and light by the rapid combining of oxygen with a substance

219

**FIGURE 15-1** Burning wood is an example of combustion.

**FIGURE 15-2** A hand warmer is an example of an exothermic reaction that takes place without combustion.

splitting one gram of Uranium 235 produces about 239,000 kilocalories or almost 1 million Btus!

Not all chemical reactions give off heat, but rather absorb it from the environment. These types of reactions are called **endothermic reactions**. An endothermic reaction takes in thermal energy from its surroundings. An example of an endothermic reaction is the combination of hydrogen and oxygen to form water, $H + 2O + 68$ kilocalories $= H_2O$. Here, the reaction of forming 18 ml of liquid water from gaseous hydrogen and oxygen requires nearly 270 Btus.

This concept is what makes water a carrier of energy, because the energy absorbed to make the water molecule can be released later when it is separated. This is how the hydrogen fuel cell works (see Chapter 13). Another example of an exothermic reaction is cold packs (Figure 15-3). Cold packs are often used to treat athletic injuries, where the application of cold is needed to reduce swelling. In this case, solid ammonium nitrate is held within a plastic container within a pack of water. When the plastic container is ruptured, the ammonium nitrate mixes with the water that absorbs heat, causing the temperature of the solution to reach nearly the freezing point of water. Only 1 gram of ammonium nitrate is needed to lower the temperature of 50 ml of water by 2°C.

The study of thermochemistry also looks at the heat gained or lost during phase changes. This type of energy is known as latent heat. A phase change occurs when a substance changes its state of matter. For example, **vaporization** is the phase change when a liquid turns into a gas. The heat of vaporization is the amount of energy required to convert a liquid into a gas. In order for a molecule of liquid water to vaporize, energy must be absorbed. For one gram of water to vaporize, you must put in 0.84 kilocalories of heat energy or 3.3 Btus. This endothermic reaction is used to create steam that powers an electric turbine. In the United States, burning coal is

**endothermic reaction**—a chemical reaction that takes in thermal energy from its surroundings

**vaporization**—the phase change of a liquid turning into a gas

(A) (B)

**FIGURE 15-3** (A) A cold pack is an example of an endothermic reaction. (B) Ammonium nitrate when mixed with water within a cold pack produces an endothermic reaction.

one of the main sources used to supply the energy required for vaporization. The process of melting, the phase change from a solid to a liquid, is also endothermic. In this case, less energy is required per gram of water to melt it. It takes only 0.08 kilocalories of heat energy to melt one gram of ice to water. Although it takes energy to melt or vaporize water, condensation and freezing are exothermic, because energy is released during these phase changes. When water vapor condenses, it releases the same amount of energy that was gained from vaporization. **Condensation** is the process of a gas turning into a liquid. The same is true for freezing because energy must be taken out of liquid water for it to turn into a solid. The gain or loss of energy as a result of the change in phase associated with vaporization and condensation is used in geothermal heat pumps as a source of heating or cooling (see Chapter 12).

**condensation**—the phase change of a gas turning into a liquid

## ELECTROCHEMISTRY

**electrochemistry**—the study of the chemical reactions that produce electricity

**oxidation**—when an atom or molecule loses an electron

**reduction**—when an atom or molecule gains an electron

**electrolyte**—a solution that contains ions

**ion**—an atom or molecule with an electric charge

**cation**—an ion with a positive electric charge

**anion**—an ion with a negative electric charge

**Electrochemistry** is the study of the chemical reactions that produce electricity. An electric current involves the movement of electrons through a conductor (see Chapter 13). The movement of electrons in a solution occurs as a result of an oxidation/reduction reaction. The term **oxidation**, when applied to electrochemistry, simply means that a substance is losing an electron. The opposite of this is known as **reduction**, that occurs when a substance gains an electron. A simple way to remember this is by using the following acronyms: LEO and GER. LEO means Lose Electrons Oxidation, and GER, Gain Electrons Reduction. It is the loss and gain of electrons in an oxidation/reduction reaction that moves electrons and produces an electric current. The chemical reactions used to produce electric currents in batteries are known as voltaic cells. A voltaic cell is simply an electrochemical reaction that generates an external electric current. This occurs as a result of the interactions of two different metals within an **electrolyte**. An electrolyte is a solution that contains ions. An **ion** is an atom or molecule that has an electric charge. There are two types of ions, one with a positive charge called a **cation**, and one with a negative charge called an **anion**. In a voltaic cell, the metal electrode used to produce the

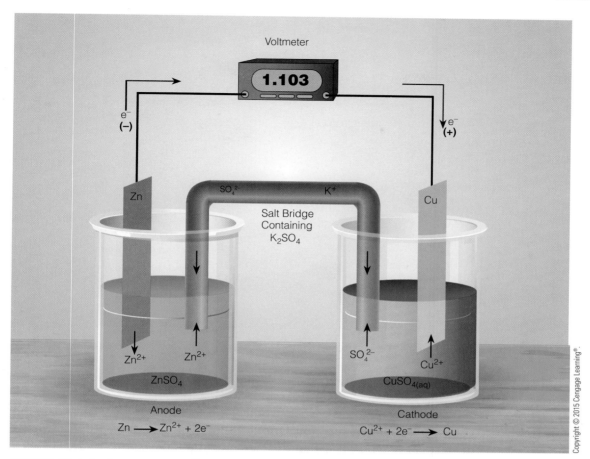

**FIGURE 15-4** A zinc-copper cell.

electrons is called the **anode**, and the metal electrode that gains the electrons is called the **cathode**.

An example of a simple electrochemical cell is the zinc-copper cell, also known as a voltaic cell (Figure 15-4). A piece of zinc metal (the anode) is placed in electrolyte solution of zinc sulfate ($ZnSO_4$). In another container, a piece of copper (cathode) is placed in an electrolyte solution of copper sulfate ($CuSO_4$). The two electrodes are connected by a wire, and the electrolyte solutions are joined by a salt bridge. A salt bridge is a "U" shaped tube filled with potassium nitrate, and plugged at each end with steel wool. The bridge is then placed in both containers, connecting them. The bridge is needed to allow the exchange of ions that build up near each electrode. Once the circuit is connected, the metallic zinc anode is oxidized by the zinc ions given off by the $ZnSO_4$ electrolyte, forming two electrons that travel through the wire to the cathode. At the cathode the copper ions in the $CuSO_4$ electrolyte are reduced as they take up the two electrons produced at the anode. This forms solid copper metal that is deposited on the cathode and the electric current is produced. The anode metal in an electrochemical cell is also known as the negative terminal of the battery and the cathode the positive terminal. As the reaction continues to produce an electric current, the zinc metal begins to lose mass as a result of its electrons being stripped away. Eventually the reaction will stop when either the zinc metal or copper ions in solution have been exhausted.

**anode**—an electrode that produces electrons

**cathode**—an electrode that gains electrons

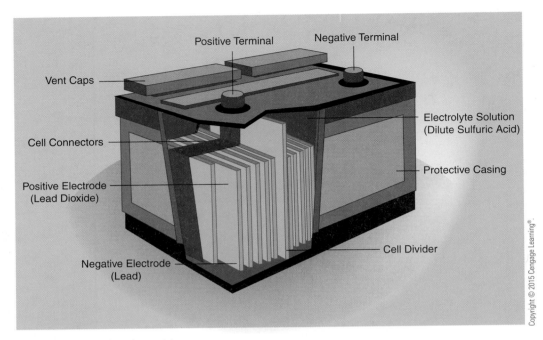

**FIGURE 15-5** A lead-acid battery.

The many batteries we use to power our technology are all based on the voltaic cell chemical reaction. One common type of battery, especially used to start car engines, is called the lead–acid battery (Figure 15-5). This is a popular battery because not only can it produce a powerful electric current, but it can also be recharged. A lead-acid battery has a cathode made of lead dioxide and an anode made of lead. The electrolyte solution is sulfuric acid. Electrons flow from the lead anode to the lead dioxide cathode producing an electric current. When power is applied to the lead–acid battery, the flow is reversed, sending the electrons collected at the cathode back to the anode. Lead–acid batteries can slowly degrade overtime as the lead oxide formed from the reduction reaction flake off and settle to the bottom of the battery compartment. This reduces the amount of charge that can be stored in the battery. Electrochemical cells that use liquid electrolyte solutions are known as wet cell batteries.

Dry cell batteries can also be used. These batteries use an alkaline paste, usually potassium hydroxide (KOH), as the electrolyte. Many dry cell batteries use zinc as the anode and manganese dioxide ($MnO_2$) as the cathode. Today, many newer, lighter, rechargeable batteries are used like the lithium–ion battery (Figure 15-6). This popular battery uses lithium cobalt oxide ($LiCoO_2$) as the anode, carbon as the cathode, and a polymer gel containing lithium salts dissolved in an organic solvent as the electrolyte.

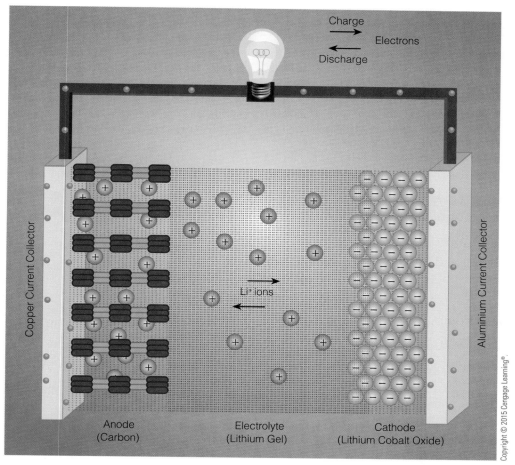

**FIGURE 15-6** Lithium ion battery.

## REVIEW OF KEY CONCEPTS

1. An exothermic reaction is a chemical reaction that gives off thermal energy. An example of an exothermic reaction includes the burning of a fossil fuel to produce heat.

2. The heat of combustion is the energy given off when a substance is burned in the presence of oxygen.

3. An endothermic reaction is a chemical reaction that takes in energy from the surrounding environment. The combination of hydrogen and oxygen gas to form liquid water is an example of an endothermic reaction.

4. The change in the phase of a substance involves the production or consumption of energy. The heat of vaporization is the amount of heat required to change a liquid to a gas. Melting is the change from a solid to a liquid.

5. It takes more energy to vaporize liquid water than it does to melt ice.

6. Oxidation refers to the losing of electrons by a substance and reduction is the gaining of electrons by a substance.

7. A simple chemical battery consists of an anode that supplies the electrons,

a cathode that accepts the electrons, and an electrolyte solution containing ions.

8. The electrons flow from the anode to the cathode. Some batteries can be recharged by applying an electric current to the anode and cathode that reverses the flow of electrons. Common batteries include the lead–acid, alkaline, and lithium ion types.

## CHAPTER REVIEW

### Short Answer

1. What is an exothermic reaction and describe one example of an exothermic reaction used to produce electricity?

2. What is meant by the heat of combustion?

3. Define the term endothermic reaction and provide one example of an endothermic reaction.

4. How is energy related to phase changes?

5. Which change in phase of water requires the greatest amount of heat?

6. What is an oxidation/reduction reaction?

7. How does a simple voltaic cell produce an electric current?

8. What are the three main parts of a battery?

9. What is an example of a common wet cell battery and a dry cell battery used today?

### Energy Math

1. How many Btus are produced by the combustion 0.25 pounds of methane gas?

2. Determine how many kilocalories you would need to vaporize 250 grams of water.

### Multiple Choice

1. Energy given off by a chemical reaction is known as:
   a. the heat of vaporization
   b. endothermic
   c. latent heat
   d. exothermic

2. Energy absorbed by a chemical reaction is known as:
   a. electrochemical
   b. endothermic
   c. latent heat
   d. exothermic

3. The rapid combining of oxygen with a substance to produce heat and light is:
   a. combustion
   b. reduction
   c. latent heat
   d. vaporization

4. The energy given off or taken in during a phase change is called:
   a. potential energy
   b. latent heat
   c. the first law of thermodynamics
   d. the second law of thermodynamics

5. The change in phase from a liquid to a gas is known as:
   a. condensation
   b. melting
   c. freezing
   d. vaporization

6. The change in phase from a gas to a liquid is known as:
   a. condensation
   b. melting
   c. freezing
   d. vaporization

7. How much energy is required to vaporize water:
   a. 1,000 Btus
   b. 0.84 kilocalories
   c. 50 Btus
   d. 0.08 kilocalories

8. When a substance loses electrons it is known as:
   a. reduction
   b. ionization
   c. electrolyte
   d. oxidation

9. When a substance gains electrons it is known as:
   a. reduction
   b. ionization
   c. electrolyte
   d. oxidation

10. The electrode that gives up electrons in a voltaic cell is called the:
    a. anode
    b. cathode
    c. electrolyte
    d. ion

11. The electrode that takes up electrons in a voltaic cell is called the:
    a. anode
    b. cathode
    c. electrolyte
    d. ion

12. The negative electrode of a battery is also known as the:
    a. anode
    b. cathode
    c. electrolyte
    d. ion

## Matching

*Match the terms with the correct definitions*

a. thermochemistry
b. exothermic reaction
c. combustion
d. endothermic reaction
e. vaporization
f. condensation
g. electrochemistry
h. oxidation
i. reduction
j. electrolyte
k. ion
l. cation
m. anion
n. cathode
o. anode

1. _____ The change in phase of a substance from a liquid to a gas.

2. _____ A metal electrode that produces electrons.

3. _____ The study of the heat absorbed or produced by chemical reactions.

4. _____ The gaining of electrons by a substance.

5. _____ A type of chemical reaction that absorbs energy.

6. _____ A positively charged ion.

7. _____ The loss of electrons from a substance.

8. _____ A solution containing ions.

9. _____ A chemical reaction that produces thermal energy.

10. _____ A negatively charge ion.

11. _____ The rapid combination of oxygen to a substance that creates heat and light.

12. _____ The study of the electricity produced by chemical reactions.

13. _____ A metal electrode that absorbs electrons.

14. _____ The change in phase of a substance from a gas to a liquid.

15. _____ An atom or molecule with a net electric charge.

## CHAPTER 16

# Work and Machines

## KEY CONCEPTS

*After reading this chapter, you should be able to:*

1. Define the term force.
2. Explain the relationship between work and energy.
3. Define the terms power, joule, and watt.
4. Explain the mechanical advantage of a machine.
5. Identify the six types of simple machines.
6. Define the term compound machine.
7. Describe how the efficiency of a machine can be determined.

## TERMS TO KNOW

force
joule
newton
power

watt
machine
mechanical advantage
resistance force

effort force
compound machines

# INTRODUCTION

The use of energy in our modern society drives many machines and technological processes that improve our standard of living. The definition of energy tells us work is being done. But what is work and how can it be described? Much of the work once done by human labor has now been replaced with machines; how is this possible? The use of machines made many processes more efficient, but also involves the use of energy to power them. Believe it or not, much of the work done today in our society is based on six simple machines that form the basis of many of our modern technologies. Together, energy and machines are used to perform work that was unthinkable one hundred years ago, greatly improving the quality of life.

## WORK

In Chapter 15, energy was defined as the ability to perform work or cause change. In physics, the term work involves the use of a **force** to produce this change. The change that occurs is usually expressed as movement over a specific distance. A simple example of work, involves picking up a pencil and moving it to a different location on your desk. In order to do work, energy must be expended. This means work and energy are related. Whenever work has been done, energy has been used. This relationship between work and energy is a fundamental aspect of physics.

The connection between energy and work was researched by the English physicist James Prescott Joule during the 1800s. He recognized the connection between heat and mechanical work, now known as the work–energy theorem. This states the change in the kinetic energy of an object is equal to the work done on the object. He realized whenever work was being done, energy had to be expended. Joule's discoveries also helped establish the first law of thermodynamics (see Chapter 15). His research has been honored by the establishment of a unit of energy named for him called the **joule**. One joule is the amount of energy required to apply the force of one newton on an object to move it a distance of one meter. A **newton** is a measurement of force roughly equal to the force of gravity pulling on an object with the mass of a baseball, so one Joule is equal to the amount of energy required to move an object with a mass of a baseball, by one meter (Figure 16-1).

Work is usually calculated as the amount of force being applied to an object and the direction it is being moved. As work is related to energy, it is related to time as well. The relationship between work and time is known as **power**. If you move your pencil from one side of your desk to the other, whether it over a time period of two seconds or ten seconds, the same amount of work is being done, but there is a difference in the amount power being applied. Power is defined as the rate at which work is being done. You use less power to move your pencil over a time period of ten seconds than over two seconds. A common unit of measure for power is the **watt**. One watt is the use of one joule of energy over the time period of one second. Therefore if you move a baseball a distance of one meter in one second, you

**force**—an influence that changes the motion of an object, or produces motion of a stationary object

**joule**—the unit of energy required to apply the force of one newton on an object to move it a distance of one meter

**newton**—a measurement of force roughly equal to the force of gravity pulling on an object with the mass of a baseball

**power**—the amount of work done over a specific amount of time

**watt**—a unit of energy defined as the use of one joule over the time period of one second

229

**FIGURE 16-1** Tossing a baseball one meter in the air to demonstrate one joule of energy.

have used one watt of power. If the time to move the baseball is cut in half, then you have used two watts. The relationship between work, energy, and power is fundamental in understanding how our energy resources are used by many technological processes today.

## SIMPLE MACHINES

**machine**—a device used to increase or change the direction of a force in order to perform a specific task

A **machine** is a device used to increase or change the direction of a force in order to perform a specific task. It is extremely difficult to imagine a world without machines. Even the simplest tasks are made easier by using a machine. In the simplest of terms, a machine is a device that transmits, or transforms energy to do work. Most all machines are variations or combinations of the six simple machines: the lever, inclined plane, wheel and axle, wedge, pulley, and screw. Many of these machines operate by either altering the direction of the force being applied to do work, making it easier, or by changing the distance the work is being done.

**mechanical advantage**—the use of a machine to do work

**resistance force**—the force produced by a machine

**effort force**—the force being applied to a machine

The use of a machine to do work is called the **mechanical advantage**. The mechanical advantage is the ratio of **resistance force** to **effort force**. The effort force is the force being applied to the machine while the resistance force is the force produced by the machine. Most every machine has a mechanical advantage greater than one, meaning the force produced is greater than the force applied. We use machines because they increase the force we apply to them, and amplify the work that can be done. A machine with a mechanical advantage of one means the force put into the machine equals the resistance force produced by it. This is usually not an advantage. The higher the mechanical advantage, the more effective the machine will be. Simple machines do not increase the amount of energy applied to do work, but increase force output compared to the force input.

## The Lever

The lever is a simple machine that uses a board or pipe resting on a pivot point known as a fulcrum. An object to be moved is placed on the resistance end of the lever, and the effort force is applied on the opposite end (Figure 16-2). Increasing the distance between the effort force and the fulcrum, while decreasing the distance between the resistance force and the fulcrum, amplifies the work done by the lever. Examples of simple levers include pliers, a pry-bar, and a bottle opener. Using a wrench with a long handle is also an example of a lever. In this case the force applied to turning a bolt is known as torque that takes linear movement and converts it into rotational force. The human body also uses levers in the form of arms, legs, and fingers to perform work.

## The Inclined Plane

An inclined plane is a type of simple machine that uses a flat surface with one end higher than the other, like a ramp (Figure 16-3). The inclined plane allows you to raise an object to a higher location by applying the force perpendicular to the direction it is to be raised. Examples of useful inclined planes include stairs and ladders. Exit ramps leading off and on highways are also inclined planes.

## The Wheel and Axle

The wheel and axle is a type of simple machine that greatly reduces friction by allowing an object to be rolled (Figure 16-4). This decreases the amount of force needed to move an object. Cars, trucks, and trains use wheels and axles to allow them to move large amounts of weight much more efficiently.

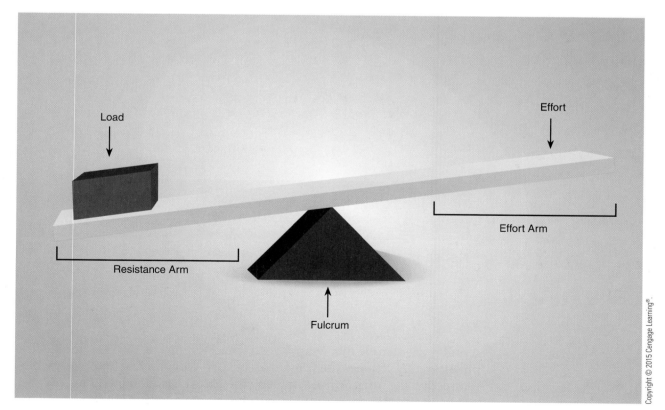

**FIGURE 16-2** Example of a lever.

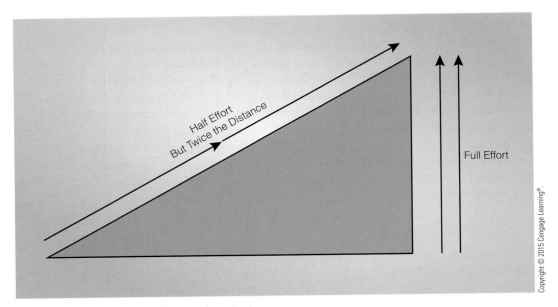

**FIGURE 16-3** Example of an inclined plane.

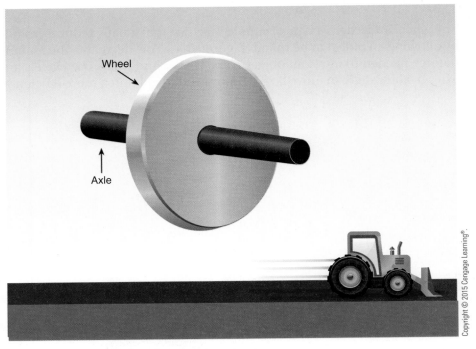

**FIGURE 16-4** Example of a wheel and axle.

## The Pulley

A pulley uses a rope that wraps over a wheel (Figure 16-5). Like the wheel, a pulley greatly reduces the friction associated with moving an object, while redirecting the force applied, making the object easier to lift. Elevators and construction cranes use pulley systems to make them able to lift heavy weights with much less effort.

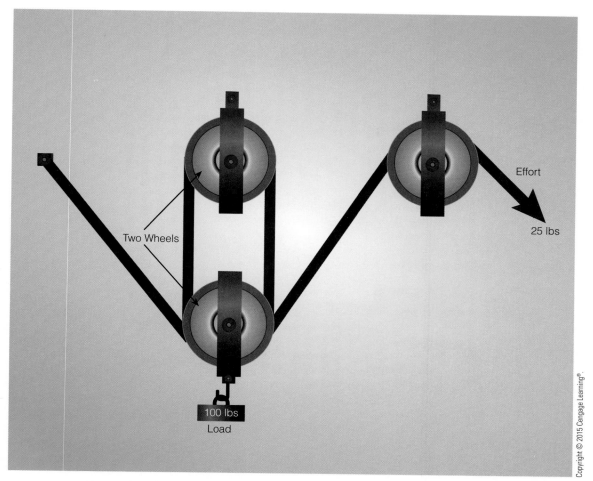

Two Wheels

Effort

25 lbs

100 lbs
Load

**FIGURE 16-5** Example of a pulley.

## The Wedge

The wedge is shaped like a triangle, that helps convert the force applied to its flat end into perpendicular forces that push outward (Figure 16-6). Wedges are often used to split objects apart or raise them upwards. Steel or iron wedges are commonly used to split apart rocks and wood.

## The Screw

The last type of simple machine is the screw. The screw transfers rotational force into linear force. This is accomplished by making spiraled grooves within a cylinder, also known as threads that work much like an inclined plane (Figure 16-7). When the screw is turned, the grooves move the cylinder forward. Metal nut and bolt fasteners employ the use of the screw to tightly hold many types of objects securely together.

## Putting it all Together

Most machines today are comprised of many moving parts that may include different types of simple machines. These are known as **compound machines**. A compound machine is a machine made up of two or more simple machines. One of the simplest forms of compound machines is the

**compound machine**—a machine made up of two or more simple machines.

**FIGURE 16-6** Example of a wedge.

**FIGURE 16-7** Example of a screw.

wheelbarrow (Figure 16-8). This machine combines the use of a lever and a wheel and axle. A bicycle also is a compound machine using the wheel and axle along with a pulley, and a lever if it has hand operated brakes (Figure 16-9).

The ability of a machine to perform work can be expressed by determining its efficiency. The efficiency of a machine is calculated by dividing the energy output of the machine, by its energy input. If you input 80 watts of power into a machine that produces 70 watts of power, the machine is 88% efficient, i.e., $(80/70) \times 100 = 88\%$. Because of the second law of thermodynamics, it is impossible to have a machine that is 100% efficient. This is because the operation of any machine will lose energy to friction or heat. Some of the most efficient machines include electric generators that can have efficiencies of more than 90%. The internal combustion engine used in most cars has an average efficiency of about 20%, which is poor. This means 80% of the money you spend on gasoline to run the car actually goes to heating the atmosphere, while the remaining 20% moves it down the road! This low efficiency is clearly not a good option when using nonrenewable fuel sources such as gasoline and diesel fuel. The ability to increase the efficiency of machines we use for our daily lives is an important step in reducing our energy consumption while also increasing our energy supplies.

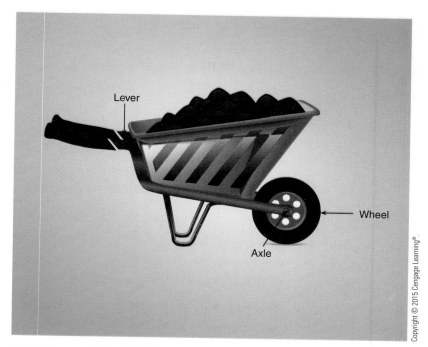

**FIGURE 16-8** Wheel barrow consists of a wheel and axle and a lever.

**FIGURE 16-9** A bicycle consists of a wheel and axle, pulleys, and a lever.

# REVIEW OF KEY CONCEPTS

1. A force is an influence that changes the motion of an object or produces motion of a stationary object.

2. Work is defined as a change caused by a force in a specific direction. The change is usually expressed by movement of some kind. The force exerted to do work expends energy.

3. The amount of energy required to do work is measured by the unit known as a joule. One joule is the amount of energy required to move an object with the force of one newton, by one meter. Power is the rate at which work is done. The common unit of power is the watt, and is equal to using one joule of energy over the time period of one second.

4. A machine is a device used to increase or change the direction of a force in order to perform a specific task. This makes doing work much easier and more efficient. The use of a machine to amplify the work being done is called the mechanical advantage.

5. Most all machines are based on six simple concepts known as simple machines. These include the lever, wedge, inclined plane, wheel and axle, pulley, and the screw.

6. Many machines in use today are considered compound machines. A compound machine is made up of two or more simple machines.

7. The efficiency of a machine is a measure of the ratio of output to the input.

# CHAPTER REVIEW

## Short Answer

1. What is work?

2. Explain the relationship between work and energy.

3. Define the terms power, joule, and watt.

4. Explain the mechanical advantage of a machine.

5. What are the six types of simple machines?

6. What is a compound machine?

7. Describe how the efficiency of a machine can be determined.

8. Describe three modern examples of the use of simple machines.

## Energy Math

1. How many joules does it require to raise three baseballs a height of four meters?

2. Using your answer from above, how many watts of power did it require to raise the three baseballs over a period of two seconds using the formula: watts = energy/time?

3. If a coal-fired steam turbine burns 18 million Btus of coal to do 8.46 million Btus of work, how efficient is the turbine?

## Multiple Choice

1. The use of a force to produce change is known as:
   a. power
   b. work
   c. the joule
   d. a newton

2. A measurement of force roughly equal to the force of gravity pulling on an object with the mass of a baseball is called:
   a. power
   b. work
   c. the joule
   d. a newton

3. The rate at which work is being done is called:
   a. power
   b. work
   c. the joule
   d. a newton

4. The use of one joule of energy over the time period of one second is known as:
   a. power
   b. watt
   c. the joule
   d. a newton

5. A simple machine that uses a fulcrum is called a(n):
   a. screw
   b. wedge
   c. inclined plane
   d. lever

6. A ramp is an example of what type of simple machine?
   a. A screw
   b. A wedge
   c. An inclined plane
   d. A lever

7. A triangular shaped object used to redirect force is called a(n):
   a. screw
   b. wedge
   c. inclined plane
   d. lever

8. A pair of pliers is an example of a(n):
   a. screw
   b. wedge
   c. inclined plane
   d. lever

9. What simple machine helps reduce friction?
   a. A screw
   b. A wheel and axle
   c. An inclined plane
   d. A wedge

10. What is the approximate efficiency of the internal combustion engine?
    a. 100%
    b. 75%
    c. 25%
    d. 5%

## Matching

*Match the terms with the correct definitions*

a. force
b. joule
c. newton
d. power

e. watt
f. machine
g. mechanical advantage

h. resistance force
i. effort force
j. compound machines

1. _____ The force put into a machine.

2. _____ The amount of energy required to move an object with a force of one Newton over one meter.

3. _____ A device used to increase or redirect a force in order to do work.

4. _____ A device made up of two or more simple machines.

5. _____ The rate at which work is being done.

6. _____ The ratio of a machine's resistance force to its effort force.

7. _____ A unit of power equivalent to the use of one joule over the time period of one second.

8. _____ The force output of a machine.

9. _____ A unit of force equal to moving an object with a mass of a baseball by one meter.

10. _____ An influence that changes the motion of an object or produces motion of a stationary object.

## CHAPTER 17

# Heat Engines

## KEY CONCEPTS

*After reading this chapter you should be able to:*

1. Explain the basic concept of a heat engine.

2. Describe the four basic steps in the operation of a Rankine Cycle heat engine and provide an example of the use of a Rankine engine.

3. Explain how a Brayton Cycle engine operates and provide one example of a Brayton engine.

4. Describe the process of how a four-stroke internal combustion engine produces power.

5. Explain the difference between spark ignition and compression ignition internal combustion engines.

6. Describe how a heat pump moves heat.

## TERMS TO KNOW

heat engine

hot reservoir

cold reservoir

heat exchanger

condenser

compression

heat pump

evaporator

# INTRODUCTION

Much of the use of our energy resources, especially fossil fuels like coal and crude oil, go to the production of heat in order to perform work. This transfer of thermal energy into mechanical energy can be accomplished by using an engine. Engines are used mostly to convert the potential energy stored within fuels into motion. Since the use of steam as a power source, many types of engines have been developed in order to power our modern society, and understanding how they operate is an important step in the ability to improve their efficiency.

## THE CARNOT CYCLE

A **heat engine** is a device that converts thermal energy into mechanical energy by a cyclic process. The ideal concept of a heat engine is best explained as energy flowing from a source to a sink while performing work along the way. In most heat engines the source of heat, also known as the **hot reservoir**, is the thermal energy produced by combustion. The sink for most heat engines, also called the **cold reservoir**, is the atmosphere, where the energy is dissipated as lost heat. The Earth is sometimes referred to as a heat engine that allows life to flourish on our planet and drives many non-living processes. In this case the hot reservoir is the sun, work is performed by the living and nonliving systems of the planet and the cold reservoir is outer space. Because of the laws of thermodynamics, no heat engine can be 100% efficient. At some point in the heat-engine cycle, energy will be lost in the process of doing work. This occurs as both friction and heat loss to the surrounding environment. The ideal example of an industrial heat engine is known as the Carnot Cycle that was devised by French engineer, Sadi Carnot in 1824.

The Carnot Cycle heat engine uses heat to expand gas within a chamber to move a piston up and down (Figure 17-1). Although the Carnot Cycle engine is mostly a theoretical look at how efficient a heat engine can be, its basic design is the basis for many heat engines widely used today.

## THE RANKINE CYCLE ENGINE

The Rankine Cycle engine is a type of heat engine mostly used to produce electricity. This type of engine has been in operation for more than one hundred years. The Rankine Cycle uses heat energy to change the phase of a working fluid within a closed system in order to perform work. Water is often used as the working fluid in these engines. Because the combustion of the fuel that produces the heat occurs outside the system, these engines are also known as external combustion engines. Most Rankine Cycle engines have four basic steps used to convert heat energy into mechanical energy (Figure 17-2). The first step involves pumping a working fluid into a **heat exchanger**. A heat exchanger is a device that transfers external heat energy to a working fluid. In the heat exchanger the fluid is vaporized. The vapor then enters a chamber where it rapidly expands, forcing a turbine to turn to produce mechanical power. The vapor leaves the turbine and enters a **condenser** that removes the

**heat engine**—a device that converts thermal energy into mechanical energy by a cyclic process

**hot reservoir**—the source of energy for a heat engine

**cold reservoir**—the destination or sink, for the energy used in a heat engine

**heat exchanger**—a device that transfers external heat energy to a working fluid

**condenser**—a device that cools a working fluid in order to return it to a liquid state

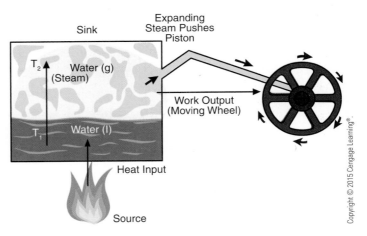

**FIGURE 17-1** A simple heat engine.

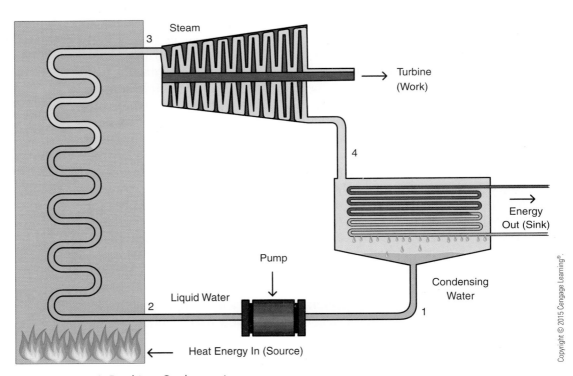

**FIGURE 17-2** A Rankine Cycle engine.

heat from the fluid, returning it to a liquid. A condenser is a device that cools a working fluid in order to return it to a liquid state. The fluid is then pumped back into the heat exchanger, thus completing the cycle. Because a Rankine engine is a heat-engine, the heat exchanger acts as the hot reservoir, while the turbine performs the work and the condenser acts as the cold reservoir. Today, both coal-fired and nuclear power plants use the Rankine Cycle to power electrical generators that produce electricity. The typical efficiency of a coal-fired Rankine Cycle engine is about 31%. This low efficiency is because of the heat loss required to condense the working fluid, along with the heat loss within the heat exchanger. Although inefficient, Rankine engines are powerful. One of the largest coal-fired power plants in the United States using a Rankine Cycle engine produces more than 3,500 megawatts of power.

# THE BRAYTON CYCLE ENGINES

Engineers recognized the inefficiencies associated with the Rankine heat engine and designed a different system that works more efficiently. This is known as the Brayton Cycle. The Brayton Cycle is different from the Rankine Cycle in a few ways (Figure 17-3). First it does not use a working fluid to transfer the energy and secondly it is based on an open system. The Brayton Cycle compresses air and mixes it with a combustible fuel. The fuel–air mixture is then ignited within a combustion chamber. The resulting superheated gas is forced through a turbine that transfers the heat energy into mechanical energy. The exhaust gas then leaves the turbine. In this case the hot reservoir is the superheated gas created by combustion, the turbine does the work, and the cold reservoir is the atmosphere. Brayton Cycle engines are also known as gas turbines or jet engines used to propel airplanes. The efficiencies of gas turbines range between 35–40%.

A new type of electric power plant combines the Brayton and Rankine cycle engines to form a hybrid system known as a Combined Cycle engine. The Combined Cycle engine uses the hot exhaust gas from a gas turbine as the heat source for a Rankine engine (Figure 17-4). Power plants using combined

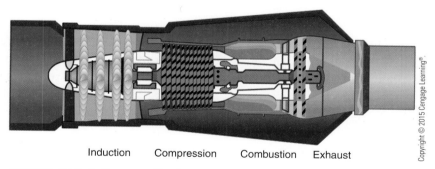

Induction     Compression     Combustion     Exhaust

Copyright © 2015 Cengage Learning®.

**FIGURE 17-3** A Brayton Cycle engine.

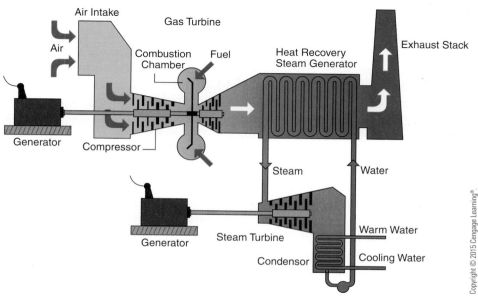

Copyright © 2015 Cengage Learning®.

**FIGURE 17-4** A Combined Cycle engine.

cycle technology have increased the efficiency of the process to about 60%. The waste heat generated by the hot exhaust gas that normally goes into the atmosphere is used to vaporize the working fluid in the Rankine engine.

## THE INTERNAL COMBUSTION ENGINE

The engine used to power our cars and trucks is known as the internal combustion (IC) engine. The heat source that drives the engine is produced internally, unlike the external combustion used by Rankine and Brayton engines. There are two basic types of IC engines; the spark ignition internal combustion engine (SI–IC), and the compression ignition internal combustion engine (CI–IC).

The spark ignition IC engine was developed by Nikolas Otto in 1876 for use in the first gasoline powered automobiles. It is also known as the Otto Engine or four-stroke engine. This type of engine uses a piston within a cylinder to transfer heat energy into mechanical energy. The term "four-stroke" describes the piston within the engine that goes through four strokes, or movements, in its cycle to produce power (Figure 17-5). These strokes include: intake, compression, combustion, and exhaust. The process begins when a gasoline–air mixture is drawn into the cylinder by the movement of a piston. Next, the piston compresses the vapor. **Compression** is the reduction in the volume of a substance as a result of an increase in pressure. A spark plug ignites the combustible vapor that causes it to explode within the cylinder. The super-heated gases produced by the explosion forces the piston to move, generating power that is transferred from the piston to a rotating shaft. A valve is then opened and the piston forces out the exhaust gases that go into the atmosphere. Finally the piston retracts, drawing more combustible fuel into the cylinder repeating the cycle. Today, modern SI–IC engines commonly have four, six, or eight cylinders that provide an incredible amount of power.

**compression**—the reduction in the volume of a substance as a result of an increase in pressure

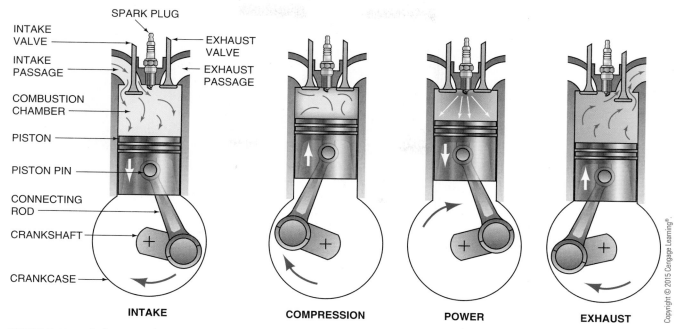

**FIGURE 17-5** A four-stroke engine.

The other popular type of internal combustion engine is known as a Compression Ignition engine, or diesel engine. This engine was developed by Rudolph Diesel beginning in 1893. The diesel engine is similar to the SI engine because it is considered a four-stroke engine, but it does not use spark ignition (Figure 17-6). The process begins when the piston retracts within a cylinder, drawing in the diesel fuel–air mixture. Next, the piston compresses the combustible vapor until it reaches its ignition point that causes it to explode. The explosion moves the piston back, creating power that is then transferred to a rotating shaft. The piston pushes out the exhaust gas, and the cycle starts again.

The CI-IC engine employs the physical concept that gases increase their temperature when compressed. The compression within the cylinder caused by the piston creates enough heat to exceed the flash point of the diesel fuel–air vapor, thus creating an explosion. The lack of a spark plug ignition in a diesel engine often makes it more difficult to start these engines in cold weather. Both types of IC engines have low efficiencies that range between 15–25%.

Another type of IC engine used on a much smaller scale is known as the two-stroke engine. These engines are commonly used for small applications like powering lawn mowers, chainsaws, and small recreational vehicles like boats, four-wheelers, motorcycles, and snowmobiles. These IC engines differ from four-stroke engines because they only have two strokes of the piston in their cycle; the compression cycle and the exhaust cycle (Figure 17-7). The two-stroke cycle begins when the piston retracts. This simultaneously draws the fuel–air vapor into the cylinder while also compressing vapor within the crankcase below the cylinder. Next, the piston moves up, compressing the vapor within the cylinder, the spark plug ignites and the resulting combustion of the fuel moves the piston down, transferring its power to a rotating shaft. The downward movement of the piston opens the exhaust port and the

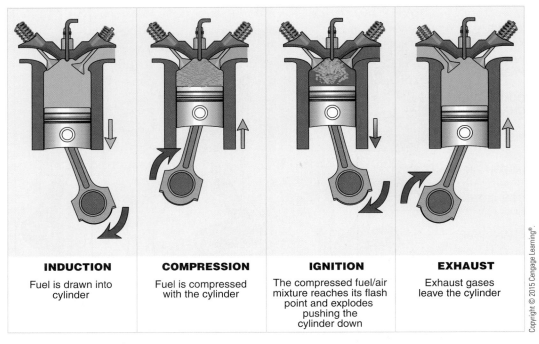

| **INDUCTION** | **COMPRESSION** | **IGNITION** | **EXHAUST** |
|---|---|---|---|
| Fuel is drawn into cylinder | Fuel is compressed with the cylinder | The compressed fuel/air mixture reaches its flash point and explodes pushing the cylinder down | Exhaust gases leave the cylinder |

**FIGURE 17-6** Diesel engine.

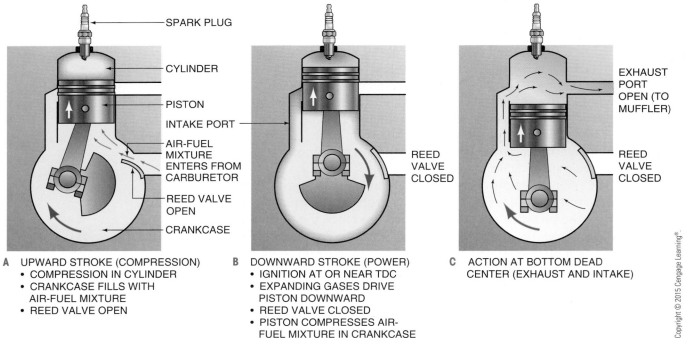

A   UPWARD STROKE (COMPRESSION)
• COMPRESSION IN CYLINDER
• CRANKCASE FILLS WITH AIR-FUEL MIXTURE
• REED VALVE OPEN

B   DOWNWARD STROKE (POWER)
• IGNITION AT OR NEAR TDC
• EXPANDING GASES DRIVE PISTON DOWNWARD
• REED VALVE CLOSED
• PISTON COMPRESSES AIR-FUEL MIXTURE IN CRANKCASE

C   ACTION AT BOTTOM DEAD CENTER (EXHAUST AND INTAKE)

**FIGURE 17-7** Two-stroke engine.

pressurized vapor held within the crankcase flows into the cylinder; driving out the exhaust. The two-stroke cycle is then complete. The reduction in strokes makes this engine much simpler and lighter than four-stroke engines. Because the crankcase of the engine also acts as a compression chamber, oil must be added to the fuel in order to lubricate the piston. The oil mixes with the fuel and causes the engine to produce a smoky exhaust that creates more pollution than a four-stroke engine. Because the exhaust stroke uses the force of the pressurized fuel vapor to expel the exhaust gases from the previous cycle, some of the unburned fuel also escapes into the atmosphere. This wastes fuel, reduces the efficiency of the engine, and adds to the pollution the engine produces. A two-stroke engine loses about 25–30% of its unburned fuel to the environment because of this. As a result of the negative environmental impacts caused by two-stroke engines, they are becoming less popular for use in recreational vehicles. In some states, boats that have two-stroke engines have been banned for use on certain waterways to prevent fuel and oil from polluting the water. The Environmental Protection Agency is also increasing the emission standards of small engines, that is affecting the use of two-stroke engines on newer vehicles.

## THE HEAT PUMP

The **heat pump** is a type of heat engine used to move heat, typically from a cooler region to a warmer region, using mechanical energy. Heat pumps are used to supply heat for buildings or as a means of refrigeration. Actually a heat pump is simply a Rankine Cycle engine operating in reverse. The heat pump, like a Rankine engine, uses a working fluid within a closed loop. The process begins when the working fluid is rapidly expanded in an **evaporator**, that turns it into a vapor. An evaporator is a device that causes a fluid to rapidly change phase from a liquid to a gas. The process of vaporization absorbs heat from the surrounding environment through a heat exchanger. The gas then enters a compressor, that causes its temperature to increase. The heated gas is then passed through a condenser that extracts the heat from the vapor, causing it to return to a liquid. The heat from the condenser is transferred through a heat exchanger to the atmosphere and the liquid is returned to the expander and the cycle repeats itself. The working fluid in a heat pump is typically a refrigerant gas with a low vaporization temperature. In the case of a geothermal heat pump, used to supply heat to a home, the evaporator absorbs heat from within the ground that is then transferred to the condenser to supply heat indoors (Figure 17-8). For refrigeration,

**heat pump**—a type of heat engine used to move heat, typically from a cooler region to a warmer region using mechanical energy

**evaporator**—a device that causes a fluid to rapidly change phase from a liquid to a gas

**CAREER CONNECTIONS**

### Agricultural Mechanic

An Agricultural Mechanic is a skilled worker who repairs and maintains all types of agricultural equipment. This career involves knowledge of repairing both diesel and gasoline motors that power the many machines that perform work on a farm. This job requires skill with a variety of tools and knowledge of the operation and repair of different farm-based machines. These machines include tractors, harvesters, irrigation equipment, balers, hydraulic systems, and many other types of specialized farm equipment. Agricultural Mechanics can find work at a farm equipment dealership or can be self-employed. The ability to be a good problem solver is a necessary skill for this job because many jobs require on-the-spot repairs along with knowledge of welding. Training at a vocational high school or a two-year technical college can prepare you for this career.

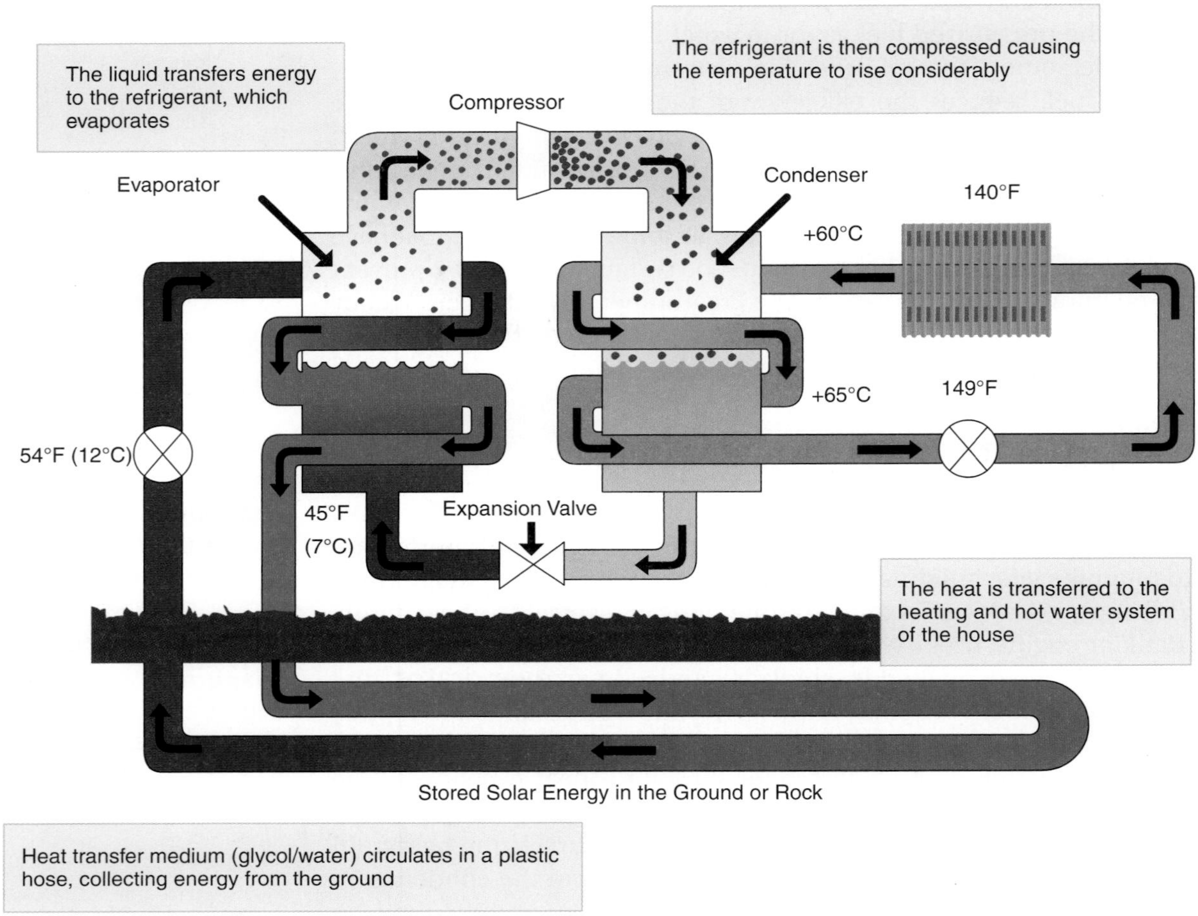

The liquid transfers energy to the refrigerant, which evaporates

Compressor

The refrigerant is then compressed causing the temperature to rise considerably

Evaporator

Condenser

140°F

+60°C

+65°C    149°F

54°F (12°C)

45°F (7°C)    Expansion Valve

The heat is transferred to the heating and hot water system of the house

Stored Solar Energy in the Ground or Rock

Heat transfer medium (glycol/water) circulates in a plastic hose, collecting energy from the ground

**FIGURE 17-8** A geothermal heat pump.

the evaporator takes heat from within the refrigerator and cools it and the heat is transferred to the condenser that expels it into your kitchen. Heat pumps require an input of energy in the form of electricity to operate the compressor. The use of a geothermal heat pump as a source of indoor heat is an efficient way of moving heat stored in the ground to your house.

# REVIEW OF KEY CONCEPTS

1. A heat engine is a device that converts thermal energy into mechanical energy. The most basic heat engine uses heat flowing from a hot reservoir to a cold reservoir with work performed along the way.

2. A Rankine Cycle heat engine, known as an external combustion engine, uses heat energy to vaporize a working fluid in order to power a turbine. This is a common method used to produce electricity by burning coal or using nuclear fission.

3. The Brayton Cycle engine, also called a gas turbine, uses hot exhaust gases from the combustion of a fuel to power a turbine. A Combined Cycle engine uses the exhaust gases from a Brayton Cycle engines as a heat source for a Rankine engine, greatly improving the efficiency of the system.

4. The internal combustion engine is another commonly used heat engine that powers our cars and trucks. This type of engine is also known as a four-stroke engine because the piston that provides the power goes through four stages, or strokes, within a cylinder. These include intake, compression, combustion, and exhaust.

5. There are two types of internal combustion engines: the spark ignition and compression ignition. The spark ignition uses a spark plug to set off the combustion within the cylinder of a gasoline engine.

The compression ignition uses the heat of a compressed vapor to initiate the combustion process. Compression ignition engines are also known as diesel engines.

6. A heat pump is a type of heat engine that moves heat from a colder area to a warmer area. Heat pumps operate much like a Rankine engine, but in reverse. These engines use refrigerant gases that have low vaporization temperatures as the working fluid to transfer the heat. Heat pumps are used as a source of heat during winter in homes and as a way to cool refrigerators and air conditioners.

## CHAPTER REVIEW

### Short Answer

1. What is a heat engine?

2. Describe the four basic steps in the operation of Rankine Cycle heat engine and provide an example of the use of a Rankine engine.

3. How does a Brayton Cycle engine operate? Provide one example of a Brayton engine.

4. Describe the process of how a four-stroke internal combustion engine produces power.

5. What is the difference between spark ignition and compression ignition internal combustion engines?

6. How does a heat pump move heat?

7. Briefly describe the process a coal-fired power plant uses the Rankine Cycle to produce electricity.

### Energy Math

1. If a Rankine Cycle engine is 31% efficient, how much work output in Btus can it produce by powering it with the combustion of 33,000 Btus of coal?

2. How much work output would a combined cycle heat engine produce by burning 33,000 Btus of coal, assuming it is 60% efficient?

3. If your car engine has an efficiency of 25%, and gasoline costs $3.50 per gallon, how much of your money per gallon goes to heating the atmosphere?

### Multiple Choice

1. In most heat engines, the source of heat is also known as the:
   a. condenser
   b. hot reservoir
   c. evaporator
   d. cold reservoir

2. Another name for a sink in a simple heat engine is a(n):
   a. condenser
   b. hot reservoir
   c. evaporator
   d. cold reservoir

3. The heat engine that is used to propel airplanes is known as a(n):
   a. Rankine Cycle engine
   b. heat pump
   c. Brayton engine
   d. internal combustion engine

4. Which heat engine is commonly used to produce electricity from the combustion of coal?
   a. Rankine Cycle engine
   b. Heat pump
   c. Brayton Cycle engine
   d. Internal combustion engine

5. Which part of the Rankine Cycle produces work?
   a. Compression
   b. Vaporization
   c. Condensation
   d. Heating

6. Another name for a Compression Ignition Internal Combustion engine is a:
   a. diesel engine
   b. two-stroke engine
   c. heat pump
   d. gas turbine

7. A refrigerator uses which type of heat engine?
   a. Rankine Cycle engine
   b. Heat pump
   c. Brayton engine
   d. Internal combustion engine

8. Which type of engine is commonly associated with water pollution?
   a. A diesel engine
   b. A two-stroke engine
   c. A four-stroke engine
   d. A gas turbine

10. What is the approximate efficiency of an internal combustion engine?
    a. 100%
    b. 75%
    c. 25%
    d. 5%

## Matching

*Match the terms with the correct definitions*

a. heat engine
b. hot reservoir
c. cold reservoir
d. heat exchanger
e. condenser
f. compression
g. heat pump

1. _____ A device that transfers heat from one substance to another.

2. _____ A type of heat engine used to move heat, typically from a cooler region to a warmer region, using mechanical energy.

3. _____ The source of energy in a heat engine.

4. _____ A device that removes heat from a gas, causing it to turn into a liquid.

5. _____ A device that converts thermal energy into mechanical energy by a cyclic process.

6. _____ The location where energy is dissipated in a heat engine, also known as a sink.

7. _____ The reduction in the volume of a substance as a result of an increase in pressure.

## CHAPTER 18

# Energy Flow through Living Systems

## KEY CONCEPTS

*After reading this chapter, you should be able to:*

1. Identify from where most energy on the Earth comes.

2. Describe the photosynthesis and chemosynthesis chemical reactions.

3. Differentiate between autotrophic and heterotrophic organisms.

4. Describe the process of primary production.

5. Explain the relationship among producers, primary consumers, secondary consumers, and decomposers in an ecosystem.

6. Define the terms herbivore, carnivore, omnivore, and detritivore.

7. Draw a diagram of a simple food chain.

8. Describe the transfer of energy from one trophic level to another in a food pyramid.

## TERMS TO KNOW

| | | |
|---|---|---|
| autotrophs | primary production | omnivores |
| photosynthesis | biomass | detritivores |
| chemosynthesis | herbivores | decomposers |
| heterotrophs | carnivores | food chain |
| respiration | | |

# INTRODUCTION

The interactions that occur in the world's ecosystems involve the exchange and movement of energy and matter between the living and nonliving worlds. An ecosystem is the term used to describe the interactions between the living and non-living factors within a specific area. The source of all the energy used in most of the world's ecosystems is the sun. The pathways solar energy moves through once it strikes the planet links all living things together. Every living thing on the Earth depends on the flow of energy through ecosystems for survival. Understanding the interactions between energy and life on the planet reveals how life on the Earth has flourished for more than 3 billion years.

## PHOTOSYNTHESIS AND CHEMOSYNTHESIS

All life on Earth requires energy to live and the way organisms' gain energy is vital for its survival. Organisms that derive energy from sunlight or from chemical reactions are called **autotrophs**, meanings "self-feeder." Autotrophs use two different processes to gain energy.

The first is the process of **photosynthesis**. Photosynthesis is the chemical reaction used by organisms to transform light energy from the sun into stored chemical energy (Figure 18-1). This is accomplished by creating sugars and starches. Photosynthesis is one of the most important chemical reactions on Earth. The photosynthesis reaction takes radiant energy from sunlight and combines it with carbon dioxide and water to form glucose (sugar), with oxygen as a byproduct. The oxygen is released into the atmosphere. All the oxygen that exists within the atmosphere is the result of photosynthesis. Glucose sugars produced by photosynthesis are joined together to form long, chain-like molecules called starches. It is the starch molecules that provide stored chemical energy for life. Green plants and phytoplankton are the two main organisms that use photosynthesis. Amazingly the photosynthesis reaction is an incredibly inefficient energy conversion process. The global average efficiency of photosynthesis is about 0.3%! This means only 0.3% of the sunlight that strikes the plant is converted into stored chemical energy!

**Chemosynthesis** is the other method autotrophs gain energy. Chemosynthesis is the process used by organisms to convert the chemical energy stored in sulfur compounds into organic compounds like carbohydrates (Figure 18-2). Commonly chemosynthesis uses compounds that contain sulfur—like hydrogen sulfide. This is how bacteria residing in deep sea hydrothermal vent communities located on the bottom of the ocean gain their energy. These communities are unique on the Earth because they do not derive their energy from sunlight as most ecosystems do.

## RESPIRATION

Another method that organisms use to gain their energy on Earth is the consumption of other organisms. These living things are called **heterotrophs**, that means "other feeder." Heterotrophs use the chemical energy

**autotrophs**—"self–feeders" are organisms that derive energy from sunlight or from chemical reactions

**photosynthesis**—the chemical reaction that takes radiant energy from sunlight and combines it with carbon dioxide and water to form glucose and oxygen as a byproduct

**chemosynthesis**—the process used by organisms to convert the chemical energy stored in sulfur compounds into carbohydrates

**heterotrophs**—"other feeders" or organisms that derive energy from consuming other organisms

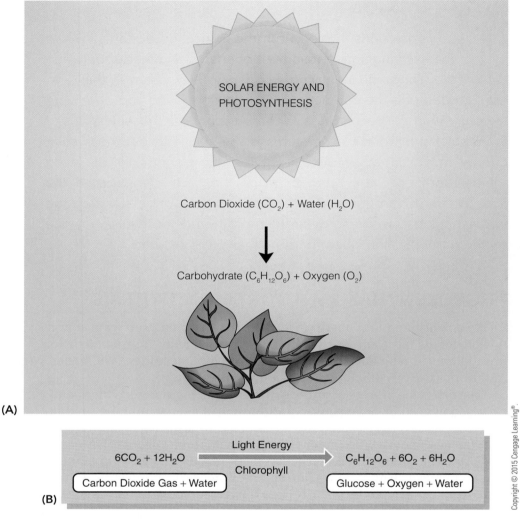

(A)

6CO$_2$ + 12H$_2$O     Light Energy     C$_6$H$_{12}$O$_6$ + 6O$_2$ + 6H$_2$O

Chlorophyll

Carbon Dioxide Gas + Water       Glucose + Oxygen + Water

(B)

**FIGURE 18-1** Photosynthesis.

**respiration**—the chemical reaction that takes the energy stored in sugars and produces carbon dioxide and water

stored in the bodies of other organisms such as plants and animals, to gain energy by respiration. **Respiration** is the opposite chemical reaction of photosynthesis. Respiration takes the energy stored in sugars like glucose and produces carbon dioxide and water as a byproduct. Human beings are heterotrophs that use respiration to gain energy from consuming food.

## PRIMARY PRODUCTION

Although it is important to understand how individual organisms gain their energy, it is the flow of energy through entire ecosystems that enables life to exist as it does on our planet. Autotrophic organisms such as plants and algae form the basis of energy in all ecosystems (Figure 18-3). With the exception of the deep sea hydrothermal vent communities, all energy flow in both aquatic and terrestrial ecosystems begins with green plants and algae. These autotrophs convert radiant energy from the sun into stored chemical energy. This forms a vital link between the sun and the rest of the organisms on Earth. Photosynthesizing autotrophs are also called producers

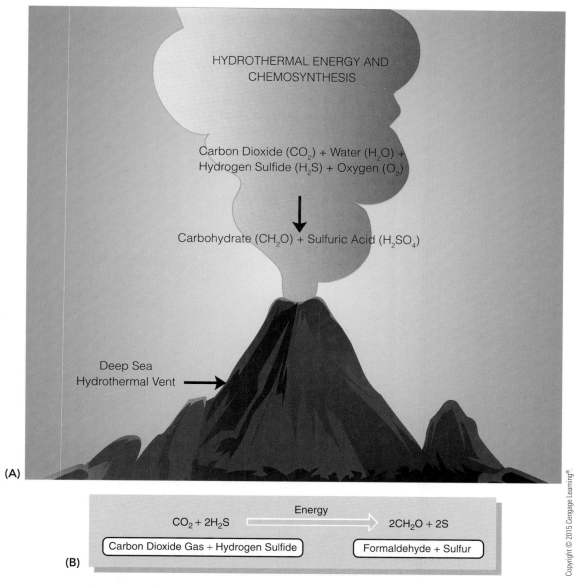

FIGURE 18-2 Chemosynthesis.

because they make the energy for the entire ecosystem. The amount of chemical energy an autotroph converts from solar energy by the process of photosynthesis is called **primary production**. Primary production is measured by determining the amount of **biomass** a plant has. Biomass is a short term for biological mass, and is the total dry weight of an organism.

## PRIMARY AND SECONDARY CONSUMERS

The heterotrophic organisms that consume the primary producers are called primary consumers. These include all organisms that eat plants or algae as a source of energy. Known as **herbivores**, or "plant eaters," primary consumers come in all shapes and sizes in a specific ecosystem. Many insects are primary consumers in ecosystems around the world. Fish and larger animals such as deer, moose, and cows are also primary consumers, or herbivores, that only consume plants (Figure 18-4).

**primary production**—the amount of chemical energy an autotroph converts from solar energy by the process of photosynthesis

**biomass**—a short term for biological mass, that is the total dry weight of an organism

**herbivores**—"plant eaters" or organisms that eat plants or algae as a source of energy

**FIGURE 18-3** Example of green plants that are producers for an ecosystem; they convert radiant energy from the sun into stored chemical energy.

**FIGURE 18-4** Cattle are examples of primary consumers, also called herbivores, who eat the producers in an ecosystem.

© PhotoLink/Photodisc/Getty Images

**FIGURE 18-5** A brown bear is an upper-level consumer in an ecosystem, also known as an omnivore, who eats plants and animals.

Organisms that consume the herbivores are known as secondary consumers. These organisms are sometimes called **carnivores**, or "meat eaters." Tigers, killer whales, and hawks are all secondary consumers. Sometimes an organism can act as both a primary and secondary consumer by eating both meat and plants. These organisms are known as **omnivores**, or "all eaters." Humans are omnivores, as are bears and dogs (Figure 18-5).

**carnivores**—"meat eaters" or organisms that consume the herbivores

**omnivores**—"all eaters" or organisms that eat both plants and animals

Some ecosystems contain higher level consumers who eat secondary consumers. These types of organisms are called tertiary consumers such as eagles or killer whales. The producers form the vital link between the sun and the primary consumers in an ecosystem; however, there exists another important category of organisms that helps recycle nutrients in a community. These organisms are called **detritivores**, or **decomposers**, that eat dead organisms. Although considered gruesome, detritivores perform a valuable service to the ecosystem by breaking down or consuming the remains of dead organisms. This allows valuable nutrients to be recycled in an ecosystem. If not for the detritivores, things would be quite messy here. Decomposers are like nature's garbage collectors. Common detritivores include vultures, many types of insects, fungi, and bacteria (Figure 18-6).

**detritivore**—organisms that eat dead things, also known as decomposers

**decomposer**—organisms that break down dead organisms by consuming them

## FOOD CHAINS AND WEBS

The specific way energy flows through a community of organisms in an ecosystem is called a **food chain**. The food chain is a series of eating processes that energy and nutrients flow from one organism to another.

**food chain**—a series of eating processes by which energy and nutrients flow from one organism to another

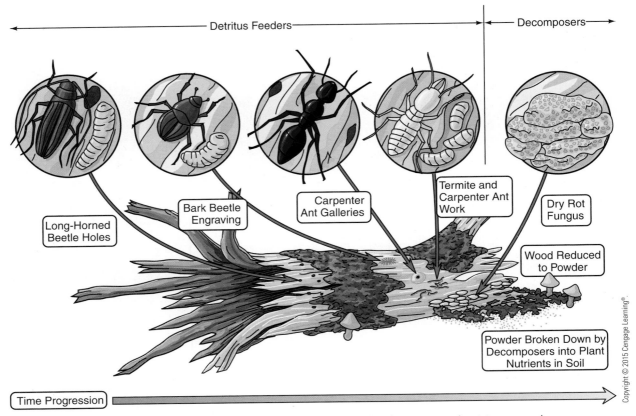

**FIGURE 18-6** A variety of organisms act as decomposers, also known as detritivores, who consume dead organisms.

A food web is complex interaction of food chains within a specific ecosystem (Figure 18-7). All ecosystems have their unique food chains or webs that sustain all the living things in an ecosystem's community. All food chains or webs always contain their own producers, primary consumers, secondary consumers, and decomposers.

## CAREER CONNECTIONS

### Conservation Biologist

A Conservation Biologist is a person who works to restore or maintain habitats or species of organisms being affected by human activity. Their work involves the study of ecosystems, plants and animals, and how best to prevent them from being disturbed by modern society. Conservation Biologists look at the complex relationships that exist between an organism and its environment. They use this knowledge to devise a plan on how to maintain this relationship in order to prevent the extinction of species. Conservation Biologists do much of their work outdoors and require a four-year degree or more in biology and natural resource management.

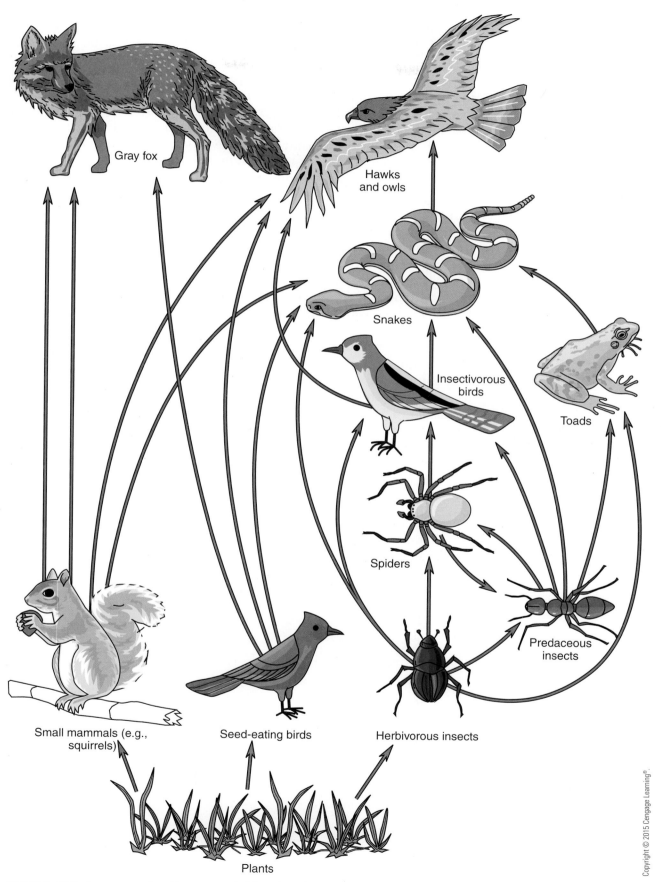

**FIGURE 18-7** An example of a terrestrial food web.

# THE ENERGY PYRAMID

The movement of energy through an ecosystem can also be illustrated by the energy pyramid. The energy pyramid is a visual representation of the way energy moves through a food chain (Figure 18-8). Each step in the food chain, illustrated in the energy pyramid, is called a trophic or feeding level. The base of the pyramid is formed by the producers who take in radiant energy from the sun and transform it into chemical energy in the form of sugar and starches by the process of photosynthesis. The next level in the pyramid is occupied by the primary consumers that eat the producers. This is the second trophic level. When a primary consumer eats a producer, approximately 10 percent of the total energy contained by the producer is gained by the primary consumer. Almost 90% of the total chemical energy contained in the producer is given off as heat when the organism digests, uses, and stores the chemical energy derived from the first trophic level. This "loss" of energy is explained by the second law of thermodynamics and is an important part of the movement of energy through a food chain. The energy is not lost, but it is no longer available for use by organisms in the food chain. Typically when an organism eats something, approximately 90% of the energy is given off to the atmosphere as heat. The third trophic level is occupied by the secondary consumer. Once again, when the secondary consumer eats the primary consumer, it only gains approximately 10% of the total chemical energy of the second trophic level. The purpose of the energy pyramid is to show how energy moves through a food chain and, more important, how much of the energy in an ecosystem is lost to the atmosphere as heat. Each successive step of a trophic level in a food chain results in a large loss of energy. This is why upper level consumers must eat large quantities of food to survive. The other important aspect of the energy pyramid is how the sun forms

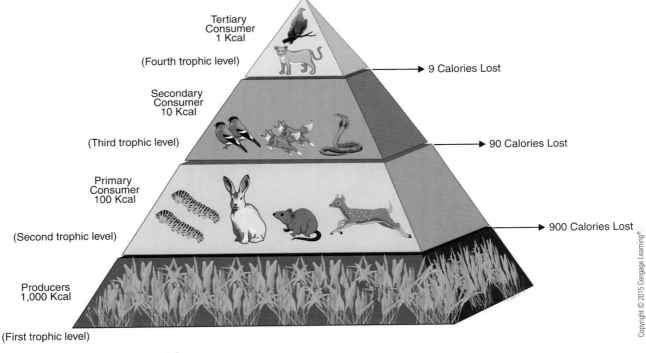

Tertiary
Consumer
1 Kcal

(Fourth trophic level)

→ 9 Calories Lost

Secondary
Consumer
10 Kcal

(Third trophic level)

→ 90 Calories Lost

Primary
Consumer
100 Kcal

(Second trophic level)

→ 900 Calories Lost

Producers
1,000 Kcal

(First trophic level)

**FIGURE 18-8** Energy pyramid.

the base for all energy in an ecosystem. It also shows how the producers act as the important link between the sun and all other organisms on the Earth. When the energy pyramid is applied to production agriculture, the efficiencies of plant or meat-based diets are revealed. Consuming only plants uses sunlight more efficiently than consuming meat, although meat is more nutritionally dense.

## REVIEW OF KEY CONCEPTS

1. Organisms derive energy in two principal ways, from photosynthesis and chemosynthesis.

2. Photosynthesis uses solar energy and carbon dioxide to create sugars and oxygen as a byproduct. All green plants and algae use photosynthesis to gain energy. Photosynthesis is the way by that most ecosystems on the Earth derive their energy. Chemosynthesis is a chemical reaction that gains energy from the breakdown of specific molecules that usually contain sulfur. This is the way deep sea hydrothermal vent communities derive their energy without access to sunlight.

3. Organisms that produce their own energy by these two processes are called autotrophs, or "self-feeders."

4. The amount of chemical energy created by autotrophs is called primary production. It is measured in the amount of biomass produced by an organism. Another way organisms derive energy on the Earth is by consuming other organisms. These types of living things are called heterotrophs, or "other feeders."

5. Autotrophs are known as producers in an ecosystem because they produce food. Organisms that consume the producers are called primary consumers. Other organisms that eat the primary consumers are called secondary consumers. Some ecosystems have tertiary consumers that eat the secondary consumers; these are also known as upper-level consumers. An important aspect of an ecosystem is the decomposers.

6. The primary consumers are also called herbivores because the eat plants. Secondary consumers are animals known as carnivores because they eat meat. Organisms that consume both plants and animals, such as humans, are called omnivores. The decomposers consume waste and dead organisms and are called detritivores.

7. The specific movement of energy through an ecosystem by a series of eating processes is called a food chain.

8. Many interrelated food chains are known as food webs. An energy pyramid is used to illustrate the way energy is used in a specific food chain. The pyramid is composed of a series of feeding levels called trophic levels. The movement of energy from one trophic level to another results in the loss of energy from a food chain as heat to the atmosphere. Energy pyramids also reveal the important link between the sun and all living things within an ecosystem.

## CHAPTER REVIEW

### Short Answer

1. Define the term autotroph and provide one example.

2. Describe the difference between chemosynthesis and photosynthesis.

3. Define the term heterotroph and provide one example.

4. Define the terms producer, primary consumer, and secondary consumer and explain how energy moves up through a food chain.

5. Provide an example for each of the following: herbivore, carnivore, omnivore, and detritivore.

6. Draw a simple food chain.

7. Approximately how much energy is lost between each trophic level in an energy pyramid?

8. Explain the differences and similarities between a naturally occurring food web and a human-created agricultural food chain.

### Energy Math

1. If the first trophic level in a food chain contains 5,000 units of energy, how many units of energy are transferred to the third trophic level?

### Multiple Choice

1. Most energy on the Earth is derived from:
   a. ecosystems
   b. producers
   c. the sun
   d. autotrophs

2. An organism that makes its own food is called a(n):
   a. secondary consumer
   b. autotroph
   c. heterotroph
   d. decomposer

3. Bacteria in hydrothermal vent communities deep in the ocean use which process to gain energy?
   a. chemosynthesis
   b. primary production
   c. photosynthesis
   d. nitrification

4. Combining carbon dioxide with water and sunlight to produce glucose sugar is an example of:
   a. chemosynthesis
   b. respiration
   c. photosynthesis
   d. nitrification

5. The amount of biomass that an ecosystem produces is called:
   a. chemosynthesis
   b. primary production
   c. photosynthesis
   d. nitrification

6. An organism that consumes only plants is called a(n):
   a. omnivore
   b. herbivore
   c. carnivore
   d. decomposer

7. The bacteria, fungi, and insects that eat waste and dead organisms are known as:
   a. omnivores
   b. herbivores
   c. carnivores
   d. decomposers

8. Approximately how much energy is given off to the atmosphere as heat between trophic levels in a food pyramid?
   a. 10%
   b. 30%
   c. 50%
   d. 90%

9. The movement of matter between the living and nonliving world is called:
   a. the food chain
   b. biogeochemical cycling
   c. the energy pyramid
   d. a food web

**10.** The source of most carbon in the carbon cycle is:
a. the hydrosphere
b. rocks in the Earth's crust
c. the atmosphere
d. fossil fuels

**11.** The process of gaining energy by breaking down sugars in the presence of oxygen is called:
a. photosynthesis
b. respiration
c. chemosynthesis
d. nitrification

## Matching

*Match the terms with the correct definitions*

a. autotrophs
b. chemosynthesis
c. heterotrophs
d. primary production
e. biomass

f. herbivores
g. carnivores
h. omnivores
i. detritivores

j. decomposers
k. food chain
l. respiration
m. photosynthesis

**1.** ____ A classification of organisms that must consume other organisms to gain energy, such as humans.

**2.** ____ A classification of organisms that eat both plants and animals.

**3.** ____ A type of organism that produces its own food by photosynthesis, such as a plant or algae, or by chemosynthesis, such as certain bacteria.

**4.** ____ A method of deriving energy from the breakdown or formation of organic compounds.

**5.** ____ The model pathway energy and matter takes through an ecosystem by a series of eating processes.

**6.** ____ A type of animal that only eats plants.

**7.** ____ An organism that breaks down and decays dead organisms or waste.

**8.** ____ The chemical process where carbohydrates are broken down in the presence of oxygen to derive energy and produce carbon dioxide.

**9.** ____ Meat-eating organisms.

**10.** ____ The process by which plants use photosynthesis to convert solar energy into chemical energy that is stored in plant material.

**11.** ____ A type of organism that consumes dead and decayed organisms or waste.

**12.** ____ The total dry weight of an organism.

**13.** ____ The chemical reaction that takes radiant energy from sunlight and combines it with carbon dioxide and water to form glucose and oxygen as a byproduct.

# Solid Bioenergy Fuel from Agriculture

---

## TOPICS TO BE PRESENTED IN THIS UNIT INCLUDE:

- Solid fuel biomass energy resources

- Solid biofuel energy production

- Environmental impacts of solid biomass fuels

## OVERVIEW

The use of solid biomass as a source of fuel has been employed by human beings for millions of years. Once a principle energy resource for society, solid biomass was slowly replaced by nonrenewable forms of fossil fuels such as coal, oil and natural gas. Today, because of the concern over the decline in these nonrenewable resources, biomass is once again being used as a primary fuel source. The ability for agriculture to produce renewable solid fuels to provide heat and power from plants, also known as bioenergy, will play an important role in our sustainable energy future.

## CHAPTER 19

# Solid Fuel Biomass Energy Resources

## KEY CONCEPTS

*After reading this chapter, you should be able to:*

1. Define the terms biomass and feed-stock.

2. Explain how the heating value of coal compares to that of biomass.

3. Describe the three different categories of woody residue and provide one example of each.

4. Explain in what form woody biomass can be burned.

5. Describe two crop residues that can be used as a solid biomass fuel.

6. Define the term energy crop.

7. Identify two woody biomass and one perennial grass solid biofuel crop.

## TERMS TO KNOW

biomass
cellulose
feedstock
woody residue

timber
agroforestry
board-foot
corn stover

energy crop
energy plantation
coppicing

# INTRODUCTION

**P**lants have long used the photosynthesis reaction to convert the sun's radiant energy into stored chemical energy to produce their own food. The use of biomass as a fuel source is just another way of storing solar energy for use later. Today, agriculture is employing this process by producing solid fuel crops that are considered replacements for traditional, nonrenewable fossil fuels such as coal, oil, and natural gas. Cornell University professor, Dr. Jerry Cherney commented on the notion of why we should use biomass as a fuel: "It (only) takes 70 days to grow a crop of grass pellet fuel, but takes 70 million years to grow of crop of fossil fuel." This straight forward approach to solving our energy requirements makes solid biomass an attractive source of renewable heat and an important aspect of future agricultural production.

## SOLID BIOMASS FEEDSTOCKS

**Biomass** is defined as the total dry weight of a plant. It is the product of photosynthesis that takes radiant energy from the sun, along with carbon dioxide and water to produce glucose (sugar) and oxygen as a byproduct. The glucose is then assembled into starches; the stored food within plants. The use of biomass as a fuel has long been employed by human beings. Naturally occurring biomass in the form of wood, grass, and animal dung has been burned to supply heat for warmth, cooking, and tool making by humans and their ancestors for tens of thousands of years (see Chapter 1). The starches that make up plant material come mostly in the form of **cellulose** and lignin. Cellulose is a main component of plant cell walls that is a polymer of glucose. Both of these compounds are carbohydrates, meaning they contain carbon, hydrogen, and oxygen (Figure 19-1).

Once the water content of plant material is reduced, the plant becomes a readily combustible fuel. The ratio, or number of hydrogen to carbon atoms (H/C) that exist in hydrocarbon fuels, provides an estimate of its potential heating value. The heating value is a measure of the amount of kilocalories of heat energy per gram of fuel that is generated during its combustion. The hydrocarbon with the highest heating value is methane, and can produce approximately 13 kcal per gram of fuel. Methane also has H/C ratio of 4.0; the highest of all hydrocarbon fuels. The average H/C ratio of plant biomass falls in between 1.2–1.7, giving it an average heating value of about 4.3 kcal/gram; slightly lower than that of coal. This illustrates the potential for using biomass as a renewable fuel source comparable to the current nonrenewable fossil fuels we use.

One important aspect of biomass is its water content. Because all living things contain water, it is necessary to remove it from biomass in order to make it easier to burn. The amount of water within biomass is known as its moisture content. The heating value of biomass is increased as the moisture content is lowered. The moisture content of solid biofuels varies

**biomass**—a short term for biological mass; the total dry weight of a plant

**cellulose**—a main component of plant cell walls that is a polymer of glucose

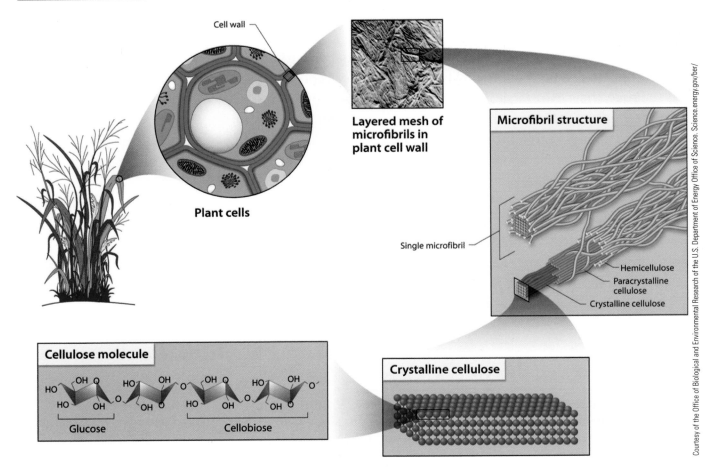

FIGURE 19-1 Illustrations of glucose and cellulose.

greatly depending on the type of plant being used. Woody plants that are "seasoned", meaning they have been air dried for a long period of time after cutting, and have a moisture content of less than 20% for optimal burning. Fresh cut plant material, referred to as "green" can have moisture contents more than 60% that make it inefficient to burn. Grasses and other non-woody plants are much easier to dry than woody plants like trees, and should have moisture contents below 10%. Basically, the lower the moisture content, the more efficient the biomass will be in producing heat energy. Biomass used specifically for energy production is known as

## CAREER CONNECTIONS

### Forester

A career in forestry involves the management of forests for both public and private use. Forestry covers many aspects of maintaining forests for timber, recreation, watershed, and wildlife management. The work of a Forester involves the study of forest ecosystems and how best to maintain them for a specific function or multiple use. To do this job you must have a love of working outdoors in all kinds of weather. Typically a Forester needs a background in biology and requires a four-year college degree in forestry management.

a biomass **feedstock**. A feedstock is a raw material used to supply energy or an industrial process. Currently there are three agriculturally based categories of biomass feedstock that can be used as a source for solid fuel; they include wood residue, crop residue, and energy crops.

**feedstock**—a raw material used to supply energy or an industrial process

## WOODY RESIDUE

**Woody residue**, also known as woody biomass, is the left-over biomass that results from the production, harvest, and processing of **timber**. Timber is defined as wood used for producing a product. The agricultural practice of growing timber is known as **agroforestry**. The forest industry in the United States produces timber for a variety of uses. Timber in America is used for lumber, paper, furniture, and a multitude of other wood-based products. Approximately 45% of all the timber produced in the United States goes to the construction of homes. Timber production is usually recorded in the amount of board-feet of lumber produced by sawmills. A **board-foot** is the unit of measure for lumber equal to the size of a piece of wood being 1 foot long, by 1 foot wide, and 1 inch thick. One board-foot is also equivalent to 144 cubic inches. In 2012, timber production in the United States produced more than 24 billion board-feet of lumber. The residue left over from this production can be used as a biomass energy feedstock. It's estimated for every 12,000 board-feet of timber harvested, more than 2.2 tons of woody residue is produced. The types of woody residue that can be used as a solid biomass fuel are often categorized as being primary, secondary, or tertiary.

**woody residue**—the left over biomass that results from the production, harvest, and processing of timber

**timber**—wood used for producing a product

**agroforestry**—the agricultural practice of growing timber

**board-foot**—the unit of measure used for lumber equal to the size of a piece of wood being 1 foot long, by 1 foot wide, and 1 inch thick

### Primary Woody Residue

Primary woody residue comes in the form of slash that includes the branches and trimmings from trees (Figure 19-2). Managing forests for timber production involves many activities that produce slash. Timber stands are trimmed to encourage the growth of desired trees. The stands

**FIGURE 19-2** Forest slash.

© Christopher Kolaczan/Shutterstock.com

are also cleared of debris and other vegetation that can build-up on the forest floor, increasing the potential for forest fires. Both of these practices produce slash that was traditionally disposed of in landfills or composted. Gathering and processing this waste for fuel makes timber production more efficient and cost effective.

## Secondary Woody Residue

Secondary woody residue includes the waste produced from wood processing facilities (Figure 19-3). This includes biomass in the form of bark, undesirable pieces of wood, and sawdust produced at lumber mills. Wood wastes from manufacturing facilities that make wood products are also considered as a secondary woody residue.

## Tertiary Woody Residue

Tertiary woody residue, known as urban residue, includes any wood waste produced from landscaping and construction (Figure 19-4). All of these residues can be dried and chopped into woodchips that can then be burned as fuel.

## Economics of Woody Residue

The cost and heat content of woody residue biomass is comparable to that of coal. The average energy content of wood residue biomass is approximately 8,750 Btus per pound, while bituminous coal ranges between 8,000–13,000 Btus per pound. Along with turning the woody residue into woodchips, this biomass feedstock can also be pelletized. Biomass pellets are small cylinders of biomass about 1 inch long, and the width of a pencil. They are produced by chopping the wood into fine pieces, that are then fed through a press that produces the pellets. Pellets are more consistent in size, and

**FIGURE 19-3** Sawmill waste being used by a power plant.

**FIGURE 19-4** Collection of urban landscape waste to be converted into combustible biomass.

are easier to manage within a combustion unit. The Department of Energy reports primary and secondary wood residue from the forest products industry accounts for 2% of the energy we produce in the United States each year. Predictions about expanding the use of these timber waste resources could potentially supply more than 225 million tons of solid biomass fuel annually representing about 25% the amount of coal we use each year.

## CROP RESIDUE

Crop residue solid biomass fuel is the remains of certain agricultural crops not used directly in our food system. They consist of leaves, stems, and other unused parts of plants grown in farm fields. Currently, soybeans, corn, and small grains like wheat are the largest crops grown in America. The residue left over from their harvest and processing could supply biomass for use as a solid fuel. Most corn is used for food, and leaves the other parts of the plant (the stalks, leaves, and corncobs) in the fields during harvest. This material is known as **corn stover** (Figure 19-5). Corn stover has an energy content of about 7,560 Btus per ton; slightly lower than bituminous coal that ranges between 8,000–11,000 Btus. You must burn about 16% more corn biomass to equal the average heat output of coal. Corn production in the United States makes about 75 million tons of corn stover each year and represents the largest amount of crop waste. Wheat straw is another crop residue that could be used as a biomass fuel. Straw has a lesser heating value than corn stover, producing about 6,840 Btus per ton. There is approximately 11 million tons of wheat straw produced annually. Other crops can produce residue as well. These include soybeans, cotton, potatoes, oats, rice, and sorghum. Together, the total amount of crop residue

**corn stover**—the left over biomass that results from the production, harvest, and processing of corn

Courtesy of National Renewable Energy Laboratory/Photo by Jim Yost.

**FIGURE 19-5** Corn stover.

grown today amounts to about 113 million tons of dry biomass. These residues can be bailed or chopped and made into pellets in a similar way as woody biomass is processed into pellet fuel.

## ENERGY CROPS

**energy crop**—an agricultural crop grown specifically for use an as energy source

An **energy crop** is an agricultural crop grown specifically for use an as energy source. These crops are also known as dedicated feedstocks. An ideal energy crop is one that is not considered a food crop. Energy crops are usually a better option when it comes to using agriculture to produce energy because they eliminate the problems that arise when a traditional food crop is used as both an energy resource and a food resource. This competition often results in the inflation of the price of the crop that can raise the cost of food. For example, corn is used for many food products in the United States and as feed for cattle.

Approximate 50% of the corn grown in America is used to feed to cattle. Corn can also be used to make ethanol that is a biofuel. In 2007, the United States government mandated 24% of the corn grown be used to make ethanol that is added to gasoline. This demand has caused the price of corn to increase. From August, 2010 to January, 2011 corn prices have more than doubled. That is good news for corn growers because they are getting paid more for their crop, but bad news for dairy farmers and cattle growers who buy the corn as feed and reduces their profits. Other factors have lead to the rise in corn prices besides ethanol production, such as lowered production rates, increasing imports, weather–related losses of corn crops in some countries, and the lowered value of the dollar. One way to alleviate this competition between food and fuel is to grow crops specifically used for fuel that do not compete with the food market. This is what an energy crop is. Although there are different types of energy crops that can be grown to produce solid biomass fuels, choosing the right crop for the specific climate and growing conditions that exist in a region is crucial for their success.

In the United States there are a few energy crops identified as being especially beneficial as a renewable fuel. These include shrub willows, hybrid poplars, switchgrass, and *Miscanthus* (Figure 19-6). All energy crops that are being developed tend to be perennial, drought resistant, and high yielding. Another benefit to growing energy crops is they can be produced on marginal croplands not presently good enough to sustain food crops. It is important that energy crops be non-food crops so they do not directly affect the price of food.

One of the challenges facing energy crops is how to efficiently transport the feedstock to the processing plant or power plant. Studies show increasing the density of the biomass in the field at harvest is key to lowering production costs because it makes the transport of the biomass more efficient. Research is being conducted on how to create dense pellets or cubes of biomass during harvest in the field. Having energy crops located close to the processing plant is important. The closer the feedstock is to the plant, the greater the efficiency and the greater the reduction in costs. Another challenge involves the potential long-term storage of the biomass prior to its use. Making sure the biomass does not begin to spoil and breakdown is

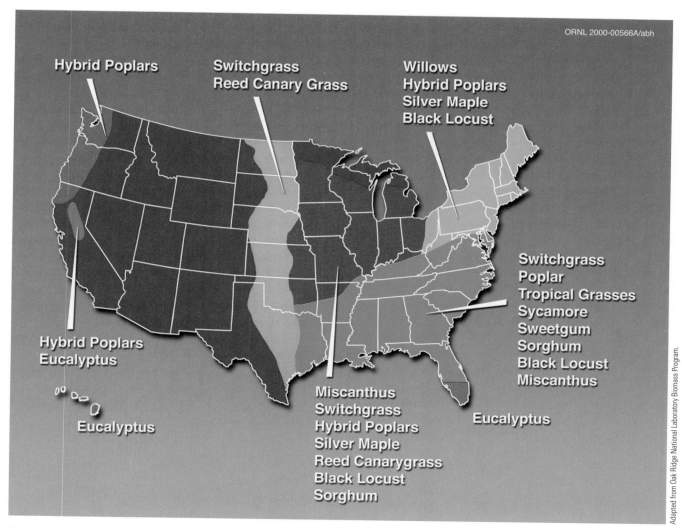

**FIGURE 19-6** Map of energy crop growing regions.

crucial for improving its efficiency. The creation of high-density biomass cubes that are resistant to spoilage may also aid in the success of bioenergy crops.

## Shrub Willow

The shrub willow is a perennial species of woody plant similar to the species *Salix purpurea* that grows to a height of about 15–20 feet. There are more than 450 different species of shrub willow that grow in North America and Eurasia. Researchers at the State University of New York's College of Environmental Science and Forestry and at Cornell University have created more than 200 hybrid species of shrub willow that are an ideal fuel crop. These woody plants reach full height in three years and are resistant to insects, disease, and droughts (Figure 19-7). The trees are harvested and chopped into woodchips that can be burned in furnaces, boilers, or used in gasifiers. The shrub willow was once a major crop grown in New York that was used for basket making and furniture. Today, shrub willow is being grown specifically as a source of solid biomass fuel. The growing of a perennial dedicated biofuel feedstock is also known as an **energy plantation**.

The cultivation of shrub willow begins with the planting of willow cuttings. Once the soil has been tilled, cuttings are placed in the ground

**energy plantation**—growing a perennial dedicated biofuel feedstock

Courtesy of Lawrence Smart, Cornell University

**FIGURE 19-7** Shrub willow.

about 2–3 feet apart in double rows. This arrangement produces between 5,000–6,000 trees per acre. During the first year of growth, weeds need to be suppressed either mechanically or chemically. New trials on the use of cover crops planted between rows of young willows are being researched as a means of weed suppression. The use of nitrogen fixing clovers as a cover crop can act not only as a weed suppresser but also as a source of plant available nitrogen.

After the first growing season, the willows should grow to a height between 3–8 feet tall. Once they drop their leaves in the fall, the willows are then **coppiced**. Coppicing is the practice of cutting the stem of the sapling to spur the growth of multiple stems. Typically shrub willows are coppiced about two inches above the soil surface. This allows the tree to create a lush canopy the following year.

> **coppicing**—the practice of cutting the stem of the sapling to spur the growth of multiple stems

After the second growing season, the double row of tree plantings should grow to the point where their canopies join together, blocking sunlight from reaching the ground. This reduces the need to suppress weeds. The plants are finally harvested at the end of the third year of growth, just after they enter dormancy. This usually occurs during November in the Northeast. The plants are harvested using a combine that can chop stems as thick as six inches. The willows are simultaneously harvested and chopped into woodchips in the field (Figure 19-8). One acre of shrub willow can produce between 4–6 tons of dry biomass per acre each year. Several shrub willow plantations are currently producing biomass in the Northeast for use as solid biofuels. The East Lycoming School District in Pennsylvania is in the process of developing a 60 acre shrub willow plantation to supply the school's woodchip boiler. This project will produce 60–70% of the campus' fuel needs.

The energy balance for shrub willow solid biomass fuel is about 1–13. This means for every unit of energy put into producing shrub willow biomass yields about 13 units of energy. The current energy balance for most

Courtesy of Lawrence Smart, Cornell University.

**FIGURE 19-8** Harvesting willow.

fossil fuels (coal, oil, and natural gas) is about 1–1, although gasoline actually has a negative energy balance, meaning it takes more energy to produce it than it gives.

## Hybrid Poplar

Another form of solid biomass fuel is the hybrid poplar (Figure 19-9). Poplars are a quick growing tree species known as cottonwoods. One common hybrid poplar grown as an energy crop (*Populus x canadensis*) is a cross between the Eastern Cottonwood and the Back Cottonwood. Poplar plantations plant cuttings spaced 8 to 10 feet apart. The trees are ready to harvest after the first 5–6 years of growth, reaching heights of 60 or more feet. The average production of biomass after 6 years is approximately 17 dry tons per acre. Typically, polar plantations can produce between 1–6 dry tons of biomass annually. The trees are harvested by cutting them down. They can then be chopped into woodchips. Once a poplar stand is established, there is no need to replant after harvest because the trees will re-sprout shoots from their stumps.

## Other Types of Woody Biomass Crops

Besides poplars and shrub willows, other woody biomass fuel crops include yellow pine and eucalyptus trees. The eucalyptus tree is one of the fastest growing trees in the world with growth rates of approximately six feet per year. Both of these species can only be grown in warmer climates. All of these woody biomass fuels are known as short rotation woody crops, and have a great potential for becoming a renewable solid fuel replacement for coal.

Courtesy of Oak Ridge National Laboratory.

**FIGURE 19-9** Poplar.

## Switchgrass

Besides woody biomass, perennial grasses are also good candidates for solid biomass fuels. One crop being touted as a productive perennial bio-fuel is Switchgrass (*Panicum virgatum*) (Figure 19-10). Switchgrass is a native prairie grass to North America. It is drought resistant and grows to a height of about 8 feet. Switchgrass fields are established by plant-ing 10 pounds of seed per acre in a prepared field. It takes two years for the grass to establish itself, with about 10 to 20 plants growing within one square meter. The plants are then harvested by cutting them at a height of about six inches above the soil surface. A mature field of Switchgrass in the Northeast can yield about 3–7 dry tons per acre, and between 5–10 tons in the Midwest. During the first two years of growth, weed suppression is nec-essary to establish the field. Once harvested and dried, the Switchgrass can be bundled and burned like wood logs, or chopped and converted into pel-lets. Grass pellets produce about 5% less heat per pound of fuel than wood fuel and produce more ash. The energy balance for Switchgrass is about 1:15, meaning every unit of energy put into the fuel to grow it, yields about 15 units of energy.

## *Miscanthus*

Another potential energy crop is *Miscanthus giganteus* (Figure 19-11). This perennial grass is native to Asia and has shown to be a promising energy crop that can be grown in the Midwest and the Northeast. *Miscanthus* has demonstrated yields of up to 25 tons per acre and has a higher energy input/output ratio than Switchgrass. This perennial grass can reside in a field for up to 30 years before it needs to be replanted, and grows to a height of 8–12 feet. Twenty tons of dried *Miscanthus* provides about the same amount of heat as 12 tons of coal. In Europe, *Miscanthus* is used primar-ily as a combustible fuel in power plants by co-firing it with coal. Co-firing

Roy Kaltschmidt, Lawrence Berkley National Laboratory.

**FIGURE 19-10** Switchgrass.

**FIGURE 19-11** *Miscanthus.*

Courtesy of S. Long, University of Illinois.

means burning two or more fuels at once. Other perennial grasses are being explored as potential solid biomass fuel crops. These include big bluestem, reed canary grass, and eastern gamagrass.

## REVIEW OF KEY CONCEPTS

1. Biomass is a term that describes the total amount of dry plant material produced by photosynthesis. Biomass is mostly in the form of cellulose and lignin that are polymers of glucose sugar.

2. Biomass can be harvested, dried, and burned much like coal. The heat content of biomass compared to bituminous coal is just slightly less.

3. Woody residue is a type of biomass that comes from the timber industry. It includes primary sources like slash and trimmings from forest stands, secondary sources from sawmills and wood processing facilities, and tertiary sources from landscaping and construction.

4. The cost and heat content of woody residue biomass is comparable to that of coal. The average energy content of wood residue biomass is approximately 8,750 Btus per pound, while bituminous coal ranges between 8,000–13,000 Btus per pound. Along with turning the woody residue into woodchips, this biomass feedstock can also be pelletized.

5. Crop residue can be a source of solid biomass fuel. This includes the leftover parts of plants discarded during harvesting or processing. Corn stover, the stalk, leaves, and corncobs of the corn plant and wheat straw are two large sources of crop residue.

6. A feedstock is a raw material used to supply energy or an industrial process. Energy crops are dedicated crops grown solely for use as a fuel feedstock.

7. Short rotation woody crops are wood-based plants grown to quickly produce biomass. Shrub willows, hybrid poplars, and eucalyptus are examples woody biomass crops that are grown on energy plantations. Perennial grasses can also be used a source of solid biomass. Switchgrass, *Miscanthus giganteus*, big bluestem, reed canary grass, and gama-grass are all varieties of grasses that can be harvested, dried, and burned directly or turned into fuel pellets. The heat content of solid biomass fuels is similar to that of coal.

## CHAPTER REVIEW

### Short Answer

1. Define the terms biomass and feedstock.

2. How does the heating value of coal compare to that of biomass?

3. What are the three different categories of woody residue and provide one example of each?

4. Explain in what form woody biomass can be burned.

5. What are two crop residues that can be used as a solid biomass fuel?

6. What is an energy crop?

7. Identify two woody biomass and one perennial grass solid biofuel crops.

8. Explain why the use of biomass as a solid fuel source is just another form of solar energy.

### Energy Math

1. If the United States burned 936 million tons of coal in 2009 to produce electricity, how many acres of shrub willow would we need to replace coal?

### Multiple Choice

1. The total amount of dry weight of a plant is called:
   a. photosynthesis
   b. respiration
   c. biomass
   d. feedstock

2. The highest moisture content biomass should have in order to burn efficiently is:
   a. 60%
   b. 40%
   c. 20%
   d. 5%

3. Wood wastes from manufacturing facilities that make wood products or saw mills are also known as:
   a. primary woody residue
   b. secondary woody residue
   c. tertiary woody residue
   d. food processing waste

4. Urban residue, including any wood waste produced from landscaping and construction is known as:
   a. primary woody residue
   b. secondary woody residue
   c. tertiary woody residue
   d. food processing waste

5. Wood waste from forest slash and trimmings is called:
   a. primary woody residue
   b. secondary woody residue
   c. tertiary woody residue
   d. food processing waste

6. Which crop residue represents the largest potential source of biomass from crop production?
   a. Wheat straw
   b. Corn stover
   c. Soybean plants
   d. Switchgrass

7. Which potential woody biomass fuel can be harvested every 6 years?
   a. Hybrid poplars
   b. Yellow pine
   c. Eucalyptus
   d. Shrub willow

8. Which biomass fuel is coppiced to increase its yield?
   a. Hybrid poplars
   b. Yellow pine
   c. Eucalyptus
   d. Shrub willow

9. Which biomass crop is harvested every three years?
   a. Hybrid poplar
   b. Switchgrass
   c. Eucalyptus
   d. Shrub willow

10. Coppicing is the process of:
    a. weed control
    b. cutting to stimulate growth
    c. harvesting biomass
    d. drying to remove moisture

## Matching

*Match the terms with the correct definitions*

a. biomass
b. feedstock
c. woody residue
d. timber
e. board foot
f. agroforestry
g. corn stover
h. energy crop
i. energy plantation
j. coppicing

1. ____ Wood used for producing a product.

2. ____ The growing of a perennial dedicated biofuel feedstock.

3. ____ The unit of measure used for lumber that is equal to the size of a piece of wood being 1 foot long, by 1 foot wide, and 1 inch thick.

4. ____ The agricultural practice of growing timber.

5. ____ An agricultural crop grown specifically for use an as energy source.

6. ____ The left over biomass that results from the production, harvest, and processing of timber.

7. ____ The total dry weight of a plant.

8. ____ The unused portion of a plant that includes the stem, leaves, and cob.

9. ____ The raw material that is used to supply energy or an industrial process.

10. ____ The process of cutting a tree to spur on the growth of multiple shoots.

## CHAPTER 20

# Solid Biofuel Energy Production

## KEY CONCEPTS

*After reading this chapter, you should be able to:*

1. Define the term biopower.

2. List five sources of biomass that can be used to supply biopower.

3. Describe the process of direct combustion of solid biomass.

4. Explain co-firing and describe two benefits of using co-firing.

5. Describe the process of combined cycle biopower.

6. Explain the benefits of modular combined cycle biopower.

7. Identify three methods of home heating that can be derived from biomass.

## TERMS TO KNOW

biopower

biosolids

co-firing

combined cycle biopower

pyrolysis

syngas

combined heat and power

hydronic heat

# INTRODUCTION

The birth of the era of electricity during the early part of the twentieth century was driven by the use of coal. Coal is the prime producer of heat that powers thousands of electric turbines across the nation by converting water into steam. Today, because of the negative effects coal has on the environment, new, renewable sources of heat are being explored. The use of biomass as a means to create heat by combustion to produce steam and electricity is known as biopower. Today, the use of biopower is the second largest producer of renewable electricity in the United States behind hydropower. The use of biomass as a source of heat for homes and businesses is also becoming more widespread. The traditional use of fossil fuels like oil and natural gas to provide warmth in colder climates can be replaced with sustainable biomass energy resources produced by American farmers.

## BIOPOWER

**Biopower** is the use of the combustion of biomass to produce electricity. Today it is becoming a more attractive alternative to using coal because it is a renewable energy resource. Common sources of biomass used for biopower include wood residues, crop residues, energy crops, food processing waste, and **biosolids**. Biosolids are the dried, solid remains collected from wastewater treatment plants and animal manures. Today in the United States there are approximately 300 biopower plants that use solid biomass fuels. Together these power plants produce more than 6,000 megawatts of electricity. This is the approximate amount of electricity required to power more than 4 million homes. About 86% of biopower electrical plants use wood as the solid biomass fuel, 13% using agricultural residues, and 2% paper pulp manufacturing residues known as black liquor. Black liquor is composed of mostly hemicellulose and lignin carbohydrates that are byproducts of making paper. There are currently three main methods of using solid biomass to generate electricity: direct combustion, co-firing, and combined cycles, also known as gasification.

**biopower**—the use of the combustion of biomass to produce electricity

**biosolids**—the dried, solid remains collected from wastewater treatment plants and animal manures

### Direct Combustion

Direct combustion biopower uses the same basic technology as traditional coal-fired power plants. Biomass fuel is burned in a combustion chamber that generates the heat needed to run a Rankine Cycle engine. Rankine Cycle heat engines ultimately convert the heat of combustion into steam that drives a turbine generator. Biomass typically generates 7,000–9,000 Btus per pound of fuel, that falls in between the two types of coal commonly used to produce electricity. Bituminous coal ranges between 10,000–15,000 Btus per pound, and sub-bituminous 8,000–13,000 Btus per pound. The majority of biopower electrical plants use direct combustion technology. There are two types of direct combustion boilers; Stoker Boilers and Fluidized Bed Boilers.

## Stoker Boilers

Stoker combustion uses a traveling grate, called a stoker, to move fuel through the combustion chamber (Figure 20-1). Air and heat moves up from below the grate, burning the fuel and creating the high temperature heat of around 2,000°F (3,632°C), then used by a heat exchanger to produce steam. Fluidized bed boilers inject the biomass fuel into the combustion chamber with high-pressure air. The biomass burns in a mixture of hot air that creates a fluid-like, high temperature combustion gas with temperatures around 1,500°F (2,732°C). This heat is then passed through a heat exchanger that creates steam.

## Fluidized Bed Boiler

The biomass must be chopped into fine pieces for use in a fluidized bed boiler. This results in less unburned fuel within the combustion chamber than in a stoker boiler. Because of this, fluidized bed combustion is more efficient compared to the stoker. The fluidized bed generally emits less nitrogen oxides because it burns at a lower temperature. A 49 megawatt, direct combustion power plant in Northern California consumes wood waste from mills, orchards, and logging operations to produce electricity. The plant uses three traveling grate stoker-fired boilers to burn 450,000 tons of biomass residue each year.

## Co-firing

**co-firing**—the process of burning biomass that is mixed with coal or natural gas

**Co-firing** is the process of burning biomass mixed with coal, oil, or natural gas. Biomass is used as a supplement to burning coal in power plants, and

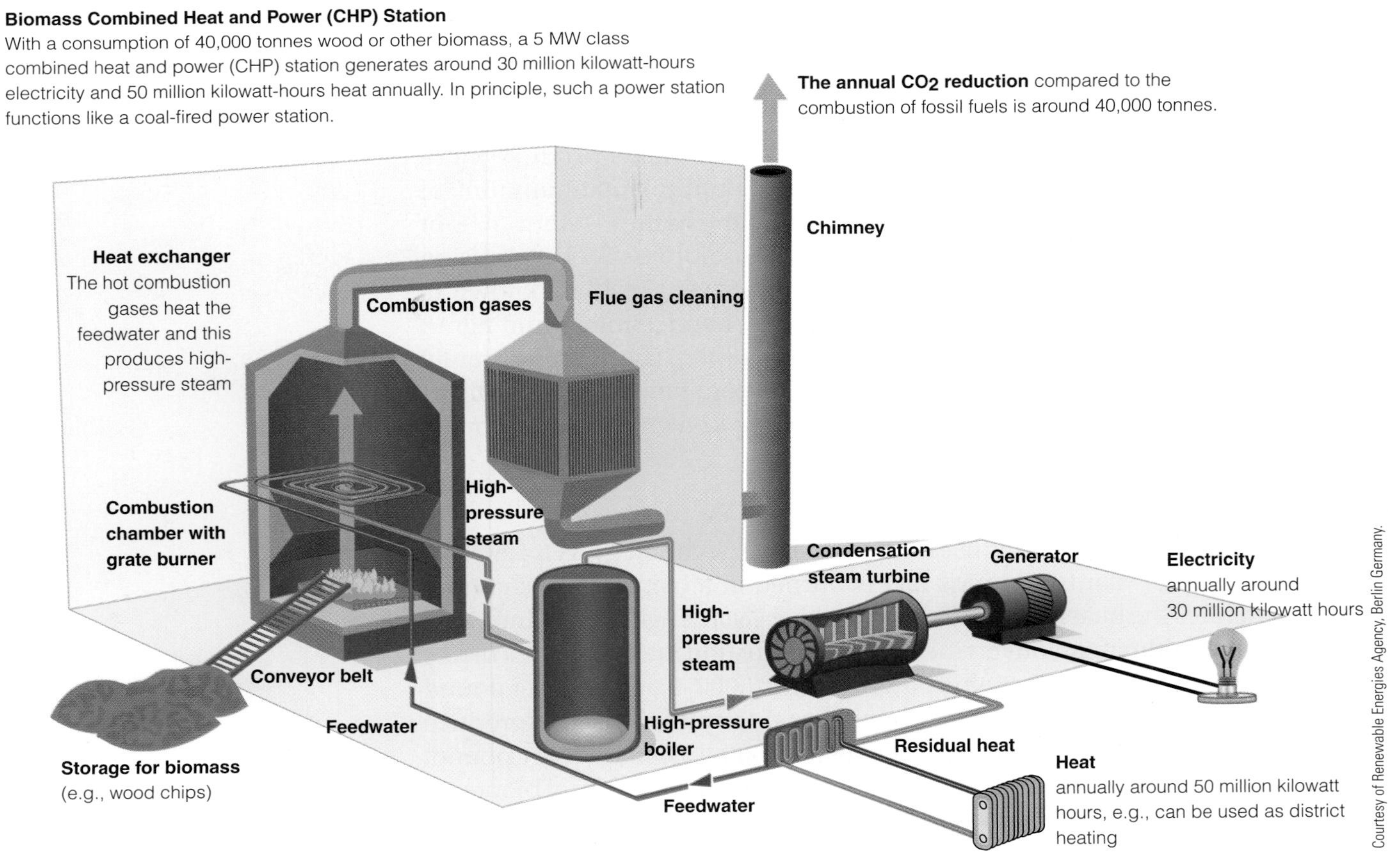

**Biomass Combined Heat and Power (CHP) Station**

With a consumption of 40,000 tonnes wood or other biomass, a 5 MW class combined heat and power (CHP) station generates around 30 million kilowatt-hours electricity and 50 million kilowatt-hours heat annually. In principle, such a power station functions like a coal-fired power station.

**The annual CO2 reduction** compared to the combustion of fossil fuels is around 40,000 tonnes.

**Heat exchanger** The hot combustion gases heat the feedwater and this produces high-pressure steam

**Combustion gases**

**Flue gas cleaning**

**Chimney**

**Combustion chamber with grate burner**

**High-pressure steam**

**Condensation steam turbine**

**Generator**

**Electricity** annually around 30 million kilowatt hours

**Conveyor belt**

**Feedwater**

**High-pressure steam**

**High-pressure boiler**

**Residual heat**

**Heat** annually around 50 million kilowatt hours, e.g., can be used as district heating

**Storage for biomass** (e.g., wood chips)

**Feedwater**

Courtesy of Renewable Energies Agency, Berlin Germany.

**FIGURE 20-1** Direct combustion biomass power plant.

represents about 10%–20% of the total fuel burned. Biomass is injected into the boilers with the coal, creating heat used to produce steam. The use of biomass reduces the emissions of both nitrogen oxides and sulfur oxides. Also, because biomass represents about 10%–20% of the fuel burned, there is a reduction in the amount of carbon dioxide produced. This is because biomass represents a renewable source of carbon. Co-firing is an easy-to-implement strategy to use biomass as an energy source because it can be used directly in most every existing coal-fired power plant in operation today. It requires retrofitting of equipment to existing boilers in power plants in order to maximize its efficiency. The Greenridge power plant in Dresden, New York, uses wood residue to co-fire with coal. Currently the plant burns between 600–1,200 tons of biomass monthly, representing about 10% of the energy produced by the 108 megawatt capacity plant.

## Combined Cycle Biopower and Biomass Gasification

**Combined cycle biopower**, also known as biomass gasification, is the process of converting solid biomass fuel into a combustible gas that can be burned to produce electricity. Biomass gasification use **pyrolysis**, the heating of a hydrocarbon fuel in a low oxygen environment. This converts the solid biomass into carbon monoxide and hydrogen **(syngas)**; a combustible gas. The syngas produced from the pyrolysis of biomass is the same as that produced from the destructive distillation of coal.

The process of combined combustion begins when the biomass is chopped into woodchips or made into pellets (Figure 20-2). The biomass

**combined cycle biopower**— also known as biomass gasification, the process of converting solid biomass fuel into a combustible gas that can be burned to produce electricity

**pyrolysis**—the heating of a hydrocarbon fuel in a low oxygen environment

**syngas**—a combustible gas composed of carbon monoxide and hydrogen

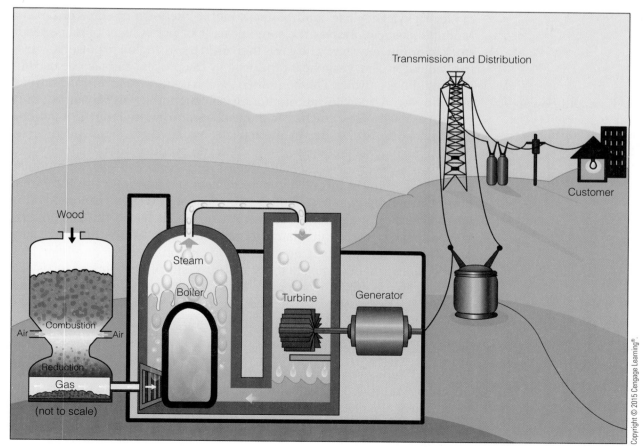

**FIGURE 20-2** Biomass gasification power plant.

fuel is then fed into a combustion chamber, where it is heated at high temperature, usually exceeding 1,400°F (760°C), in a low oxygen environment. This converts the solid biomass into syngas. Gasification converts approximately 85% of the biomass into combustible gas, 10% solid biochar, and 5% bio-oil. Biochar is a byproduct of biomass gasification that makes an excellent soil additive. Particulates of biomass and other impurities must be separated from the gas. This is done using cyclone separators, wet scrubbers, or dry filters. Once the gas has been cleaned, it is fed into a gas turbine where it is combusted. The turbine powers a generator that produces electricity.

The most efficient biomass gasification systems use a combined cycle. This process uses the hot exhaust gas produced from the gas turbine to drive a Rankine Cycle engine that produces steam to drive an electric turbine. The combination of the gas turbine with the Rankine engine increases the efficiency to about 60%. The average efficiency of a direct combustion power plant is between 35%–40%. This technology is being applied to use many different sources of biomass. A biopower plant in Burlington, Vermont uses biomass gasification to produce eight megawatts of electricity. This consumes 200 tons of woodchips each day. This gasification process adds to the total output of the plants 50 megawatts, and is enough to supply electricity to the City of Burlington. The other 42 megawatts is produced by the direct combustion of biomass in the form of wood residue.

Although most biomass gasification systems are used for large-scale electricity production, there is also the possibility to use this technology on a smaller scale. These systems are known as modular biomass gasification units. These stand-alone systems can produce between 5–50 kilowatts of electricity. They use solid biomass fuel in the form of chips or pellets. Modular gasification uses the same process of pyrolysis of biomass feedstock to produce syngas. The gas is then used by an engine to power an electric generator. The waste heat produced by the process can also be used to heat buildings or water. This combined system is known as **combined heat and power**. A 100 kilowatt modular gasification unit will consume 2 tons of biomass each day. Smaller, 5 kilowatt systems need about 100 pounds per day. These units are an attractive alternative for small businesses, manufacturing facilities, and schools that can produce both electricity and heat using a variety of renewable biomass resources. A 15 kilowatt modular gasification unit was installed to provide heat and power for the vocational horticulture program at North Park High School in Walden, Colorado. This unit is operated by agriculture students, who feed between 120–240 pounds of wood pellets into the unit daily to supply all of their heating and electricity requirements to run their greenhouse operation. Another modular biomass gasifier is being tested at the State University of New York's Morrisville campus. This biomass gasifier produces 100 kilowatts of both heat and power and is designed to use the bedding from horse stalls as its fuel source.

**combined heat and power—** the process of using the waste heat from a power plant to heat nearby buildings

## BIOMASS HEAT

The use of biomass as a renewable fuel can be used to supply heat during the colder months. Traditionally, oil and natural gas are the primary fuels burned to heat homes during winter. Wood stoves also provide many homes with supplemental heat. All of these can be replaced by using more efficient, sustainable solid biomass fuels. The combustion of biomass pellets produces less ash and particulate matter than wood, along with producing more complete combustion creating virtually no smoke. The efficient burning of biomass pellets also creates more heat than traditional wood stoves. Pellet stoves are self-feeding and can be operated by a thermostat can turn them on and off automatically to sustain a desired amount of heat output. They can also be filled with enough pellets for one or more days depending on the size of the hopper. There are two basic types of automated systems that can use solid biomass as a heat source; free-standing stoves or pellet furnaces.

Free-standing biomass pellet stoves can be used as space heaters within a building or home. These stand-alone units can produce between 5,000 and 50,000 Btus, which is enough heat for an average room or an entire house. They are self-starting and self-feeding. They typically use a blower to distribute heat produced by the stove. Stand-alone pellet stoves are designed to burn a variety of biomass fuels. These include corn, grains, fruit pits, wood pellets, and grass pellets. Some pellet stoves burn only one type of biomass while others can use different fuels, so it is important to find out the capabilities of the specific type of pellet stove you are considering for use. A typical stand-alone pellet stove will burn between 40 and 80 pounds of pellets each day, depending on the outside temperature.

Pellet furnaces are designed to work like traditional fossil-fuel furnaces installed within homes and businesses today (Figure 20-3). Different fuel sources, usually in the form of biomass pellets, are burned to produce heat. Corn and wood pellet furnaces are widely used, along with newer models that can also be fueled by grass pellets. The pellets can be delivered and stored in individual 40 pound bags, or large deliveries can be made to a storage hopper from a truck. The delivery trucks dispense the pellets using large diameter hoses. Pellet furnaces can create forced air heat that is distributed by a blower and ventilations ducts. They also can be used to provide hydronic heat.

**Hydronic heat** uses circulating hot water at temperatures between 130°F–180°F (54°C–82°C). Biomass fuel is burned within a combustion chamber that heats the water. The hot water is then circulated by a series of pumps to radiators to distribute the heat. Hydronic fluid can also be used in radiant floor heating. In this application, hot water is circulated through pipes installed within flooring. This creates an efficient source of indoor warmth. Biomass furnaces can provide a source of heat to produce domestic hot water. Although most pellet stoves and furnaces in use today use wood as their biomass source, new grass pellets are being developed (Figure 20-4). Grass is a quick growing biomass crop that has a heat value similar to wood pellets. The only drawback to using grass pellets is they produce more ash than wood pellets that requires frequent cleaning of the ash pan compared to burning hardwood pellets. Also, not all pellet stoves and furnaces are designed to use grass pellets (Figure 20-5).

**hydronic heat**—the use of circulating hot water heated to temperatures between 130°F–180°F (54°C–82°C) to heat a building or home

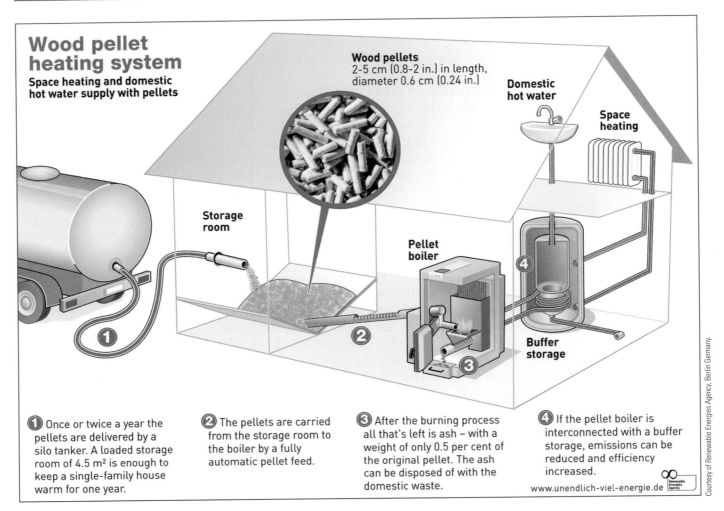

# Wood pellet heating system

**Space heating and domestic hot water supply with pellets**

**Wood pellets**
2–5 cm (0.8–2 in.) in length, diameter 0.6 cm (0.24 in.)

**Domestic hot water**

**Space heating**

**Storage room**

**Pellet boiler**

**Buffer storage**

**①** Once or twice a year the pellets are delivered by a silo tanker. A loaded storage room of 4.5 m² is enough to keep a single-family house warm for one year.

**②** The pellets are carried from the storage room to the boiler by a fully automatic pellet feed.

**③** After the burning process all that's left is ash – with a weight of only 0.5 per cent of the original pellet. The ash can be disposed of with the domestic waste.

**④** If the pellet boiler is interconnected with a buffer storage, emissions can be reduced and efficiency increased.

www.unendlich-viel-energie.de

**FIGURE 20-3** Pellet boiler system.

**FIGURE 20-4** Grass pellets.

**FIGURE 20-5** Pellet stove.

## REVIEW OF KEY CONCEPTS

1. Biopower is the use of biomass to supply electricity. Currently there are three methods to convert solid biomass fuel into electrical power: direct combustion, co-firing, and combined cycling.

2. Five sources that can be used to supply biopower are wood residues, crop residues, energy crops, food processing waste, and biosolids.

3. Direct combustion uses a combustion chamber to burn the biomass that has been chipped. This is done in two types of boiler systems; the stoker boiler and fluidized bed combustion.

4. Co-firing is the process of burning solid biomass fuel with coal; typically in a mixture of 5%–10% biomass and 90%–95% coal. Co-firing reduces the emissions of sulfur oxides, nitrogen oxides, and carbon monoxide.

5. Combined cycle biomass combustion is the process that solid biomass is converted into gas. This is accomplished by pyrolysis, the process of heating biomass in a low oxygen environment. This produces syngas, a combustible mixture of hydrogen and carbon dioxide gas. The syngas is then burned in a gas turbine that powers an electrical generator. The hot exhaust gas from the gas turbine can then be used to heat water into steam to turn another turbine. This combined cycle makes the system efficient.

6. Smaller, combined cycle units, called modular units can be used to create both heat and electricity on lesser scale. These units use solid biomass fuels to produce between 5–50 kilowatts of electricity. The use of biomass to create both heat and electricity is called combined heat and power.

7. Solid biomass can be used to produce heat for buildings and homes during winter. This can be done using stand-alone pellet stove space heaters or large pellet furnaces. Biomass fuels like corn grain or wood pellets can be used to create heat from their combustion. This energy can be used in forced hot air or hydronic heating systems. Grass pellets are also being developed, that have a similar heat content to that of woody biomass.

## CHAPTER REVIEW

### Short Answer

1. What is biopower?

2. List five sources of biomass that can be used to supply biopower.

3. What is the process of direct combustion?

4. Explain co-firing and describe two benefits of using co-firing.

5. What is combined cycle biopower?

6. What are the benefits of modular combined cycle biopower?

7. Identify three methods of home heating that can be derived from biomass.

8. Briefly discuss the advantages that using solid biomass has over traditional fossils fuels used to produce heat and power:

### Energy Math

1. If a traditional coal-fired power plant use 3,200 tons of coal per day to produce electricity, how much coal could be saved each day if 10% of the plant's fuel was replaced by co-firing with biomass?

2. If a 15 kW modular combined power unit consumes 180 pounds of biomass each day, how many acres of shrub willow would you need to grow to supply fuel to power it for one year, if one acre of willows yields 5 tons of dry biomass?

### Multiple Choice

1. The use of biomass to produce electricity is called:
   a. biosolid
   b. biopower
   c. pyrolysis
   d. syngas

2. Which of the following terms is defined as the conversion of biomass into a combustible gas used to power a gas turbine?
   a. Direct combustion
   b. Co-firing
   c. Combined cycle
   d. Hydronics

3. Which of the following terms is defined as the combustion of solid biomass in a fluidized bed boiler?
   a. Direct combustion
   b. Co-firing
   c. Combined cycle
   d. Hydronics

4. The process of burning biomass together with coal is called:
   a. direct combustion
   b. co-firing
   c. combined cycle
   d. hydronics

**5.** The process of using circulating hot water as a heat source is called:
   a. direct combustion
   b. co-firing
   c. combined Cycle
   d. hydronics

**6.** What is the approximate efficiency of a combined cycle biomass power plant?
   a. 5%
   b. 10%
   c. 35%
   d. 60%

## Matching

*Match the terms with the correct definitions*

a. biopower
b. biosolids
c. co-firing

d. combined cycle biopower
e. pyrolysis
f. syngas

g. combined heat and power
h. hydronic

**1.** ____ The heating of biomass in a low oxygen environment to produce syngas.

**2.** ____ The process of using circulating hot water as a source of heat.

**3.** ____ Dried biomass in the form of animal manure and sludge from wastewater treatment plants.

**4.** ____ A combustible mixture of hydrogen and carbon monoxide gases.

**5.** ____ The process of supplying thermal energy and electricity from biomass.

**6.** ____ The production of electricity from solid biomass.

**7.** ____ The process of burning biomass with coal or natural gas.

**8.** ____ The process of converting solid biomass into a combustible gas that powers a turbine.

## CHAPTER 21

# Environmental Effects of Solid Biomass Fuels

## KEY CONCEPTS

*After reading this chapter, you should be able to:*

1. Define the term sustainable.

2. Explain four benefits solid biomass fuels have on the environment.

3. Identify two legumes and explain why they are important for agriculture.

4. Explain what it means to be carbon-neutral.

5. Describe two benefits farmers gain from growing a perennial energy crop.

6. Indentify three negative effects of using solid biomass as a fuel source.

7. Explain the relationship between using crop residue for fuel and hypoxia.

## TERMS TO KNOW

sustainable

ecosystem

legume

runoff

acid precipitation

condensable particulate
  matter

carbon-neutral

monoculture

cultural eutrophication

hypoxia

sediment pollution

watershed

# INTRODUCTION

Solid biomass fuels are an important renewable fuel option. Today they represent the second most widely used renewable energy resource behind that of hydropower. Solid biofuels can be produced by established agricultural practices in a variety of climates to supply a sustainable source of fuel. They can also be derived from other sources, such as manufacturing and processing wastes, making these practices more efficient. Like any energy resource though, solid biomass energy has an impact on the environment. Understanding the positive and negative aspects of biomass fuel will help decide where it may best be applied to assure a safe and sustainable energy future.

## POSITIVE EFFECTS OF SOLID BIOMASS FUELS

The use of solid biomass fuels for generating heat and power have many benefits over using traditional fossil fuels such as coal, oil, and natural gas. First and foremost is their renewability. Solid biomass is a renewable, **sustainable** form of solid fuel. Unlike coal that is not sustainable, biomass derived from a dedicated energy crop like shrub willow, hybrid poplar, switchgrass, or *Miscanthus* can be sustained over time. This then provides an inexhaustible energy resource. Engineering an agricultural system that provides the required amount of biomass feedstock to supply a proposed power plant is basic to its sustainable operation. A 30 megawatt biomass facility consumes approximately 110,000 tons of wood per year. To meet the fuel requirements of the biomass power plant, surrounding farmers would need to plant 55,000 acres of shrub willows. This would produce a constant supply of biomass using a three year harvest scheme, assuming a yield of 6 dry tons of biomass per acre. Ideally, the energy crops should be grown near the power plant to reduce the cost of transporting the biomass. Transportation costs are an important factor for the cost effectiveness of any feedstock. Simply put, the farther you have to transport something, the higher the cost. In order to make solid biomass fuels successful for both the farmer and the power plant operator, a crop production plan must be designed to produce an annual harvest to sustain the biomass feedstock needed to supply a steady source of energy.

Typically biomass power plants are in the 30–80 megawatt range (Figure 21-1). This is much smaller than traditional coal-fired plants that are usually around 500–1,200 megawatts. Biomass power plants are smaller because it is not cost effective to transport biomass solid fuels over long distances because of the reduced energy density that biomass has compared to fossil fuels. Energy density refers to the amount of energy contained per unit volume of a substance. Therefore, in order to make efficient use of local agricultural and forestry resources, the power plants are much smaller.

**sustainable**—to maintain at a constant level

**FIGURE 21-1** Woodchips outside a biomass power plant.

## Less Impact on Ecosystems

Studies have shown the growth of perennial, short rotation crops like switchgrass, *Miscanthus,* poplar, and shrub willow, also have a lesser environmental impact on surrounding **ecosystems** when compared to traditional row crops grown for food.

Studies on short rotation energy crops also revealed an increase in the diversity of animals within the perennial energy plantation as well (Figure 21-2). Both birds and mammals benefit from the habitat created within these perennial fields. Woody biomass crops can act as corridors between forest ecosystems that allow the migration of certain species. The trees and grasses also are prime habitat for the reproduction of birds.

## Soil Health

The establishment of perennial energy crops has demonstrated a positive effect on the health of soil. Once the crop is established, there is a large reduction in the rate of erosion of topsoil from the fields. This is the result of the development of a perennial root system that holds soil in place. Also, the soil surface is eventually covered by the growing vegetation. This greatly reduces soil loss compared to that of traditional annual row crops. The use of cover crops planted between rows of wide row trees like shrub willow and poplar can prevent soil erosion during early growth, and if the cover crop is a **legume** then it can also act as a source of nitrogen for the crop. A legume is a family of plants that have seedpods and also house tiny

**ecosystem**—the interaction between the living and non-living components in a specific area

**legume**—a family of plants that have seedpods and house tiny nodules in their root system that is home to nitrogen producing bacteria

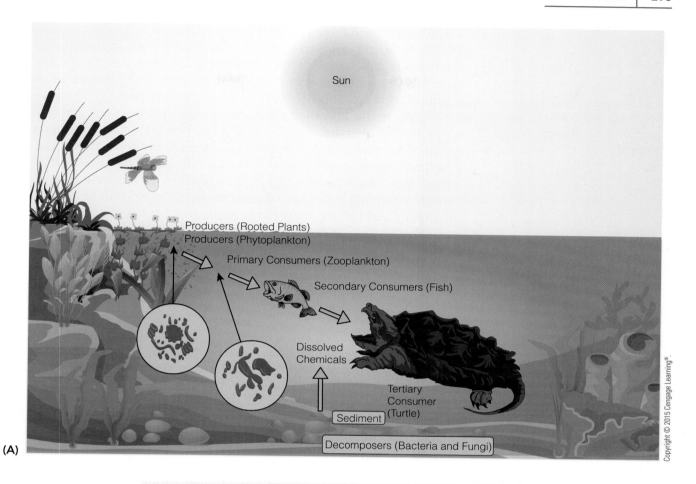

Sun

Producers (Rooted Plants)
Producers (Phytoplankton)
Primary Consumers (Zooplankton)
Secondary Consumers (Fish)

Dissolved
Chemicals

Tertiary
Consumer
(Turtle)

Sediment

Decomposers (Bacteria and Fungi)

(A)

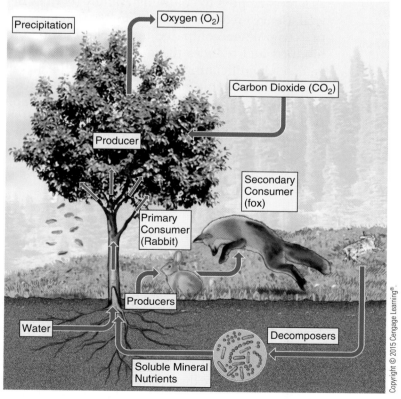

Precipitation

Oxygen (O$_2$)

Carbon Dioxide (CO$_2$)

Producer

Secondary
Consumer
(fox)

Primary
Consumer
(Rabbit)

Producers

Water

Decomposers

Soluble Mineral
Nutrients

(B)

**FIGURE 21-2** (A) Aquatic ecosystem. (B) Terrestrial ecosystem.

nodules in their root system that is home to nitrogen producing bacteria. The bacteria add nitrogen to the soil that plants need for healthy growth. Common legume cover crops include alfalfa, clover, and vetch. After the establishment of the energy crop, the organic material content of the soil begins to increase. This helps prevent soil loss to **runoff**.

Runoff is the loss of soil sediments to surface water as a result of water flowing over the ground surface. Organic material within soil binds soil particles together that reduces erosion. Excess organic material from perennial energy crops results from the buildup of dead vegetation and annual leaf drop from the plants. This leads to an increase of beneficial carbon within the soil. There is a decrease in the amount of nitrogen fertilizer required for the energy crop compared to that of a traditional row crop. This ultimately reduces the potential for the runoff of nitrogen into surrounding surface waters that is the basis for algae blooms. A reduction in the use of chemical pesticides is a benefit because many hybrid short rotation energy crops have been bred to be insect and disease resistant.

## Reduces Emissions

Another benefit of using woody biomass as a solid fuel is the reduction in emissions that result when they are burned. The combustion of biomass in power plants leads to significant reductions of nitrogen oxides (~60%), and an almost complete reduction of sulfur oxides (~99%) when compared to the combustion of bituminous coal. Both of these air pollutants contribute to the formation of **acid precipitation**. There is also a decrease in **condensable particulate matter** (PM–CON) emissions. The burning of biomass reduces PM–CON by over 85%. Burning biomass as a replacement to coal substantially reduces heavy metal emissions as well. Heavy metals are toxic elements that include arsenic, selenium, and mercury. When compared to the combustion of bituminous coal, solid biomass fuel reduces these heavy metal emissions by over 60%. There is also a significant impact on carbon dioxide emissions. Because biomass is a renewable source of carbon, it is considered **carbon-neutral**. Carbon-neutral means same amount of carbon produced by its combustion is then absorbed from the atmosphere by the growth of the plants that comprise the energy crop. This leads to a balance of carbon between the combustion and the growth of the biomass fuel crop; therefore there is no long-term gain of carbon dioxide in the atmosphere as a result of burning biomass (Figure 21-3).

## Increased Opportunities and Efficiencies for Farmers and Other Industries

The growth of short rotation, perennial energy crops has the potential to bring fallow or marginal agricultural fields back into production, creating a cash crop for farmers, while providing a source of renewable solid fuel and having a positive impact on the environment. Not all biomass used to generate heat and power is from a dedicated energy crop. Many biomass plants use woody residue from forestry, manufacturing, and landscaping. Much of this waste was once discarded within landfills. Redirecting this resource for use as a solid, renewable fuel makes many of these processes more efficient and environmentally friendly.

**runoff**—the carrying away of soil and sediments by the action of water moving across a surface

**acid precipitation**—rain or snow with a pH below 5.0

**condensable particulate matter**—fine particles produced from combustion that are captured in a liquid during a stack test

**carbon-neutral**—the amount of carbon dioxide produced by combustion is removed by plants, negating the amount of carbon dioxide entering the atmosphere

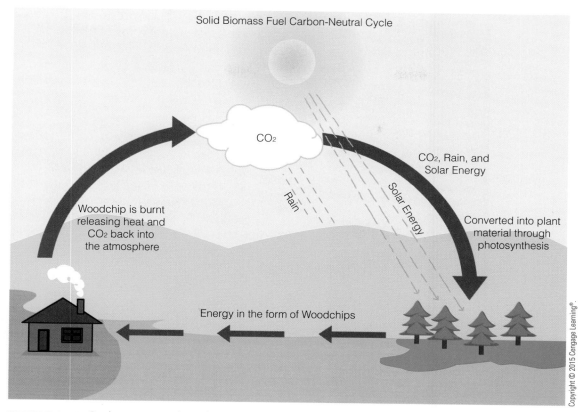

Solid Biomass Fuel Carbon-Neutral Cycle

CO₂

CO₂, Rain, and Solar Energy

Rain

Solar Energy

Woodchip is burnt releasing heat and CO₂ back into the atmosphere

Converted into plant material through photosynthesis

Energy in the form of Woodchips

**FIGURE 21-3** Carbon-neutral cycle.

# NEGATIVE EFFECTS OF SOLID BIOMASS FUELS

Although there are many positive aspects of using solid biomass as a source of fuel, there are a few negative aspects associated with it as well.

## Runoff and Soil Quality

The initial establishment of a perennial crop can accelerate soil erosion during the first years of planting the crop (Figure 21-4). The tilling, planting, and early growth of energy crops exposes fields to runoff, that lead to the pollution of surrounding surface waters. After the first year, the growth of the crop begins to minimize this because the plants grow large enough to cover the soil surface and reduce erosion. During this phase of growth, herbicide and fertilizer use is necessary to establish the crop. The use of these chemicals may impact local ecosystems as a result of their incorporation into the food chain. The harvesting of dedicated perennial energy crops also exposes soil to runoff and erosion. The cutting of woody biomass or grasses opens up the soil for more exposure to erosion.

Harvesting biomass can lead to soil compaction from the heavy machinery used in forests and fields. Some biomass power plants consume agricultural waste in the form of corn stover or wheat straw. This biomass was traditionally chopped and blown back on farm fields during harvest. This crop residue was then plowed under, to put organic material and nutrients back into the soil. Redirecting this crop residue away from fields and burning it can have an impact on soil fertility. This may also increase the use of chemical fertilizers, especially nitrogen fertilizers made from natural gas.

Courtesy of U.S. Department of Agriculture, Natural Resources Conservation Service.

**FIGURE 21-4** Runoff.

## Energy Production versus Food Production

Concerns have been raised regarding the impact of using land to grow energy crops instead of food crops and how that will affect our food supply. This issue can be addressed by the use of marginal farmlands and bringing fallow farmland back into production in the form of an energy plantation.

## Insect Populations and Disease

**monoculture**—the growth of one species of crop

Large-scale energy plantations may establish **monocultures** that can become susceptible to insects and diseases. A monoculture is the growth of one species of crop. The development of hybrid strains of plants may address this potential problem. Another way to eliminate a monoculture is to plant more than one type or variety of energy crop. This practice is used widely in current production agriculture in the form of strip cropping. Strip cropping divides farm fields into strips of different crops.

## Water Quality

**cultural eutrophication**—the increase of nutrients within water by human activity that leads to the rapid growth of aquatic plants and algae

**hypoxia**—low levels of oxygen

An increase in fertilizer to replace lost nutrients in the soil has an impact on water quality as well. Nutrients in the form of nitrates and phosphates in water can lead to **cultural eutrophication**. Cultural eutrophication is the increase of nutrients from human activity in water that leads to the rapid growth of aquatic plants and algae (Figure 21-5). The algae then die and sink to the bottom, where they are decomposed by bacteria. The bacteria remove oxygen from the water in order to decompose the excess algae that causes a drop in dissolved oxygen levels. Low levels of dissolved oxygen in water are known as **hypoxia**. This can cause death to aquatic organisms that need oxygen to live. Poor forestry management techniques when harvesting trees and forest debris for use as biomass fuel have significant effects on water quality

Courtesy of the National Oceanic and Atmospheric Administration.

**FIGURE 21-5** Algal bloom on Lake Erie.

as well. The practice of clear-cutting forests exposes large amounts of forest soils to runoff and erosion that create **sediment pollution** in surface water. This sediment disrupts aquatic ecosystems by blocking sunlight and reducing photosynthesis, clogs fish gills, disrupts fish reproduction, and introduces toxic metals into water. Forests act as an important part of the **watershed** by absorbing water and preventing runoff. Proposals to use slash and trimmings produced from timber management can remove beneficial ground cover that reduces runoff and acts as a source of nutrients within forests.

**sediment pollution**—the addition of soil and rock particles to water

**watershed**—the total land area that drains into a specific river, river system or body of water

## Pollutants

The combustion of biomass, like coal, produces air pollutants in the form of mainly nitrogen oxides and particulate matter. Some solid biomass fuels like switchgrass also have higher amounts of chlorine and potassium compounds in their ash. Chlorine causes problems by being corrosive to the boiler, exhaust and control equipment. It can form chlorinated hydrocarbons in the exhaust gas. These have a negative impact on air quality. The combustion of grass also produces more bottom ash than wood, in addition to an increase in the amount of carbon monoxide emitted when burning solid biomass. This may be the result of higher moisture content within the biomass and the lower heat content. The average heat content of biomass is about 7,000–9,000 Btus per pound compared to bituminous coal that falls between 8,000–13,000 Btus per pound. One way to minimize these emissions is to use emission control equipment or biomass gasification. Biomass gasification converts the solid biomass into a gas that burns more cleanly than direct combustion. Tests of the use of biomass when co-fired with coal also reveal lower temperatures generated within boilers. This leads to an overall decrease in the efficiency of the power plant. The lower temperatures result from the lower heating value of biomass compared to that of coal.

# REVIEW OF KEY CONCEPTS

1. The use of solid biomass fuel is a sustainable source of heat and power, meaning it can be maintained at a constant level.

2. The production of dedicated short rotation crops like shrub willow, poplar, switchgrass, and *Miscanthus* can produce a sustainable fuel. Perennial fuel crops are especially beneficial to the environment because they reduce the problems of runoff erosion and nutrient loss that are often associated with annual row crops. Burning solid biomass crops instead of coal can cause reductions in the production of atmospheric pollutants. The combustion of biomass reduces nitrogen oxides, condensable particulate matter, and heavy metals, along with an almost complete reduction of sulfur oxides emissions. Solid biomass fuel in the form of woody residue can use the waste leftover from timber production and wood manufacturing as a source of fuel. This avoids the disposal of this waste in landfills.

3. Legume cover crops like alfalfa, clover, and vetch, planted during the early stages of establishing wide row energy crops like shrub willow and poplar, help prevent erosion as well. Legumes can also act as a source of nitrogen that enriches soil fertility.

4. The use of biomass as solid fuel is carbon-neutral; meaning the amount of carbon produced by combustion is removed by plants, therefore offsetting the amount of carbon dioxide entering the atmosphere.

5. The use of perennial energy crops acts as a habitat for many species of birds and mammals. The planting of short rotation perennial energy crops have shown to have a positive impact on wildlife.

6. The negative aspects of growing biomass to use as a solid fuel involve the loss of soil as a result of erosion. Soil exposed during the planting and harvest of energy crops lead to runoff and the loss of nutrients. Excessive runoff results in the sediment pollution of surface water. Also, the use of crop residue as a biomass fuel removes organic material and nutrients from soil that was once replaced during harvest. This can lead to an increase in runoff and the use of fertilizer that pollutes surface water causing cultural eutrophication. The growth of some energy crops can compete for space that was once used to grow food that may affect our food supply. The combustion of biomass produces more carbon monoxide emissions than coal, and also particulate matter, that negatively affects air quality. Some biomass crops like switchgrass, have higher potassium and chlorine levels cause corrosion problems in boilers and produce harmful emissions. The combustion of biomass also has a higher moisture content and lower heating value than coal, reduces the efficiency of the power plant.

7. Harvesting of crop residue for use in biomass removes nutrients from the soil that would remain if the crop residue were left to decompose. This results in the need

for an increase in fertilizer use to replace lost nutrients in the soil. Runoff from the land can cause fertilizers to leach into the water that leads to the rapid growth of aquatic plants and algae. The algae then die and sink to the bottom, where they are decomposed by bacteria. The bacteria take oxygen from the water in order to decompose the excess algae, and cause a drop in dissolved oxygen levels within water. Low levels of dissolved oxygen in water are known as hypoxia. This can cause death to aquatic organisms that need oxygen to live.

## CHAPTER REVIEW

### Short Answer

1. What is meant by the term sustainable?

2. Explain four benefits solid biomass fuels have on the environment.

3. What are legumes; explain why they are important for agriculture?

4. What is carbon-neutral?

5. Describe two benefits farmers receive from growing a perennial energy crop.

6. Indentify three negative effects of using solid biomass as a fuel source.

7. What is the relationship between using crop residue for fuel and hypoxia?

8. Briefly discuss three advantages and three disadvantages of using solid biomass fuel to produce heat and power.

### Energy Math

1. A traditional coal-fired power plant produces 1,200 pounds of sulfur oxides (SOx) each day. If this same plant burned a solid biomass fuel, how many pounds of sulfur dioxide would be produced daily?

### Multiple Choice

1. The interactions between living and nonliving things in a specific area is:
   a. photosynthesis
   b. sustainable
   c. biomass
   d. an ecosystem

2. A term that describes something that can be continually supplied is:
   a. feedstock
   b. sustainable
   c. biomass
   d. nonrenewable

3. Legumes are important for energy crops because they:
   a. have seedpods
   b. fix nitrogen in the soil
   c. can be harvested and burned
   d. provide habitat for birds and mammals

4. Which of the following is a legume?
   a. Shrub willow
   b. Poplar
   c. Switchgrass
   d. Alfalfa

**5.** Which atmospheric pollutant is almost eliminated when biomass is burned instead of coal?
   a. Carbon monoxide
   b. Nitrogen oxides
   c. Sulfur oxides
   d. Particulate matter

**6.** Which atmospheric pollutant has been shown to increase when biomass is burned instead of coal?
   a. Carbon monoxide
   b. Nitrogen oxides
   c. Sulfur oxides
   d. Particulate matter

**7.** An increase in the emission of chlorinated hydrocarbons is related to the combustion of:
   a. switchgrass
   b. coal
   c. poplar
   d. shrub willow

**8.** The combustion of solid biomass fuel:
   a. is carbon-neutral compared to coal
   b. produces less carbon dioxide than coal
   c. produces more carbon dioxide than coal
   d. produces the same amount of carbon dioxide as coal.

## Matching

*Match the terms with the correct definitions*

a. sustainable
b. ecosystem
c. legume
d. runoff

e. acid precipitation
f. condensable particulate matter
g. carbon-neutral
h. monoculture

i. cultural eutrophication
j. hypoxia
k. sediment pollution
l. watershed

**1.** _____ Polluted rain or snow that has a pH below 5.0.

**2.** _____ The introduction of soil and sediment into water as a result of runoff.

**3.** _____ The interaction between the living and nonliving things in a specific area.

**4.** _____ The growth of one species of crop on agricultural lands.

**5.** _____ The act of supplying something at a constant level.

**6.** _____ The total land area drained by a river or stream.

**7.** _____ The loss of soil by water moving over the soil surface.

**8.** _____ The amount of carbon produced by combustion is equal to that absorbed by plants.

**9.** _____ An increase in plant nutrients in water that causes an algal bloom.

**10.** _____ Microscopic droplets of liquid produced by combustion.

**11.** _____ A family of plants that fix nitrogen in the soil.

**12.** _____ When an area has low levels of oxygen.

# UNIT VI

# Liquid Bioenergy Fuels from Agriculture

---

## TOPICS TO BE PRESENTED IN THIS UNIT INCLUDE:

• Alcohol transportation fuels

• Biodiesel

• Biomass to liquids

## OVERVIEW

The use of liquid fuels by society has an advantage because it is easily stored, distributed, and transported. Today, liquid fuels from nonrenewable petroleum powers most of the world's motor vehicles. This is mainly in the form of gasoline and diesel fuel. The ability for agriculture to produce liquid fuels to replace these traditional fossil fuels holds great promise for the future. The growth of dedicated energy crops that can be converted into fuels and used by motor vehicles is not only sustainable, but can reduce the emissions of carbon dioxide that contributes to global warming.

## CHAPTER 22

# Alcohol Transportation Fuels

## KEY CONCEPTS

*After reading this chapter, you should be able to:*

1. Define the terms fermentation and distillation.

2. Describe the four steps used to produce ethanol at a biorefinery.

3. Identify how many gallons of ethanol can be produced by one acre of corn.

4. Explain the difference between E10, E15, and E85.

5. Describe the difference in the energy content of one gallon of ethanol compared to one gallon of gasoline.

6. Identify three sugar and starch-based ethanol feedstocks.

7. Define the term cellulosic ethanol.

8. Describe three positive effects that using ethanol as a liquid fuel has on the environment.

9. Describe three negative effects that using ethanol as a liquid fuel has on the environment.

10. Explain the difference between methanol and biomethanol.

11. Describe how most methanol is made today in the United States

12. Explain the process of producing biomethanol from biomass.

13. Explain the advantage of producing biomethanol.

14. Discuss the benefits of using M85 as a liquid fuel for cars.

15. Describe how a methanol fuel cell produces energy.

16. Explain three negative aspects of using methanol.

17. Explain the difference between butanol and biobutanol.

18. Discuss the process of producing biobutanol from biomass.

19. Explain the advantages of producing biobutanol.

20. Describe the difference between Bu11 and Bu100.

21. Explain how the energy content of butanol compares to that of ethanol and methanol.

## TERMS TO KNOW

ethanol

fermentation

sugar

distillation

bio-refinery

starch

polymer

cellulose

endosperm

hydrolysis

distiller's dried grain and solubles (DDGS)

dewater

biomethanol

destructive distillation

char

biomass gasification

syngas

gasifier

methanol fuel cell

biobutanol

# INTRODUCTION

Probably the most widely recognized liquid fuel produced by agriculture today is ethanol. Ethanol is a form of alcohol produced by the fermentation of sugars. Ethanol is commonly mixed with gasoline and has been touted as its possible replacement. Other alcohol-based fuels such as methanol and biobutanol are also possible renewable alternatives that can be used instead of traditional liquid fossil fuels. Debates continue about the role these alcohol fuels will play in the future and how they will impact the use of agricultural land, food production, and the environment.

# ETHANOL

**Ethanol**, or ethyl alcohol, is a colorless combustible hydrocarbon fuel ($C_2H_5OH$) commonly known as grain alcohol.

## Ethanol Production

Ethanol is produced by the **fermentation** of **sugars** by the action of yeast. Fermentation is the process of breaking down a substance by microorganisms in the absence of oxygen (Figure 22-1). The fermentation of sugars by yeasts in the genus *Saccharomyces*, commonly known as baker's yeast, converts glucose sugar into ethyl alcohol. The process involves the mixture of sugar, water, and yeast. The broth is heated to about 90°F (30°C), the optimal temperature for yeast to ferment the sugars. The byproduct of the fermentation of sugars by yeast is ethyl alcohol and carbon dioxide gas. It is estimated for every 1 gram of sugar, 0.51 grams of ethanol should be produced by fermentation.

Because the fermentation process involves a large quantity of water, the resulting fermented broth contains about 7.7% ethanol and 92.3% water. The ethanol must then be separated from the water. This is accomplished by **distillation**. Distillation is the process of separating liquids in a solution as a result of their different boiling points (Figure 22-2). The boiling point of ethanol is 172°F (78°C), while the boiling point of water is 212°F (100°C). If you heat the ethanol–water solution to 172°F, then the ethanol will vaporize, leaving the water behind. The ethanol vapor can be condensed and collected. Typically the ethanol–water solution is distilled two or three times to purify it to contain 95% ethanol.

The combustion of ethanol is an efficient, clean process that only produces carbon dioxide and water vapor: $C_2H_5OH + 3O_2 = 2CO_2 + 3H_2O$. Ethanol has been produced by humans for thousands of years. The same procedure used to produce ethanol fuel is also used to make wine, beer, and liquor. The main difference between the two processes is the concentration of ethanol after distillation. Wine contains 13% ethanol, beer 4%, while 80-proof liquor contains 40%. Ethanol used as a liquid fuel however contains 95% ethanol that would be toxic if consumed. The fermentation of sugars is also used to make bread. The desired byproduct of baking is the carbon dioxide gas

**ethanol**—a colorless combustible hydrocarbon fuel ($C_2H_5OH$) commonly known as grain alcohol

**fermentation**—the process of breaking down a substance by microorganisms, in the absence of oxygen

**sugar**—a simple carbohydrate molecule, like glucose ($C_6H_{12}O_6$)

**distillation**—the process of separating liquids in a solution as a result of their different boiling points

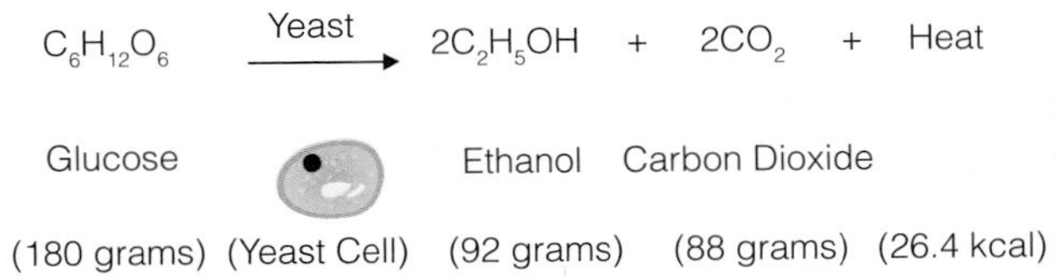

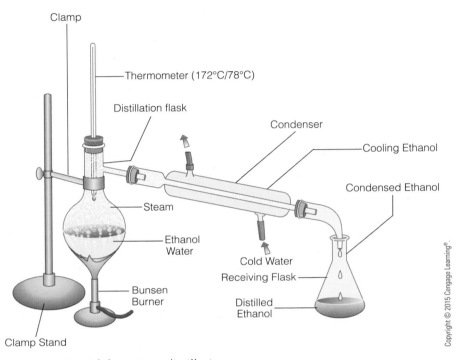

$$C_6H_{12}O_6 \xrightarrow{\text{Yeast}} 2C_2H_5OH + 2CO_2 + \text{Heat}$$

Glucose        Ethanol   Carbon Dioxide

(180 grams) (Yeast Cell) (92 grams)   (88 grams) (26.4 kcal)

**FIGURE 22-1** The ethanol fermentation reaction.

Clamp

Thermometer (172°C/78°C)

Distillation flask

Condenser

Cooling Ethanol

Condensed Ethanol

Steam

Ethanol
Water

Cold Water

Receiving Flask

Bunsen
Burner

Distilled
Ethanol

Clamp Stand

**FIGURE 22-2** A laboratory distillation apparatus.

produced by the fermentation of sugars by yeast. This causes bread to rise. The alcohol is then burned off as part of the high temperature baking process. The modern production of ethanol fuel takes place at a manufacturing plant called a **bio-refinery**, and involves four main processes: pretreatment, hydrolysis, biochemical conversion, and distillation.

## Pretreatment

The pretreatment process involves the preparation of the plant biomass to be used as the source of sugar for fermentation. This can vary depending on the type of crop, but mostly involves the grinding of the plant material into a fine powder. All plants contain sugars in the form of **starches**. These long chain molecules are known as **polymers**. Carbohydrates are molecules made up from atoms of carbon, hydrogen, and oxygen. Most starch within plants is in the form of **cellulose**, hemicellulose, and lignin, that are all known as biopolymers that form cell walls (Figure 22-3).

Cellulose is a polymer of glucose and is the most abundant form of plant material on the earth. Cellulose consists of long, straight chains of

**bio-refinery**—a modern manufacturing plant used for the production of biofuels

**starch**—a common name for carbohydrate molecules composed of long chains of sugars

**polymer**—long chains of the same molecule

**cellulose**—a polymer of glucose and the most abundant form of plant material on the Earth.

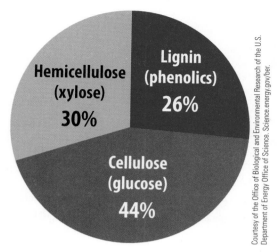

Courtesy of the Office of Biological and Environmental Research of the U.S. Department of Energy Office of Science. Science.energy.gov/ber.

**FIGURE 22-3** Composition of biomass.

glucose that are tightly held together. The pretreatment of biomass makes it easier to access the sugars locked up in starches. Today the main feedstock for ethanol production in the United States is corn. In 2012, 40% of all the corn used in the United States went to ethanol production. Corn contains starch within the **endosperm** of the corn kernel. When making ethanol from corn, pretreatment involves grinding up the corn kernels by what is known as the dry grind process. The corn kernels are cleaned and separated from impurities like dirt, metal, and plant material by blowing them through a series of metal screens. Next, the dry corn is pounded by a hammer mill into powder with each grain being between 0.5–2 mm in diameter. Once the dry grind process is complete the corn powder is then mixed with water. About 1.5 pounds of ground corn is mixed per gallon of water prior to fermentation. This mix is then sent to the next phase of ethanol production known as hydrolysis.

**endosperm**—the stored food within a plant's seed

## Hydrolysis

Because plant material has most of its sugars locked up in polymers like cellulose, it is necessary to break the chains apart into individual glucose molecules (Figure 22-4). This process is done by **hydrolysis**. Hydrolysis is the separation of a substance by water. Hydrolysis in the production of ethanol involves the use of special enzymes, acids, and heat to break down the cellulose and hemicellulose into simple sugars like glucose that can then be fermented by the yeast. Some biomass feedstocks like corn grain do not require extensive hydrolysis because the starch within the endosperm is easily broken down into sugar. Other feedstocks composed of mostly cellulose, like switchgrass and *Miscanthus*, require a more extensive hydrolysis process involving the use of enzymes and heat to release glucose from the cellulose. Current research in the fermentation of cellulosic ethanol is focusing on the best types of pretreatment to yield the greatest amount of sugars from biomass. Part of the research is focused on identifying effective cellulases. Cellulase is an enzyme that breaks cellulose down into glucose.

Another challenge to pretreatment is breaking down lignin so enzymes can access the cellulose. All plant cell walls are constructed of tough lignin that surrounds the chains of cellulose, much like the way concrete surrounds iron re-enforcing bars in a concrete structure. In order to get at the

**hydrolysis**—the separation of a substance by water

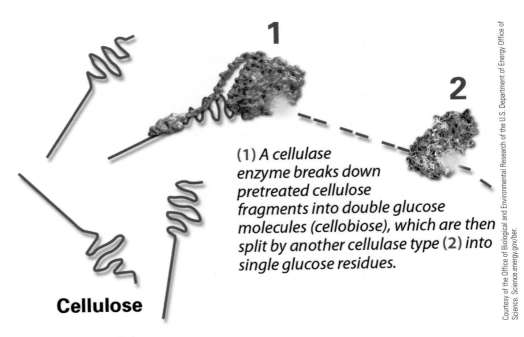

Courtesy of the Office of Biological and Environmental Research of the U.S. Department of Energy Office of Science. Science.energy.gov/ber.

**FIGURE 22-4** Cellulose.

cellulose, pretreatment methods are needed to chip away at the "concrete" lignin to expose the cellulose. Deriving ethanol from cellulose is known as cellulosic ethanol that could make the process of ethanol production extremely efficient.

The Genencor company has developed an enzyme known as Accellerase 1500 that is used as a pretreatment enzyme for ethanol produced from cellulose. This enzyme is made by a genetically modified strain of decomposing fungus. The enzyme operates at temperatures between 122°F–149°F (50°C–65°C) and at pH levels between 4 and 5. Research is also being conducted on using high pressure and heat to help break down the cellulose. Other technologies are being developed to increase the rate of enzyme reactions in order to breakdown starch using nanotechnology. Researchers at Cornell University's Biofuels Research Laboratory are experimenting with magnetic nanoparticles that house enzymes that are mixed using a magnetic field. This action has shown to improve enzyme reaction rates greatly. Once hydrolysis is complete, the solution is mixed with the yeast, and fermentation begins.

### Biochemical Conversion

The biochemical conversion process (fermentation) takes place in a heated chamber for two to three days. These large fermentation chambers range in size between 300,000–500,000 gallons and bio-refineries usually contain multiple chambers. Each biochemical conversion is done in separate batches and is not a continuous process.

### Distillation

When fermentation is completed, the solution is distilled to separate the ethanol so it is 95% pure. On average, 300–420 gallons of ethanol can be produced from one acre of corn. The remaining biomass within the solution is known as sillage or **distiller's dried grains and solubles (DDGS)**. This can be **dewatered**, the process of removing suspended solids from

**distiller's dried grains and solubles (DDGS)**—the remaining biomass left over after the fermentation process, also known as sillage

**dewatered**—the process of removing suspended solids from within a water solution

within a water solution. The DDGS can then be processed as animal feed and contains useful proteins, fats, and minerals. DDGS is often mixed with cattle feed at levels of up to 40%. The use of DDGS makes corn ethanol production more efficient and cost effective. A typical dry grain process ethanol bio-refinery can produce 15 pounds of DDGS from one bushel of corn. Eighty-five percent of the DDGS used in the United States goes to feeding dairy cows and beef cattle with the remaining 15% used for hogs and poultry. Sillage waste not used for feed can be burned as a sold biomass fuel. Figure 22-5 shows an overview of the ethanol production process and Figure 22-6 an ethanol refining plant.

Once distillation is complete, the ethanol is denatured with an additive like methanol or gasoline, so it cannot be consumed or taxed as an alcoholic beverage. One of the largest ethanol bio-refineries in the United States is in Blair, Nebraska operated by Cargill Inc. and produces 195 million gallons of ethanol per year. Another ethanol bio-refinery in the Northeast is located in Fulton, NY, and owned by Sunoco. The plant uses 41 million bushels of corn to produce 100 million gallons of ethanol annually. Overall, Iowa is the largest producer of ethanol that produced more than 3.5 billion gallons of ethanol in 2011. There are other methods of fermenting ethanol being developed that use bacteria instead of yeasts. A strain of bacteria, *Zymomonas mobili*, is being used to increase the efficiency of the fermentation process. *Z. mobili* can tolerate higher concentrations of ethanol during fermentation, making the conversion of sugars to fuel more efficient.

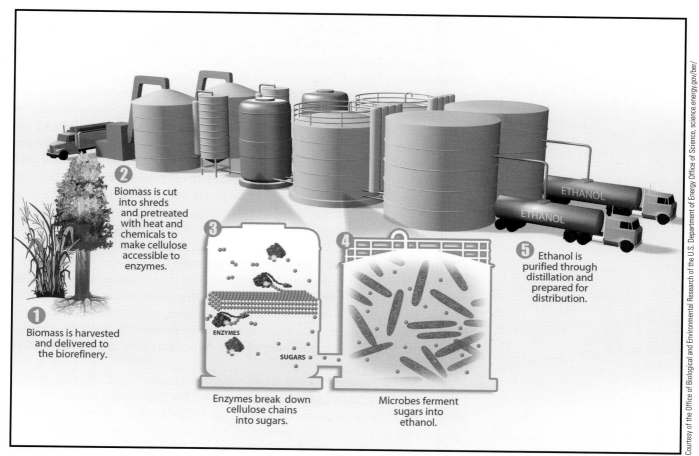

① Biomass is harvested and delivered to the biorefinery.

② Biomass is cut into shreds and pretreated with heat and chemicals to make cellulose accessible to enzymes.

③ Enzymes break down cellulose chains into sugars.

ENZYMES

SUGARS

④ Microbes ferment sugars into ethanol.

⑤ Ethanol is purified through distillation and prepared for distribution.

ETHANOL

**FIGURE 22-5** Cellulosic ethanol production.

Courtesy of the U. S. Department of Agriculture, Agricultural Research Service/photo by Steven Vaughn.

**FIGURE 22-6** Use of E85 ethanol as an automotive fuel. Ethanol plant, West Burlington, Idaho.

## Ethanol Use

In 2012, 13.3 billion gallons of ethanol were produced in the United States. Approximately 91% of this was derived from corn grain as the biomass feedstock. The cost of producing ethanol from corn fluctuates greatly. Approximately 50% of the cost of ethanol production is the cost of the feedstock. As the price of corn rises, so does the cost of ethanol. The price of producing ethanol in the United States averaged between $1.05 per gallon in January of 2005, to $2.25 per gallon in January 2011. Similarly the cost of a gallon of ethanol purchased at the pump fluctuated from $1.05–$3.10 per gallon during the same time period. In 2007, the United States government enacted the *Energy Independence and Security Act* that mandated the production of renewable fuels, mainly in the form of ethanol. By 2011, the total production of ethanol from corn must to exceed 12.6 billion gallons, and top off at 15 billion gallons by 2022. During this same time period there is to be an increase in the production of ethanol produced from crops other than corn that range from 1.35 to 21 billion gallons. These feedstocks will include cellulose-based plant material derived from crop residue and biomass from plants such as switchgrass and *Miscanthus*. The price of ethanol is expected to decrease in the future as the technology for manufacturing of ethanol is improved, the scale of production is increased, and other crops besides corn are used as a feedstock. If switchgrass can be used as the feedstock for ethanol production, it is estimated that production costs can be reduced to $1.80 per gallon.

Ethanol is currently used as an additive to gasoline. It is commonly mixed at a 10% ethanol, 90% gasoline ratio. This fuel is also known as E10, considered to be a low level blended fuel. In 2012, approximately 12.6 billion gallons of ethanol was added to gasoline in the United States. The Environmental Protection Agency (EPA) recently approved the use of E15 (15% ethanol–85% gasoline) in cars manufactured after the 2000 model year.

This is an intermediate blend of fuel that includes 15%–85% ethanol. E15 is also used by NASCAR racing engines.

There are special engines are called Flex-Fuel engines that can run on gasoline or E85 (85% ethanol–15% gasoline). E85 is considered an alternative fuel by the EPA (Figure 22-7). E100 (100% ethanol) is used by Indy race cars.

One drawback of using ethanol as a replacement for gasoline is its lower energy content. One gallon of ethanol has an energy content of 84,000 Btus. This is approximately 73% of the energy contained within one gallon of gasoline. This means you have to use 1.4 gallons of ethanol to get the same amount of mileage as one gallon of gas. Another important aspect of ethanol production is its energy balance or EROI (Energy Return on Investment). This is a measure of the energy you need to produce a resource and how much energy it ultimately gives out. Calculating an EROI takes into effect many inputs and outputs of energy during the whole process. For an agricultural-based energy resource, this includes all of the energy that goes into growing, harvesting, processing, and transporting a crop, as well as taking into account any possible byproducts derived from the process. For a fossil fuel, the EROI includes locating, extracting, transporting, and refining the resource. This is why EROI values can differ for a specific energy resource based on the range of the study. The energy balance of corn ethanol production has been debated for many years. Currently the energy balance for corn ethanol ranges between 0.7–1.3. This means for every one unit of energy put in to make ethanol from corn, you get between 0.7 and 1.3 units out. Most researchers agree an EROI of 3 or greater is desirable for a biofuel to be economically successful. In comparison, the energy balance of gasoline, that has been estimated to be between minus 1.2 and 3.0. This means every one unit put into making gasoline from oil, yields either negative 1.2 to 3 units of energy. The negative value from some studies suggest gasoline actually produces less energy than it takes to process it from crude oil, making it extremely inefficient

**FIGURE 22-7** Use of E85 ethanol as fuel.

## CAREER CONNECTIONS

### Bio-refinery

The growth in the production of biofuels like ethanol is creating many new career opportunities working in bio-refineries. Bio-refineries hold opportunities in many job fields that include plant managers, technicians, maintenance, plant operators, accounting, microbiologists, lab technicians, grain specialists, electricians, chemists, and commodity managers. The expansion of this industry will continue into the future creating many new career opportunities.

## Ethanol Feedstocks

Although 91% of the ethanol produced in the United States today comes from corn, there are other biomass feedstocks that can be used to make ethanol. Ethanol feedstocks are divided into two basic categories: sugar and starch-based and cellulose-based. Sugar and starch-based feedstocks include plants that have readily available sugar that does not need significant processing prior to fermentation. These plants are typically high in sugar content and include corn, sorghum (milo), and sugarcane (Figure 22-8). The use of corn as a feedstock typically yields 2.4 gallons of ethanol per bushel of corn.

Sugarcane is an especially productive ethanol feedstock because its stalk contains 20% sugar that is easily fermented. The energy balance for sugarcane ethanol is 1–8, much higher than corn ethanol. You can produce almost twice as much ethanol from one acre of sugarcane compared to corn. Sugarcane is a tropical grass, so it can only be grown in warmer states like Texas, Louisiana, and Florida. Brazil is the second-largest producer of ethanol in the world, growing more than 9 million acres of sugarcane that

**FIGURE 22-8A** Corn feedstock.

**FIGURE 22-8B** Sorghum feedstock.

**FIGURE 22-8C** Sugar Cane feedstock.

produces 6.5 billion gallons of ethanol annually. This equates to an average yield of 720 gallons of ethanol per acre of sugarcane. Eighty-five percent of every vehicle in Brazil is a flex fuel vehicle that runs on E85. Until December 31, 2011, the United Sates imposed an import tax on Brazilian ethanol of 54 cents per gallon. This tariff was designed to make it difficult to import of ethanol into the United States from Brazil, and allowed ethanol producers in the United States to expand production. Analysts do not believe the tax cut will have a negative effect on U.S. ethanol production because of the increasing costs of transporting ethanol from Brazil and their declining production as a result of a poor sugarcane harvest in 2011. Proponents of the tax cut suggest this will help import a greater amount of lower cost ethanol into the expanding biofuels market.

Another sugar-based plant is sorghum, also known as milo that is currently a small player in the ethanol industry, accounting for less than 1% of production. Some ethanol is produced from whey, a byproduct of cheese making, wheat, and potatoes. The production of ethanol from these traditional food crops using existing fermentation technology is known as first generation biofuel.

The other category of ethanol feedstock is cellulose-based, also known as cellulosic ethanol. Cellulosic ethanol uses the whole plant as a source of sugars. The process involves breaking down cellulose molecules into simple sugars. Currently there is no commercial production of cellulosic ethanol because of the difficulties associated with breaking down the cellulose polymers. Research is currently underway at solving the problem of using cellulose and involves the use of enzymes, heat treatments, and microorganisms to access the sugar locked within the cellulose. Theoretically any biomass can be used as a source of sugar for ethanol production if the bonds that hold starches together can be easily broken. These types of nonfood crops are known as second generation biofuels, and derive their sugar from the breakdown of cellulose. Crops being researched as sources of cellulosic ethanol include switchgrass, *Miscanthus giganteus*, and the use of corn stover (Figure 22-9).

Courtesy of Dan Putnam, UC Davis.

**FIGURE 22-9** *Miscanthus giganteus.*

Estimates show the energy balance of cellulosic ethanol could reach up to 1–36 depending on the development of pretreatment technology. This would make the cost of producing ethanol drop considerably. Estimates of using switchgrass to produce ethanol predict a yield between 600–1,200 gallons of ethanol per acre if the cellulose can be broken down into simple sugars, then fermented. This is equal to 1 ton of dry biomass yielding between 100–150 gallons of ethanol.

Third generation biofuels are also under development. These include the potential for developing crops capable of producing high amounts of sugars and low amounts of lignin.

## Environmental Impacts of Ethanol

The use of ethanol in gasoline began in the 1970s as a way to increase the octane rating of gasoline. This was done to add oxygen to the fuel and reduce the emissions of carbon monoxide from car exhaust. Today, most gasoline contains a 10% mixture of ethanol, known as E10. This burns more cleanly than pure gasoline. The use of E10 and E85 show a reduction in the emission of carbon monoxide, nitrogen oxides, particulate matter, and hydrocarbons as compared to gasoline. Also, blending ethanol with gasoline reduces the amount of oil needed to supply fuel for our cars and trucks. A 10% blend of ethanol reduces our consumption of oil by 10%. Considering we use 378 million gallons of gasoline per day in the United States, a 10% mix of ethanol can save more than 37 million gallons of gasoline each day. There is also a reduction of carbon dioxide emissions when using ethanol. Because ethanol is derived from biomass, some of the carbon produced by combustion is absorbed by plants grown to produce the next year's crop (Figure 22-10). Depending on the blend of ethanol and the fossil energy input into the production of the biomass crop, and its processing and fermentation, there can be a reduction of $CO_2$ between 19–52%. It is also estimated the use of cellulosic ethanol may someday reduce carbon emissions from cars by as much as 86%. This can decrease the impact our motor vehicles have on producing greenhouse gases and contributing to global warming.

Although there are many positive aspects of using ethanol as a liquid fuel, there are some negative aspects as well. The evaporation of liquid fuels during refueling in the warmer months can contribute to the formation of ozone (photochemical smog). The evaporation of hydrocarbons

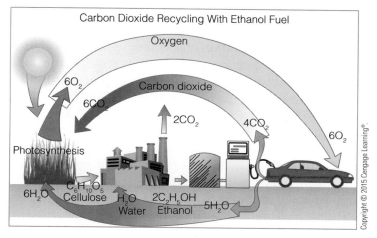

**FIGURE 22-10** Ethanol and the carbon cycle.

like gasoline and ethanol interact with sunlight and nitrogen oxides in the atmosphere to form ground level ozone and peroxyacytyl nitrate also known as PANs. These are both irritants to the eyes, nose, throat, and lungs. There is also an increase in the emissions of some volatile organic compounds (VOCs) like acetaldehyde, as compared to the combustion of gasoline.

The use of ethanol can have an impact on the price of food as well when it is made from traditional food crops like corn. Much of the corn grown in the United States is used to feed animals to produce meat or dairy. The rise in corn prices as the demand increases for its use in ethanol production can also cause a rise in the price of food. This is why a change to using a dedicated energy crop that uses biomass from a non-food crop like switchgrass is more desirable for the production of ethanol. This can reduce the competition between the cost of food and the cost of energy.

There is also a concern over the amount of agricultural land required to support an increase in ethanol production. The Department of Energy and the Department of Agriculture conducted a study to examine the feasibility of replacing 30% of our petroleum use with biomass fuels. This report concluded it may be possible to produce one billion tons of biomass to meet the goal of a 30% replacement of oil by biofuels by the year 2030. The report identified increases in crop yields by 50% and the planting of 55 million acres of energy crops would be required to meet this goal without impacting our food supply.

Today, because most ethanol is produced from corn, concerns about effects that increased cultivation have on the environment have been raised. Corn is the largest crop produced in the United States and the heaviest user of pesticides and fertilizer. Fertilizer, mainly in the form of nitrogen and phosphorus, can runoff into surface waters causing water pollution. Also, the cycling of pesticides within the environment can have negative effects on beneficial insects, and lead to the bioaccumulation of toxins within animals and birds. Bioaccumulation is the increase in the concentration of pesticides in the bodies of organisms as you move up the food chain. The higher level consumers in a food chain can build-up toxic levels of pesticides that can lead to adverse health effects. There is also a concern for increased erosion of soil as a result of growing more corn. Corn is a wide row crop that exposes a large amount of soil to the atmosphere. This leads to excessive runoff and sediment pollution of water if proper soil management techniques are not used.

The fermentation of ethanol can have negative effects on the environment. Industrial scale fermentation takes place within a specialized manufacturing plant called a bio-refinery. This facility requires the use of energy to produce heat and power for the fermentation process. If this energy is supplied in the form of fossil fuels, then emissions by coal, oil, and natural gas may lead to increases in atmospheric pollution. Also, fermentation requires large quantities of water during the process. It is estimated about 12 gallons of water is needed to produce one gallon of ethanol. Therefore 12 gallons of wastewater is produced per gallon of ethanol that must be treated at the end of the fermentation cycle.

Another problem with the use of ethanol is that it's corrosive to certain metals like aluminum, brass, and zinc along with rubber and certain plastics in higher concentrations like E85 that makes it difficult to use existing gasoline distribution systems. In order to transport and dispense ethanol, the correct equipment approved for ethanol use must be used. Ethanol also absorbs water from the atmosphere that can dilute its concentration.

# METHYL ALCOHOL

Methanol, or methyl alcohol, is a colorless, combustible hydrocarbon molecule consisting of one carbon atom, four hydrogen atoms, and an oxygen atom ($CH_3OH$). It is commonly known as wood alcohol because it was once produced from wood. Methanol is considered more toxic than ethanol because it can cause chemical burns to the skin when touched. Methanol also produces a nearly invisible flame when burned.

## Methanol and Biomethanol Production

Most all methanol used today is produced from natural gas. In a process known as steam reforming, methane gas is exposed to high temperatures and pressures in the presence of steam that forms methanol. **Biomethanol** is the same as methyl alcohol, but is produced from biomass. Methanol production from biomass involves the use of **destructive distillation**. Destructive distillation is the heating of a hydrocarbon in the absence of oxygen. This is also known as pyrolysis and results in the formation of hydrogen and carbon monoxide gases, along with tar, hydrocarbon oils, and **char**. The most efficient way to produce biomethanol is by using **biomass gasification**. A biomass gasifier uses heat, steam, and pressure in a low oxygen environment to produce hydrogen and carbon monoxide gas also known as **syngas** from biomass. Typically a fluidized bed **gasifier** is used.

A biomass gasifying system injects finely chopped biomass into the top of the gasifier, while high-pressure steam heated to 2,200°F–2,300°F (~1,200°C–1,300°C) is blown in from the bottom of the chamber (Figure 22-11). The biomass is converted into syngas as it mixes with the steam. This high temperature gasification process converts 90% of the biomass into syngas, reducing the amount of impurities like tar, char, and hydrocarbon oils produced during low temperature conversion. Syngas can be further processed into methanol by exposing it to heat and pressure within the presence of a catalyst as shown in the following reaction: $2H_2 + CO = CH_3OH$.

Efficient gasification of one ton of woody biomass can produce 186 gallons of methanol. The benefit of methanol production by gasification is it can use a variety of different biomass feedstocks including woody residue, crop residue, and energy crops. One disadvantage of the process is that it is energy intensive. The production of heat is required for the conversion of biomass to methanol.

## Methanol Use

Methanol is mainly used today to produce methyl tertiary butyl ether (MTBE), that is added to gasoline to oxygenate it and make it burn more efficiently. This reduces the production of carbon monoxide produced by vehicle exhaust. MTBE has been shown to be a pollutant in drinking water and possibly a carcinogen. A carcinogen is something that causes

**biomethanol**—methyl alcohol is produced from biomass

**destructive distillation**—the heating of a hydrocarbon in the absence of oxygen

**char**—the solid remains of partially decomposed hydrocarbons

**biomass gasification**—the use of heat, steam, and pressure in a low oxygen environment to produce syngas from biomass

**syngas**—a combustible mixture of hydrogen and carbon monoxide gas

**gasifier**—a device that uses high heat in a low oxygen environment to convert substances into a gas

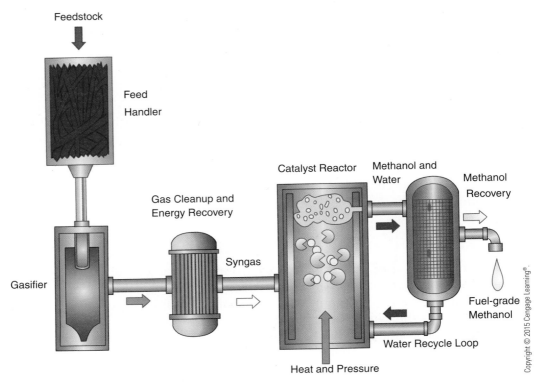

**FIGURE 22-11** Biomethanol production.

cancer. Because of the possible negative effects of using MTBE, ethanol is now being used as a gasoline additive. Methanol can be used in an internal combustion engine like ethanol. It is commonly mixed in a 10% methanol, 90% gasoline ratio known as M10, or as M85 (15% gasoline and 85% methanol). There are significant reductions of hydrocarbon, particulate matter, nitrogen oxides, and volatile organic compounds emissions when M85 is used in car engines. Methanol has an energy content of 63,000 Btus per gallon. This is 55% of the energy content of gasoline. This means you have to burn 1.8 gallons of methanol to get the equivalent mileage produced by using one gallon of gasoline. Even though methanol has a lower energy content than ethanol, it is being considered as a potential fossil-fuel replacement because any type of biomass can be converted into methanol. Also, methanol can be used in a **methanol fuel cell** to produce electricity. A methanol fuel cell, also known as the Direct Methanol Fuel Cell (DMFC) cell, is a device that uses a membrane catalyst to mix oxygen with a solution of methanol to produce carbon dioxide and water vapor, while also generating an electric current (Figure 22-12).

The development of methanol batteries to replace commonly used electrochemical batteries is under way. Potentially these batteries can provide a constant supply of electricity needed to power small appliances, such as laptop computers, cell phones, and MP3 players as long as they have a supply of methanol. This could be an advantage to the consumer because there would be no time needed to recharge them. Instead of plugging them into an electrical outlet you just refill them with methanol.

**methanol fuel cell**—a device that uses a membrane catalyst to mix oxygen with a solution of methanol to produce carbon dioxide and water vapor, while also generating an electric current

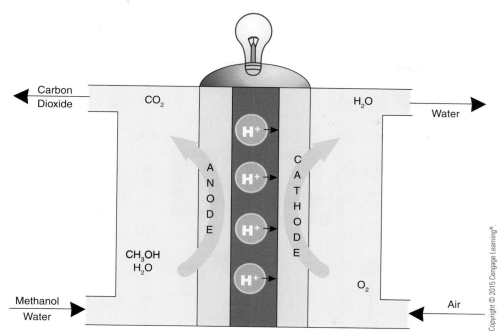

**FIGURE 22-12** Methanol fuel cell.

One of the negative aspects of using methanol as an energy source is it produces carbon dioxide when burned or used to power a fuel cell. This problem can be addressed if biomethanol is produced by using a sustainable biomass feedstock. The emissions of carbon dioxide can then be absorbed by the continual growth of the energy crop, therefore making the process carbon neutral.

## BUTYL ALCOHOL AND BIOBUTANOL

Butanol, or butyl alcohol, is a colorless, combustible hydrocarbon with the chemical formula of $C_4H_9OH$. Today most butyl alcohol is made from petroleum and is used as an industrial solvent. **Biobutanol** is butyl alcohol made from the fermentation of biomass. In a similar process to the fermentation of ethanol, the sugars in biomass are converted by microorganisms into butanol. This is known as the Acetone–Butanol–Ethanol (ABE) fermentation process (Figure 22-13). Instead of using yeast to convert the sugars to alcohol, a bacteria (*Clostridium acetobutylicum*) is used. The ABE process had been producing butanol for use as a solvent until it was replaced by a petroleum conversion method in the 1950s. Today, new advances in ABE fermentation are being developed that can convert biomass feedstocks into butanol. The same crops used to produce ethanol, like corn, sorghum, and cellulosic biomass, can also be used to produce butanol. Conversion rates of corn grain to butyl alcohol range between 1.3–2.5 gallons of butanol per bushel of corn. This represents 50% of the biomass being converted into butanol. This is slightly higher than current corn ethanol production rates. Although butanol fermentation uses a different microorganism than ethanol production, existing ethanol biorefineries can be easily converted to produce biobutanol.

**biobutanol**—a type of butyl alcohol that is made from the fermentation of biomass

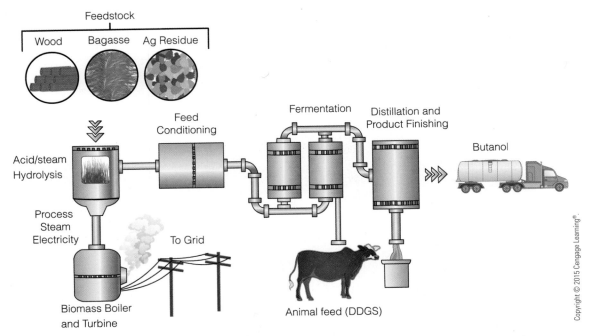

**FIGURE 22-13** Butanol fermentation.

## Uses of Biobutanol

Biobutanol has many advantages over using ethanol or methanol as a gasoline additive. One gallon of butanol contains 110,000 Btus of energy; an energy density of about 90% that of gasoline. This means average you need to burn 1.1 gallons of methanol to equal the energy in one gallon of gasoline. This is higher than the energy density of both methanol and ethanol. Currently the EPA approves an 11.5% mixture of butanol (Bu11.5) with gasoline. This improves the combustion of the gasoline by oxygenating the fuel. Studies show unmodified gasoline internal combustion engines can run on mixtures of 16% without any modification. This is higher than the current approved mixture of 10% for ethanol. These claims have yet to be approved by the EPA, but tests have also been done on using Bu100 in unmodified cars that revealed no negative effects to the engine.

The use of butanol as a gasoline additive has resulted in the reduction of carbon monoxide, hydrocarbon, and nitrogen oxide emissions. There is also a potential for a substantial reduction of carbon dioxide if the butanol is produced from sustainable biomass grown from dedicated energy crops. This has the prospect of making the use of biobutanol carbon-neutral. Another advantage to using biobutanol as a transportation fuel is it can be distributed using the existing gasoline infrastructure.

Butanol is not corrosive to current gasoline pumps, pipelines, and hoses we use today to fuel our motor vehicles. Ethanol can corrode existing gasoline fuel distribution systems and also absorb water from the atmosphere, unlike butanol. Butanol can also be mixed with diesel fuel in concentrations up to 20% without any modifications to the diesel engine. The U.S. Navy is looking at the possibility of using butanol as an additive to the jet fuel they use in planes.

# REVIEW OF KEY CONCEPTS

1. Fermentation is the process of breaking down a substance in the absence of oxygen. During fermentation, yeast consumes simple sugars like glucose to produce carbon dioxide and ethanol. Distillation is the process that ethanol is separated from water. The ethanol water mixture is heated, and separates ethanol from the water because ethanol has a lower boiling point temperature than water. The vapor is collected and condensed. This produces a 95% ethanol solution.

2. The four main processes to produce ethanol are pretreatment, hydrolysis, biochemical conversion, and distillation. The pretreatment process involves the preparation of the plant biomass to be used as the source of sugar for fermentation. Hydrolysis in the production of ethanol involves the use of special enzymes, acids, and heat to break down the cellulose and hemicellulose into simple sugars like glucose then fermented by the yeast. The biochemical conversion process (fermentation) takes place in a heated chamber for two to three days. Distillation separates out the ethanol so it is 95% pure.

3. Most all ethanol produced in the United States is made from corn. Approximately 300–420 gallons of ethanol can be made from one acre of corn.

4. Ethanol can be mixed with gasoline at concentrations of 10%, 15%, and 85%; known as E10, E15, and E85. Currently cars built after 2000 can use E10 or E15 fuels. Special Flex Fuel vehicles can use E85.

5. Ethanol has an energy content of 84,000 Btus. This is approximately 73% of the energy contained within one gallon of gasoline. This means you have to use 1.4 gallons of ethanol to get the same amount of mileage as one gallon of gas.

6. There are two types of feedstocks used to make ethanol: sugar and starch-based, and cellulose-based. Sugar and starch-based feedstocks include corn, sorghum, and sugarcane.

7. Cellulosic ethanol is made from any type of biomass. Cellulose feedstocks can be derived from any plant biomass. Plants lock up sugars in the form of complex starches like cellulose.

8. The benefits of using ethanol include the fact it is a renewable fuel, it lowers many of the emissions of pollutants produced from motor vehicles, and in the future can be produced from high yielding, sustainable biomass energy crops.

9. The problems associated with ethanol use include its corrosive nature and water absorption. Also, because most ethanol is produced in the United States comes from corn the cost of food can be affected as the demand for corn increases. The practice of growing more corn to produce ethanol can lead to increased erosion of soil, pesticide and fertilizer use, runoff, and possible water pollution.

10. Methanol, or methyl alcohol, commonly called wood alcohol, is another type of alcohol fuel. Biomethanol is methanol produced from biomass by the process of biomass gasification.

11. Most methanol produced in the United States is made from natural gas by the process of steam reforming.

12. Biomethanol is produced by biomass gasification that heats biomass feedstocks in a low oxygen environment in the presence of high temperature steam. This converts the biomass into syngas; a mixture of carbon monoxide and hydrogen. The syngas is passed through a catalyst that converts it into liquid methanol.

13. The benefit of methanol production by gasification is it can use a variety of different biomass feedstocks including woody residue, crop residue, and energy crops.

14. There are significant reductions of hydrocarbon, particulate matter, nitrogen oxides, and volatile organic compound emissions when M85 is used in car engines.

15. Methanol can be used to power a methanol fuel cell that produces an electric current. Currently methanol fuel cell batteries are being developed as replacements for rechargeable batteries in small electronics.

16. Three negative aspects of producing biomethanol are that it is more toxic, it is found to be a carcinogenic pollutant in drinking water, and it is more energy intensive to create.

17. Butanol, or butyl alcohol, is another form of alcohol produced from fermentation. Currently most butanol is made from petroleum, but it can also be made from biomass. In this case it is known as biobutanol. Just like ethanol, different forms of biomass can be potentially converted into butanol by fermentation.

18. The production of biobutanol uses bacteria instead of yeast. This is known as the ABE fermentation process.

19. Butanol has a high energy density of 90% that of gasoline, that is higher than ethanol and methanol. This means on average you need to burn 1.1 gallons of methanol to equal the energy in one gallon of gasoline.

20. Butanol can be mixed with gasoline in concentrations of up to 11.5% (Bu11.5). Studies show that this can be increased as high as 16% and that 100% butanol can be used in car engines without any modification, although this has not been proven by the EPA.

21. One gallon of ethanol has an energy content of about 84,000 Btus. Methanol has an energy content of 63,000 Btus per gallon. One gallon of butanol contains 110,000 Btus of energy.

## CHAPTER REVIEW

### Short Answers

1. What is fermentation and distillation?

2. Describe the four steps used to produce ethanol at a bio-refinery.

3. How many gallons of ethanol can be produced by one acre of corn and by one acre of sugarcane?

4. What is the difference between E10, E15, and E85?

5. Describe the difference in the energy content of one gallon of ethanol compared to one gallon of gasoline.

6. What are three sugar and starch-based ethanol feedstocks?

7. What is cellulosic ethanol?

8. What are three positive effects and three negative effects using ethanol as a liquid fuel have on the environment?

9. What is the difference between methanol and biomethanol?

10. How is most methanol made today in the United States?

11. Describe the process of producing biomethanol from biomass.

12. What is an advantage of producing biomethanol from biomass?

13. What are some of the benefits of using M85 as a liquid fuel for cars?

14. How does a methanol fuel cell produce energy?

15. What are three negative aspects of using methanol?

16. What is the difference between butanol and biobutanol?

17. How is most butanol made today in the United States?

18. Describe the process of producing biobutanol from biomass.

19. What is an advantage of producing biobutanol from biomass?

20. What are some of the benefits of using Butanol as a liquid fuel for cars?

21. How does the energy content of butanol compare to that of ethanol and methanol?

22. Briefly discuss the advantages and disadvantages of using of alcohol as a liquid transportation fuel?

23. Compare and contrast some of the properties of ethanol, methanol, and butanol.

## Energy Math

1. If the U.S. daily consumption of gasoline in 2012 was 367 million gallons per day, how many gallons of gas could be saved if cars used E15 as a fuel?

2. If 186 gallons of biomethanol can be produced by the gasification of 1 ton of biomass, how many acres of shrub willow (6 dry tons per acre)

would you need to produce to supply 100,000 gallons of biomethanol?

3. How many bushels of corn would you need to produce 367 million gallons of biobutanol using a conversion factor of 1 bushel per 2.5 gallons of butanol?

## Multiple Choice

1. Which chemical formula represents ethanol?
   a. $C_4H_9OH$
   b. $CH_3OH$
   c. $C_2H_5OH$
   d. $C_6H_{12}O_6$

2. A long chain of the same molecule, like starch is known as:
   a. sucrose
   b. glucose
   c. a polymer
   d. a catalyst

3. Which of the following is responsible for the fermentation of sugar to produce ethanol?
   a. Bacteria
   b. Gasification
   c. Pyrolysis
   d. Yeast

4. Ethanol is separated from water by:
   a. distillation
   b. gasification
   c. combustion
   d. hydrolysis

5. Which of the following chemical formulas represent glucose sugar?
   a. $C_4H_9OH$
   b. $CH_3OH$
   c. $C_2H_5OH$
   d. $C_6H_{12}O_6$

6. The process of breaking something down by using water is called:
   a. distillation
   b. gasification
   c. combustion
   d. hydrolysis

7. Which crop is the main feedstock for ethanol production in the U.S.?
   a. Switchgrass
   b. Corn
   c. Sorghum
   d. Sugarcane

8. Which crop produces more ethanol per acre?
   a. Wheat
   b. Corn
   c. Sorghum
   d. Sugarcane

9. Which fuel blend can be used by cars manufactured after the year 2000?
   a. E100
   b. E85
   c. E15
   d. None of the above

10. Which chemical formula represents methanol?
   a. $C_4H_9OH$
   b. $CH_3OH$
   c. $C_2H_5OH$
   d. $C_6H_{12}O_6$

11. Which process can be used to produce methanol?
   a. Fermentation
   b. Biomass gasification
   c. Combustion
   d. Hydrolysis

12. A mixture of hydrogen and carbon monoxide is known as:
   a. biogas
   b. methane
   c. syngas
   d. methanol

13. Which chemical formula represents butanol?
   a. $C_4H_9OH$
   b. $CH_3OH$
   c. $C_2H_5OH$
   d. $C_6H_{12}O_6$

14. Which of the following is responsible for the fermentation of sugar to produce butanol?
   a. Bacteria
   b. Gasification
   c. Pyrolysis
   d. Yeast

15. Which alcohol fuel has the highest energy density?
   a. Ethanol
   b. Methanol
   c. Butanol
   d. Gasoline

16. Which alcohol fuel can be mixed with diesel fuel?
   a. Ethanol
   b. Methanol
   c. Butanol
   d. Gasoline

17. Which alcohol feedstock can potentially produce the most fuel from fermentation?
   a. Sugar based
   b. Starch based
   c. Cellulose based
   d. Corn based

## Matching

*Match the terms with the correct definitions*

a. ethanol
b. fermentation
c. sugar
d. distillation
e. bio-refinery
f. starches
g. polymers

h. cellulose
i. endosperm
j. hydrolysis
k. distiller's dried grains and solubles (DDGS)
l. dewater
m. biomethanol

n. destructive distillation
o. char
p. biomass gasification
q. syngas
r. gasifier
s. methanol fuel cell
t. biobutanol

1. _____ The solid remains of partially decomposed hydrocarbons.

2. _____ A mixture of hydrogen and carbon monoxide gases.

3. _____ The process of removing suspended solids from a solution.

4. _____ A common name for carbohydrate molecules that are composed of long chains of sugars.

5. _____ The process of breaking apart a substance by water.

6. _____ Also known as grain alcohol.

7. _____ A heated chamber used to convert solids into a gas.

8. _____ A large fermentation plant.

9. _____ The heating of a hydrocarbon in the absence of oxygen.

10. _____ The separation of liquids as a result of the differences in their boiling points.

11. _____ A polymer of glucose that makes up cell walls.

12. _____ The stored food within a plant's seed.

13. _____ Wood alcohol produced from biomass.

14. _____ The conversion of plant material into syngas.

15. _____ The process of converting sugars into ethanol by yeast.

16. _____ A device that converts alcohol into electricity.

17. _____ The solid waste leftover from the fermentation process.

18. _____ Long chain-like molecules.

19. _____ A simple carbohydrate molecule like glucose.

20. _____ An alcohol fuel that can be produced from the fermentation of biomass by bacteria.

## CHAPTER 23

# Biodiesel

## KEY CONCEPTS

*After reading this chapter, you should be able to:*

1. Describe how a diesel engine operates.

2. Explain why Rudolph Diesel invented the diesel engine.

3. Explain reasons why pure vegetable oils are not good to use in diesel engines.

4. Briefly describe the process known transesterfication.

5. Identify the five main steps of producing biodiesel.

6. Indentify the two oil crops that produce the most biodiesel in the United States

7. Define B100, B20, and bioheat.

8. Describe two positive effects and two negative effects of using biodiesel.

## TERMS TO KNOW

ignition point

viscous

gel point

cloud point

dehulling

biodiesel

fatty acid

transesterfication

tallow

white grease

yellow grease

bioheat

# INTRODUCTION

After gasoline, diesel fuel is the second largest liquid fuel consumed in the United States. Diesel fuel powers trucks, trains, buses, ships, and other heavy equipment. A similar fuel known as Number 2 fuel oil is used to heat millions of homes. Although most of the diesel fuel we use today comes from petroleum, it can also be made from vegetable oil. The use of vegetable oils as a source of fuel has been used for thousands of years. Many plant-based oils are combustible and can actually be burned in diesel engines directly. The conversion of vegetable oil into biodiesel however is a much better option when it comes to performance. The possibility of producing a sustainable supply of liquid fuel from vegetable oils is going to play an important role in the way energy is produced and used by agriculture in the future.

## THE DIESEL ENGINE

In 1897, Rudolph Diesel built the first compression ignition internal combustion engine named in his honor. Diesel's engine was different from other internal combustion engines that used the heat from a spark plug to ignite the fuel within a cylinder that powered a piston. These engines are known as spark ignition engines that are used in most cars today. In Diesel's engine, the heat required to set off combustion was generated by the compression of the hydrocarbon vapor within the cylinder, causing the gas to reach its **ignition point**. The resulting explosion produced hot, rapidly expanding gases that move the piston and create power (Figure 23-1).

Diesel was driven to create a practical heat engine that could be used by farmers. He saw the change in agriculture, from the use of self-sustaining animal power to that of machine power that could potentially keep the famer from being self-sufficient. Diesel's plan was to invent an engine fueled by vegetable oil that could be grown by farmers. This would allow farmers to produce their own fuel for their machinery similarly to the way farmers produced feed for their work animals in the past. The result was the compression ignition engine. Diesel originally used peanut oil to fuel his early engines, but as low-cost petroleum fuels were developed, his engine was soon powered by the fossil fuel that bears his name.

The diesel engine quickly became the motor of choice for machines that required power and durability. Today diesel engines operate in much the same way as Diesel's original engine more than one hundred years ago. In fact, most all diesel engines can theoretically be run on pure vegetable oil if the weather is not too cold. This is not recommended because over time, pure vegetable oil leaves residues within the engine that can build up and clog fuel injectors, valves, and cylinders. This leads to maintenance problems and increased wear on the engine. When the temperature drops below a certain temperature, vegetable oil becomes more **viscous** and

**ignition point**—the temperature at which a gas will spontaneously combust

**viscous**—a liquid that resists flow

329

**THE DIESEL PRINCIPLE**

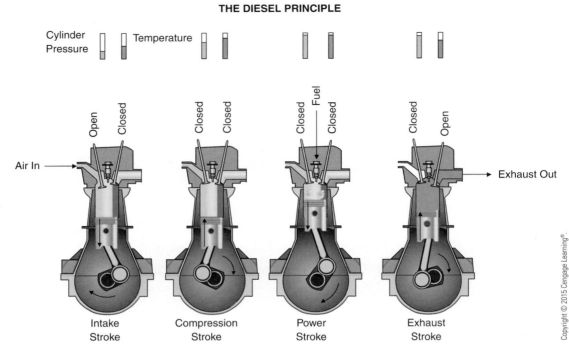

**FIGURE 23-1** Diesel engine.

**gel point**—the temperature that something begins to gel

**cloud point**—the temperature that oil starts to contain small amounts of solids that make the oil appear cloudy and more viscous

does not flow easily. If the temperature is low enough, the oil will begin to become a solid or gel. The temperature oil begins to gel is known as its **gel point**. For example peanut oil has a gel point of 37°F (2.8°C), while soybean gels at around 34°F (1°C). The gel point differs from the **cloud point.** The cloud point is the temperature that the oil starts to contain small amounts of solids making the oil appear cloudy and more viscous. The cloud point is usually higher than the gel point (Figure 23-2). The high gel points

**FIGURE 23-2** Comparing gel and cloud points.

of most vegetable oils makes the use of pure vegetable oil impractical for use in diesel engines in colder climates. This is why biodiesel is preferred.

## BIODIESEL PRODUCTION

Biodiesel is produced in the United States by more than 170 plants located all over the country. The largest plant in the United States is in Hoquiam, Washington, that produces up to 100 million gallons each year. The process of producing biodiesel involves five main steps: oil extraction, catalyst preparation, transesterfication, glycerin separation, and washing (Figure 23-3).

### Oil Extraction

Oil extraction of vegetable feedstock first involves cleaning and **dehulling**. Dehulling is important because hulls absorb oil. The seeds are then chopped into smaller pieces and washed in an oil-extracting solvent like hexane. The solvent is then separated from the seeds. The remaining seed material is dried and can be used as feed for livestock or other food products. The hexane solution is evaporated and the oil is produced. The evaporated hexane can be reclaimed and used again. Smaller batches of oil feedstock can be processed using an expeller. An expeller is a press-like device that squeezes the oil crop under high pressure (Figure 23-4). The oil seeps out of tiny sieves and is collected. This is also known as cold-pressed oil. The use of hexane solvent extraction is much more efficient than expeller extraction.

**dehulling**—the process of removing the hull or outer covering of seeds

### Catalyst Preparation

Once the oil is extracted, a catalyst is prepared for mixing. The method for producing most biodiesel in the United States is known as the base catalyzed reaction. The catalyst is typically a mixture of sodium or potassium hydroxide and methanol, known as methoxide. The proportion of methanol to sodium

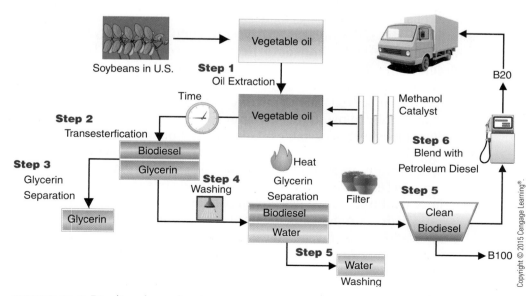

**FIGURE 23-3** Biodiesel production.

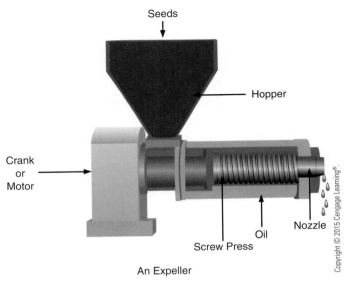

**FIGURE 23-4** An expeller used for seed oil extraction.

or potassium hydroxide is dependent on the specific oil feedstock being used. A titration procedure for determining the concentration of methoxide is done for every new batch of oil to assure complete transesterfication. The titration step involves adding small amounts of methoxide solution to the oil feedstock until a pH between 8 and 9 is reached. A typical mixture of 3.5 grams of sodium hydroxide (NaOH) or 7.0 grams of potassium hydroxide (KOH) per liter of methanol is used for most fresh oils.

## TRANSESTERFICATION

**biodiesel**—a combustible, vegetable oil-based fuel composed of fatty acids

**fatty acid**—a long hydrocarbon molecule that makes up fats and oils in animals and plants

**transesterfication**—process of mixing the triglycerides that make up vegetable oil with a methanol-catalyst solution to break them into smaller fatty acids

**Biodiesel** is a combustible, vegetable oil-based fuel composed of **fatty acids**. A fatty acid is a long hydrocarbon molecule that makes up fats and oils in animals and plants. Vegetable oils are mostly composed of triglycerides; a type of fat made up of three fatty acid molecules attached to a glycerol molecule. The arrangement of triglycerides makes vegetable oil too viscous to use in a diesel engine; therefore the process known as transesterfication is used to break the triglycerides into smaller fatty acids. This is how biodiesel is made. **Transesterfication** is accomplished by mixing the triglycerides in vegetable oil with a methanol-catalyst solution described previously (Figure 23-5). Once the catalyst has been made, it can then be added to the oil. The oil catalyst mixture is heated to about 130°F (54°C) and mixed to start the transesterfication process.

When mixed with vegetable oil, the catalyst breaks the bonds between the 3 triglycerides and the glycerol molecule. One fatty acid molecule is released from the triglyceride and joins with one methanol molecule to form a methyl ester. The hydroxyl (OH) molecule released from the catalyst solution then binds with a glycerol molecule to make glycerin. The result is a liquid hydrocarbon fuel in the form of a fatty acid methyl ester (FAME) that is less viscous than vegetable oil because fatty acid chains are broken into single chains. Approximately 80%–90% of the original vegetable oil is converted into biodiesel, with glycerin as the remaining component. The glycerin byproduct can be used to make a variety of products including soap and as an animal feed additive.

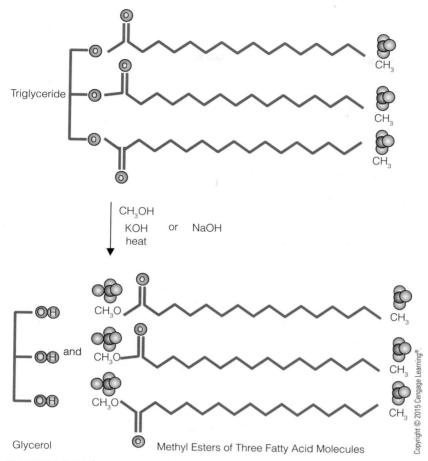

**FIGURE 23-5** The process of transesterfication.

## Glycerin Separation

Once the reaction is complete, the oil is then allowed to separate into bio-diesel and glycerin (Figure 23-6). The glycerin is denser than the biodiesel and settles to the bottom of the tank. The denser glycerin is then pumped out from the base of the mixing tank.

## Washing

The biodiesel is then washed in water to remove any impurities. Finally it is allowed to dry to vent off any remaining water and it is ready for use. Ten pounds of methoxide catalyst added to 100 pounds of oil yields approximately 90 pounds of biodiesel, and 10 pounds of glycerin. This represents a 90% conversion of oil to biodiesel. The energy balance for producing biodiesel is about 1–3.2 meaning for every one unit of energy put in to make it, yields 3.2 units of energy. This is about the same as corn ethanol but lower than sugarcane ethanol.

## OIL FEEDSTOCKS

Biodiesel can be made from any source of vegetable oil or animal fat. This includes oil crops, waste oils that are produced by the food service industry, and animal fats from meat processing. In 2012, the United States produced more than 968 million gallons of biodiesel that came from soybean oil

**FIGURE 23-6** Separation of biodiesel and glycerin.

**tallow**—rendered animal fat from processing meat

**white grease**—unprocessed fat from meat processing

**yellow grease**—used vegetable oil from the food service industry

(53%), **tallow** (5%), canola oil (10%), **white grease** (5%), **yellow grease** (8%), poultry fat (2%), and corn oil (8%) (Figure 23-7). Tallow is rendered animal fat from meat processing, and white grease is unprocessed fat from meat. Yellow grease is used vegetable oil from the food service industry. Different oil crops have varying yields of oil per acre (Figure 23-8).

Soybean typically yields 46 gallons of oil per acre. Other crops like canola (rapeseed) can yield more than 70 gallons per acre, mustard more than 50 gallons per acre, and corn about 20 gallons per acre. Other potential oil crops include palm oil (635 gallons/acre) and Jatropha, a semi-arid shrub

**U.S. Biodiesel Feedstocks 2012**

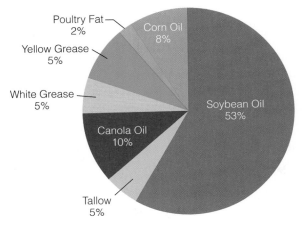

**FIGURE 23-7** Biodiesel feedstocks in the U. S.

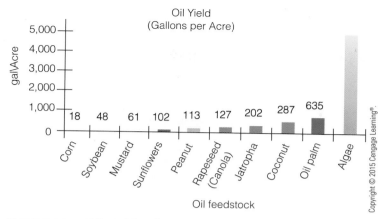

**FIGURE 23-8** Oil yields of various oil crops.

that grows on marginal land and yields about 200 gallons per acre. There is also the possibility of using algae as an oil feedstock that may yield more than 5,000 gallons of oil per acre! More than 50% of the biomass in algae is oil, making algae more than 30 times more productive than land-based oil crops. Research on the production of algae for use as an oil feedstock is ongoing and may be an important source of biomass oil in the near future (Figure 23-9).

There is also an advantage of reusing the DDGS (sillage) waste from the fermentation of corn as a feedstock for biodiesel production. The production of corn-based ethanol uses yeast to convert the sugars in corn into ethyl alcohol. The remaining solid waste, called DDGS, is high in corn oil content. Once dewatered, DDGS can be processed to extract the remaining corn oil. This could provide a feedstock for biodiesel production, making the corn-based ethanol process more efficient.

As stated before, in 2012 the annual production of biodiesel exceeded 968 million gallons; more than 168 million gallons over the EPAs target of 800 million gallons for 2011. This is a major increase in production from the 315 million gallons produced in 2010. The low production in 2010 was down from 2008 levels when 678 million gallons was produced. The reduction in production in 2010 was the result of the loss of a blender's tax incentive. A blender's tax incentive is a monetary exemption ($1.00) for every gallon of vegetable oil-based biodiesel blended into petroleum diesel. This lowered the cost of production associated with biodiesel in order to make it more competitive with petroleum fuel. The blender's tax was re-instated in late 2010, extending the credit through 2011. As a result of the tax incentive, the level of production for biodiesel exceeded 1 billion gallons. The purpose of this mandate, which is part of the Energy Independence and Security Act of 2007, is to reduce our use of foreign petroleum and cut our greenhouse gas emissions.

## USE OF BIODIESEL

Biodiesel is mainly produced today for use as an additive to diesel fuel. Commonly it is mixed at a ratio of 20% biodiesel to 80% petroleum diesel (B20). B100 is 100% biodiesel that is classified as an alternative renewable fuel. There is no real difference between biodiesel and petroleum diesel, except their energy density. Petroleum diesel has a heat content of about 18,000 Btus per pound as compared to biodiesel's average of about 15,900 Btus per pound. This equates to biodiesel having 88% the heat value of petroleum diesel. Most diesel engines manufactured after 1994 are capable of using B100, and all diesel engines can

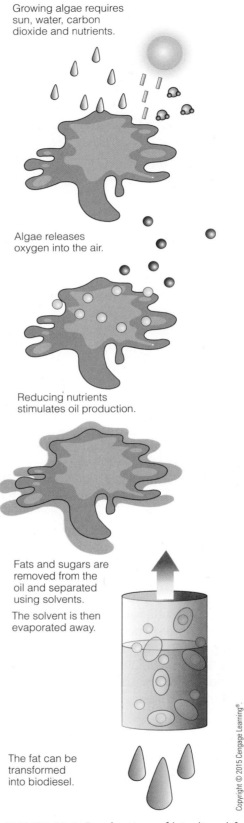

Growing algae requires sun, water, carbon dioxide and nutrients.

Algae releases oxygen into the air.

Reducing nutrients stimulates oil production.

Fats and sugars are removed from the oil and separated using solvents.

The solvent is then evaporated away.

The fat can be transformed into biodiesel.

**FIGURE 23-9** Production of biodiesel from algae.

burn B20. The main concern over the use of B100 is that biodiesel is a strong solvent. Many older hoses and gaskets in diesel engines built prior to 1994 can be corroded by biodiesel. Therefore it is necessary to assure the diesel engine you want to use biodiesel in has the correct specifications to burn B100. Another concern is the possibility of voiding your engine's warranty. Not all diesel engines are specified to burn biodiesel and you should check with the engine manufacturer before using biodiesel so as not to void your warranty. The other potential problem with using B100 is its relatively high cloud point temperature. The cloud point of B100 is around 40°F (4°C). Because of this, B20 is often used in colder climates to reduce the potential for gelling. Diesel engines used in colder climates use a fuel preheating system to decrease the viscosity of biodiesel so it flows into the engine more easily. Biodiesel can also be used as an additive or replacement for Number 2 heating oil. Number 2 heating oil is common fuel source in boilers and furnaces used to heat homes in winter. There is little difference between Number 2 heating oil and diesel fuel. The distinction between the two is often the color of the fuel. Number 2 heating oil is dyed red to differentiate it from diesel fuel used for transportation. Diesel fuel is subject to road taxes and Number 2 heating oil is not. The use of biodiesel as a source of home heating oil is called **bioheat**. Typical bioheat blends are classified as a B5 that is a 5% biodiesel and 95% Number 2 heating oil mixture.

**bioheat**—home heating oil mixed with biodiesel

The use of biodiesel has been shown to significantly reduce harmful emissions of air pollutants. The combustion of B100 reduces the emissions of carbon monoxide by 48%, unburned hydrocarbons by 67%, and particulate matter by 47%. The use of B20 also reduces these same pollutants by 25% (Figure 23-10). Because B100 burns so efficiently within a diesel engine, it produces a higher amount of nitrogen oxides compared to traditional petroleum diesel. Another drawback to using biodiesel is the amount of land and crops needed to produce it. Like any energy crop, oil seed has the potential for competing for agricultural land used to produce food. There is also the threat expanding agricultural land may pose dangers to local ecosystems. The reduction of forestland ecosystems to expanding energy crops may lead to increased erosion, runoff, and water pollution. These crops also require the use of chemical pesticides and fertilizers to combat pests and replace soil nutrients.

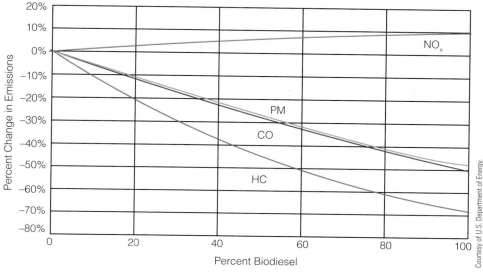

**FIGURE 23-10** Emission reductions for biodiesel mixes.

# REVIEW OF KEY CONCEPTS

1. The diesel engine is a compression ignition internal combustion engine. The ignition that starts the combustion process is caused by the heat generated from the compression of the fuel by a piston. The heat ignites the vapor and produces a hot, rapidly expanding gas that drives the piston. This is what provides the power from the engine.

2. The diesel engine was invented by Rudolph Diesel in 1893 and was originally designed to run on vegetable oil so farmers could grow their own fuel.

3. As petroleum became widely available in the early 1900s, the diesel engine was powered by petroleum-based diesel fuel. Although the diesel engine was developed for using vegetable oil, new diesel engines can experience problems when trying to use pure vegetable oil as a fuel. The buildup of residue within the engine can cause excess wear and vegetable oil has high cloud and gel points. The gel point is the temperature at which the oil becomes a solid and the cloud point is the temperature when it begins to become more viscous. This causes the use of vegetable oil to be a problem in cold weather.

4. Biodiesel is a fatty acid methyl ester fuel made from vegetable oil through a process known as transesterfication. Biodiesel is less viscous than pure vegetable oil and has lower cloud and gel points. Transesterfication is the process by which long chain triglyceride molecules in vegetable oil are broken into shorter chain fatty acid molecules by the use of a methoxide catalyst.

5. The process of producing biodiesel involves five main steps: oil extraction, catalyst preparation, transesterfication, glycerin separation, and washing.

6. In the United States most biodiesel is made from soybean oil, but other oils and animal fats can be used as a feedstock as well.

7. Biodiesel is mainly produced today for use as an additive to diesel fuel. Commonly it is mixed as a ratio of 20% biodiesel to 80% petroleum diesel (B20). B100 is 100% biodiesel, classified as an alternative renewable fuel. Biodeisel used to heat a home is known as bioheat.

8. The use of biodiesel lowers the emissions of certain atmospheric pollutants like particulate matter, unburned hydrocarbons, and carbon monoxide. Biodiesel can also be used as a substitute for home heating oil. Biodiesel experiences gelling at temperatures around 40°F, in the colder months a B20 mix is commonly used.

# CHAPTER REVIEW

## Short Answer

1. Describe the process of how a diesel engine operates.

2. Why did Rudolph Diesel invent the diesel engine?

3. Explain why pure vegetable oils are not good to use in diesel engines.

4. What is the process known as transesterfication?

5. What are the five main steps of producing biodiesel?

6. Which two oil crops produce the most biodiesel in the United States?

7. Define B100, B20, and bioheat.

8. What are two positive effects and two negative effects of using biodiesel?

9. Briefly discuss the advantages for using biodiesel as a liquid fuel compared to using petroleum diesel.

## Energy Math

1. In 2009, 152,502,000 gallons of diesel fuel was used in the United States. How many acres of rapeseed (canola) would you need to grow to produce the same amount of biodiesel?

## Multiple Choice

1. The diesel engine was originally run on:
   a. diesel fuel
   b. animal fat
   c. peanut oil
   d. tallow

2. The temperature at which biodiesel begins to form solids is called:
   a. ignition point
   b. viscosity
   c. cloud point
   d. compression

3. What is the byproduct of making biodiesel?
   a. Glycerin
   b. Methoxide
   c. Fatty acids
   d. Esters

4. What is the process called that converts oils into biodiesel?
   a. Fatty acid methyl ester
   b. Transestertfication
   c. Glycerin
   d. Methoxide

5. Rendered animal fat is also known as:
   a. yellow grease
   b. white grease
   c. tallow
   d. canola

6. Used oil from kitchens is known as:
   a. yellow grease
   b. white grease
   c. tallow
   d. canola

7. Unprocessed fat from meat used to make biodiesel is called:
   a. yellow grease
   b. white grease
   c. tallow
   d. canola

8. Which vegetable oil has the highest yield in gallons per acre?
   a. Soybean
   b. Corn
   c. Canola (Rapeseed)
   d. Mustard

9. Which vegetable oil has the lowest yield in gallons per acre?
   a. Soybean
   b. Corn
   c. Canola (Rapeseed)
   d. Mustard

**10.** Which vegetable oil produces the most bio-diesel in the United States?
   a. Soybean
   b. Corn
   c. Canola (Rapeseed)
   d. Mustard

**11.** Which atmospheric pollutant is increased by using B100?
   a. Nitrogen oxides
   b. Carbon dioxide
   c. Particulate matter
   d. Unburned hydrocarbons

## Matching

*Match the terms with the correct definitions*

a. ignition point
b. viscous
c. gel point
d. cloud point

e. biodiesel
f. fatty acids
g. dehulling
h. tallow

i. white grease
j. yellow grease
k. bioheat

**1.** ____ A combustible vegetable oil-based fuel composed of fatty acids.

**2.** ____ Unprocessed animal fat.

**3.** ____ Use cooking oil.

**4.** ____ The temperature at which a gas combusts.

**5.** ____ Rendered animal fat.

**6.** ____ The ability of a liquid to flow.

**7.** ____ The process of removing the outer seed coat.

**8.** ____ The temperature at which solids begin to form in a liquid.

**9.** ____ The temperature at which a liquid begins to gel.

**10.** ____ A long hydrocarbon molecule that makes up fats and oils in animals and plants.

**11.** ____ Home heating oil containing biodiesel.

# CHAPTER 24

# Biomass to Liquids

## KEY CONCEPTS

*After reading this chapter, you should be able to:*

1. Describe the process of fast pyrolysis and how it converts biomass into liquid fuel.

2. Define the term bio-oil.

3. Explain the process of hydrotreating.

4. Describe the Fischer–Tropsch process.

5. Discuss the benefits of converting biomass to liquid fuels.

## TERMS TO KNOW

thermochemical conversion

fast pyrolysis

bio-oil

hydrotreating

hydrocracking

Fischer–Tropsch process

alkane

# INTRODUCTION

The use of biomass to create liquid fuels similar to gasoline and diesel may provide agriculture with another way to produce energy. Research into methods of using biomass feedstocks to produce sustainable bio-oils that mimic petroleum are underway. The potential for using different forms of plant material to generate liquid fuels is yet another renewable fuel option.

## BIOMASS CONVERSION

The process of making liquid fuels from biomass involves the use of **thermochemical conversion**. Thermochemical conversion uses high heat and a catalyst to transform chemical compounds. Liquid fuels can be made from biomass by using this method. Solid biomass feedstock can be converted into a liquid fuel called bio-oil is similar to crude oil. This is done using pyrolysis. Up to 75% of the mass of the biomass feedstock can be converted into bio-oil.

**thermochemical conversion—** the use of high heat and a catalyst to transform chemical compounds

### Fast Pyrolysis

Pyrolysis is the breakdown of compounds by exposing them to high heat in a low oxygen environment (Figure 24-1). A similar process is used to convert solid biomass into syngas, a flammable mixture of hydrogen and carbon monoxide. The difference between producing a gas or a liquid from biomass involves the amount of heat applied and the duration the heat is applied. Typically biomass exposed to high heat for a longer period of time produces syngas by a process known as gasification. The use of lower heat for a shorter time is called **fast pyrolysis**, that produces a liquid byproduct. Fast pyrolysis exposes biomass to high heat for a short time. Typically the biomass feedstock is heated in the absence of oxygen to 900°F (~500°C) for approximately one second. This rapid heating converts the solid biomass into pyrolysis oil, also known as **bio-oil** or bio-crude. Bio-oil is a thick, black liquid hydrocarbon similar to crude oil (Figure 24-2). Typically, 75% of the solid biomass is converted into bio-oil by fast pyrolysis; 13% is a combustible gas and 12% a solid bio-char can be used as a soil additive. This bio-oil can be processed much like petroleum to convert it into liquid fuels like diesel and gasoline.

**fast pyrolysis—**exposing biomass to high heat for a short time in a low oxygen environment

**bio-oil—**also known as bio-crude, a thick, black liquid hydrocarbon that is similar to crude oil

### Hydrotreating

After fast pyrolysis, the bio-oil needs to be treated so its properties are more like petroleum fuels. This involves the use of **hydrotreating**. Hydrotreating is the use of hydrogen gas to lower the oxygen content of the bio-oil. This is the same process oil refineries treat crude oil.

**hydrotreating—**the use of hydrogen gas to lower the oxygen content of bio-oil

### Hydrocracking

After hydrotreatment, the bio-oil undergoes **hydrocracking**. Hydrocracking breaks down the long chains of hydrocarbon molecules that make

**hydrocracking—**the process of breaking long chain hydrocarbon molecules into shorter chains

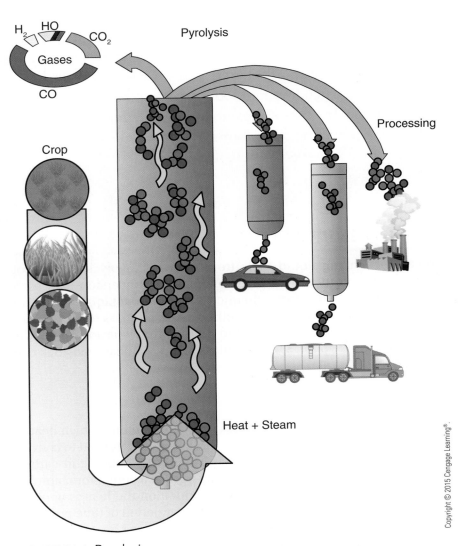

**FIGURE 24-1** Pyrolysis.

up the bio-oil and converts them into smaller chains. This allows the thick, viscous bio-oil to be converted into fuels like diesel, gasoline, and jet fuel.

## RESEARCH

The Department of Energy is conducting research into the conversion of biomass to liquids. The DoE determined it may be possible to convert solid biomass energy crops like switchgrass, shrub willow, and hybrid polar into liquid fuels but woody biomass seems to have a higher yield of bio-oil compared to that of herbaceous crops. One study showed the potential for converting one dry ton of poplar wood into 90 gallons of gasoline. This process however, involved the use of nonrenewable natural gas as the source of hydrogen needed for hydrotreatment.

## BYPRODUCTS OF CREATING BIO-OIL

The byproducts of fast pyrolysis include hydrogen and carbon monoxide gas, along with char. Char is the unburned solid remains of the biomass feedstock. This can be used as a soil additive. The gases produced by pyrolysis

**FIGURE 24-2** Bio-oil.

are flammable used to generate the heat needed for the thermochemical conversion.

## THE FISCHER–TROPSCH PROCESS

The **Fischer–Tropsch process** is a method of converting syngas into liquid fuels. Syngas is a mixture of hydrogen and carbon monoxide. It is used as a way to convert solid hydrocarbon fuels like coal and biomass into liquid fuels. The process was developed by Franz Fischer and Hans Tropsch in the early 1900s. During World War II, Germany used it to convert coal into gasoline, diesel, and aircraft fuel. The Fischer–Tropsch process uses gasification to convert solid biomass into liquid fuel (Figure 24-3). The biomass is superheated to temperatures of 2,000°F (1,200°C) within a chamber that contains a high-pressure, oxygen–steam mixture that converts it into syngas. The gasification process also creates impurities like hydrogen sulfide ($H_2S$) and carbon dioxide ($CO_2$). These gases are removed after gasification. The syngas is then exposed to a catalyst, like iron or cobalt metals that converts the syngas into liquid **alkane** hydrocarbons through the following reaction: $2n+1H_2 + nCO = C_nH_{2n+2} + nH_2O$, where n represents the number of atoms. An alkane is a hydrocarbon with the chemical formula of $CnH_{2n+2}$. If n equals 8, then the Fischer–Tropsch reaction produces octane ($17H_2 + 8CO = C_8H_{18} + 8H_2O$), an alkane that makes up gasoline. The alkanes produced by biomass can be used to produce liquid fuels.

Research suggests one ton of woody biomass can produce 47 gallons of fuel. The benefit of using the Fischer–Tropsch process is that many forms of solid biomass can be converted into liquid fuels similar to ones currently being used for transportation (Figure 24-4). These fuels would be considered renewable because they are produced from sustainable biomass.

**Fischer–Tropsch process**—a method of converting syngas into liquid fuels by exposing biomass to high heat and a catalyst

**alkane**—a hydrocarbon with the chemical formula of $CnH_{2n+2}$

1 In a low temperature gasifier, the biomass is broken down at temperatures between 400°C and 500°C into biocoke and low-temperature carbonization gas containing tar.

2 In the combustion chamber, the low-temperature carbonization gas is oxidized at temperatures exceeding 1,400°C and the biocoke is blown in.

4 The deduster removes any remaining coke dust particles.

6 The Fischer-Tropsch reactor uses catalysts to transform the gas into liquid fuel.

3 The raw gas is cooled in the heat exchanger.

5 The gas is purged of any remaining chlorides and sulfides in the washer.

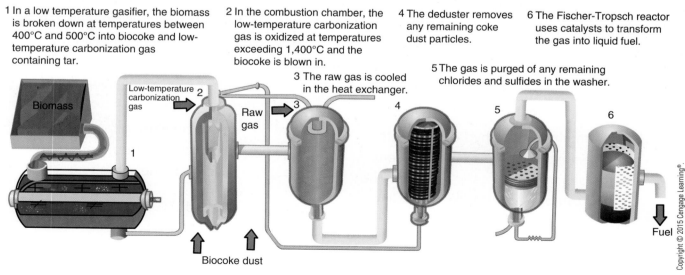

**FIGURE 24-3** Fischer–Tropsch process.

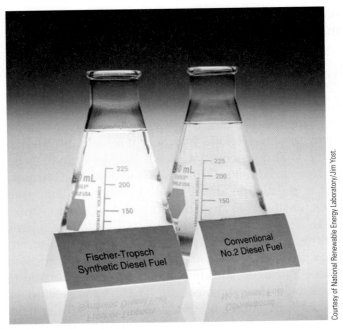

**FIGURE 24-4** Fischer–Tropsch fuel.

## REVIEW OF KEY CONCEPTS

1. The conversion of solid biomass into liquid fuels is possible by using fast pyrolysis. Fast pyrolysis exposes biomass to high heat for a short amount of time. This converts plant material into pyrolysis oil, also known as bio-oil.

2. Bio-oil is similar to crude oil that can be hydrotreated. Many biomass feedstocks can be used to produce bio-oil, including energy crops like shrub willow, hybrid poplar, and switchgrass.

3. Hydrotreament uses hydrogen gas to reduce the amount of oxygen present in the oil, making it easier to refine. The bio-oil can be converted into gasoline, diesel, and jet fuel.

4. Another biomass conversion method, called the Fischer–Tropsch process can be used to create liquid fuels. In this method, plant material is mixed with a high-pressure, steam oxygen mixture in a chamber that converts the biomass into syngas. The syngas is cleaned and exposed to a catalyst that transforms it into alkane hydrocarbons. These hydrocarbons can be refined into diesel and gasoline.

5. The benefits of converting biomass into liquid fuels include producing a form of gasoline based on renewable resources.

# CHAPTER REVIEW

## Short Answer

1. Describe the process of how fast pyrolysis converts biomass into liquid fuel.

2. What is bio-oil?

3. What is hydrotreating?

4. Describe the Fischer–Tropsch process.

5. What are the benefits of converting biomass to liquid fuels?

6. Briefly describe the processes that biomass is converted into liquid fuels.

## Energy Math

1. If one barrel equals 42 gallons, then how many barrels of bio-oil could you convert using fast pyrolysis on 6 tons of woody biomass?

## Multiple Choice

1. The use of high heat and a catalyst to transform chemical compounds is called:
   a. transesterfication
   b. thermochemical conversion
   c. hydrotreatment
   d. hydrocracking

2. A mixture of hydrogen and carbon monoxide gas is known as:
   a. syngas
   b. octane
   c. pyrolysis oil
   d. bio-oil

3. A thick, black liquid hydrocarbon that is similar to crude oil is:
   a. syngas
   b. octane
   c. biodiesel
   d. bio-oil

4. What is the process called that converts biomass into liquid fuel?
   a. Fast pyrolysis
   b. Transestertfication
   c. Hydrocracking
   d. Hydrotreatment

5. The process that mixes hydrogen with pyrolysis oil to reduce oxygen is called:
   a. fast pyrolysis
   b. transestertfication
   c. hydrocracking
   d. hydrotreatment

6. Which process converts biomass into syngas, then into a liquid hydrocarbon?
   a. Fast pyrolysis
   b. Fischer–Tropsch process
   c. Hydrocracking
   d. Hydrotreatment

7. Which of the following is an alkane?
   a. $H_2S$
   b. $CO_2$
   c. $C_8H_{18}$
   d. $C_6H_{12}O_6$

## Matching

*Match the terms with the correct definitions*

a. thermochemical conversion   d. hydrotreating            f. the Fischer–Tropsch process
b. fast pyrolysis              e. hydrocracking            g. alkane
c. bio-oil

1. _____ The combination of hydrogen with pyrolysis oil to reduce its oxygen content.

2. _____ The rapid heating of a substance to decompose it.

3. _____ A simple hydrocarbon with the formula $C_nH_{2n+2}$.

4. _____ The use of heat and a catalyst to transform a substance.

5. _____ The process of breaking long chains of hydrocarbons into smaller chains.

6. _____ The use of high temperatures, steam and oxygen to convert solids into liquid fuel.

7. _____ A thick black hydrocarbon formed from biomass that is similar to crude oil.

# UNIT VII

# Gaseous Bioenergy Fuels from Agriculture

---

**TOPICS TO BE PRESENTED IN THIS UNIT INCLUDE:**

- Biogas

## OVERVIEW

The use of agriculturally produced biogas provides the ability to transform agricultural waste into a sustainable energy source. Biogas production on the farm provides farmers with a low tech, clean source of energy that reduce the impact animal manures have on the environment. Similarly, methane produced from landfill can be a source of biogas. Both of these fuels can be used like natural gas.

## CHAPTER 25

# Biogas

## KEY CONCEPTS

*After reading this chapter, you should be able to:*

1. Describe the process of anaerobic digestion.

2. Define the term biogas.

3. Explain the five steps used to produce biogas.

4. Identify the five types of biogas digester systems.

5. Discuss the benefits of producing biogas from animal manures.

6. Discuss the ways biogas can be used.

7. Identify three negative aspects of producing and using biogas.

8. Explain how landfills produce and use biogas.

## TERMS TO KNOW

biogas
methanobacteria
anaerobic
amino acids

covered lagoon digester
complete mix digester
plug flow digester
fixed-film digester

batch digester
effluent
landfills
combined heat and power

# INTRODUCTION

Biological gas, or biogas, is a naturally occurring combustible fuel produced from organic material. Also known as swamp gas, biogas can be used just like methane; the cleanest burning fossil fuel. Methane can be produced by a variety of sources, including landfills, sewage, and animal manures.

## ANAEROBIC DIGESTION

**Biogas** is a colorless, odorless, combustible gas composed of methane and carbon dioxide. It is produced by anaerobic digestion, a form of bioconversion that involves the breakdown of organic material in the absence of oxygen to produce methane gas ($CH_4$). Biogas is also called biomethane. The conversion is done by **methanobacteria**. Methanobacteria are a class of **anaerobic** bacteria that consume organic matter and produce methane as a byproduct. The production of biogas involves three steps and includes many types of bacteria.

**biogas**—a colorless, odorless, combustible gas composed of methane and carbon dioxide

**methanobacteria**—a class of anaerobic bacteria that consume organic matter and produce methane as a byproduct

**anaerobic**—without oxygen

### Hydrolysis

The process begins when organic material is broken down by aerobic bacteria. Even though biogas is an anaerobic process, it begins by aerobic bacteria decomposing the proteins and carbohydrates that make up organic matter. The proteins are made of molecules called **amino acids**. An amino acid is a simple compound made up of hydrogen, oxygen, nitrogen, and carbon. In this first step of biogas production, water splits the proteins into individual molecules of amino acids. This is known as hydrolysis. Simple sugars and fatty acids are also produced by hydrolysis.

**amino acid**—a simple compound made up of hydrogen, oxygen, nitrogen, and carbon

### Acidogenesis

The amino acids, sugars, and fats are consumed by anaerobic bacteria to produce acids and alcohols. This is known as acidogenesis. Acidogenesis produces acetic acid, along with byproducts like hydrogen, ammonia ($NH_3$), and hydrogen sulfide gas ($H_2S$). Hydrogen sulfide is what gives decomposing organic matter that "rotten egg" odor.

### Methanogenesis

The final step is called methanogenesis; the process methanobacteria consume acetic acid to form methane and carbon dioxide, known as biogas. The methane content produced by anaerobic digestion is 60%–70%. The balance of other gases is 30%–40% carbon dioxide and trace amounts of other gases. Anaerobic digestion occurs naturally because all of the bacteria involved exist in soil. The process of biogas production can be improved by increasing the temperature. There are three categories of temperature that classify anaerobic digestion: psychrophilic that occurs at 68°F (20°C), mesophilic at 95°F–105°F (35°C–40°C), and thermophilic at 125°F–135°F (51°C–57°C). Generally, the higher the temperature, the greater the biogas

353

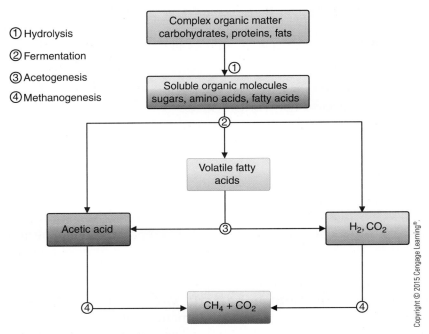

①Hydrolysis

②Fermentation

③Acetogenesis

④Methanogenesis

**FIGURE 25-1** Production of biogas.

production, but a minimum of 68°F (20°C) is necessary to create biogas. Figure 25-1 illustrates the process of biogas production.

## DIGESTER BIOGAS PRODUCTION

Biogas production involves five steps that make up the digestion process: collection system, digestion, effluent management, biogas collection/treatment, and biogas use (Figure 25-2).

### The Collection System

The collection system involves methods for collecting the organic material to be used for biogas production. Today the main sources of organic matter include animal manures, sewage sludge, and food processing waste. In theory, any organic material can be used to produce biogas as long as it has high moisture content. For biogas production, the percentage of total solids within the waste–water mixture should not exceed 15%. A collection system must be designed to easily divert the organic waste to the digester (Figure 25-3).

### Digestion

The next step is digestion. The biogas is actually produced a digester. A digester is a large, closed container designed to maximize anaerobic digestion easy collection of the biogas. Currently there are five types of anaerobic digesters: covered lagoons, complete mix, plug flow, fixed film, and batch digesters.

#### Covered Lagoon

The **covered lagoon digester** is a large outdoor pool (Figure 25-4). The bottom of the lagoon is typically uses an impermeable liner to prevent the waste from leaking into the ground. The lagoon is covered by a flexible

**covered lagoon digester—** a type of biogas generation system composed of a large outdoor lagoon covered by a flexible impermeable barrier that traps the gas

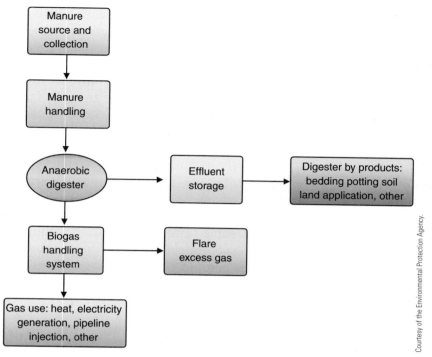

FIGURE 25-2 Biogas system.

FIGURE 25-3 Manure collection system.

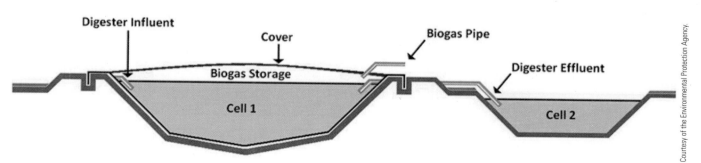

FIGURE 25-4 Covered lagoon digester.

impermeable barrier that traps the gas. Organic material mixed with water with a solid content of no greater than 3% is pumped into the lagoon. The anaerobic digestion occurs within the lagoon and pipes installed in the cover collect the gas. Lagoon digesters are best suited for warmer climates because they are difficult to heat. The average retention time for waste within a lagoon digester is 50 days. The retention time is the amount of time the waste stays in the digester to produce biogas. This is the least expensive large-scale biogas digester.

### Complete Mix

**complete mix digester**—a biogas digester that consists of a large tank containing a mechanical mixing unit that circulates organic waste

A **complete mix digester** consists of a large tank, containing a mechanical mixing unit that circulates the organic waste (Figure 25-5). The tank can be stored above or below ground. This system uses mixtures of organic waste that contain between 3%–10% solids. The waste is mixed within the tank and is heated to maximize biogas production. This system shortens the retention time to two weeks. Most complete mix digesters are operated at mesophilic temperatures (95°F–105°F) and because they are heated can be used in a variety of climates.

### Plug Flow

**plug flow digester**—a type of biogas generator that consists of rectangular tank and a mixing chamber. Organic waste is fed into the tank where it undergoes anaerobic digestion. As new waste enters, it pushes old waste through the tank

The **plug flow digester** uses a large, rectangular tank and a mixing chamber (Figures 25-6 and 25-7). Organic waste is introduced into the mixing chamber where it is blended with water to reduce its solid content to between 11%–13%. This waste is then fed into the tank where it undergoes anaerobic digestion. As new waste enters, it pushes the old waste through the tank. Gas is collected from the top of the tank by pipes. The tank is also heated to mesophilic temperatures to maximize gas production. Eventually, the waste arrives at the end of the tank where it is removed. The average

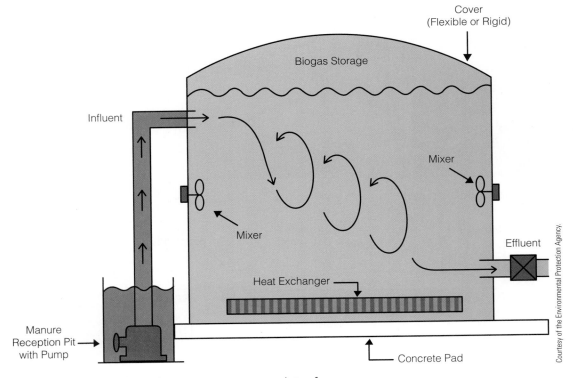

**FIGURE 25-5** Plug flow digester in use on a dairy farm.

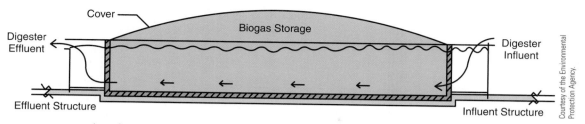

**FIGURE 25-6** Plug flow digester.

**FIGURE 25-7** Plug flow digester.

retention time for the plug flow digestion is about 15 days. This system can be used in many climates because it is heated.

## Fixed-Film Digester

The **fixed-film digester** uses a large tank filled with a plastic media with a large surface area (Figure 25-8). The plastic surface area allows a film of bacteria to grow, helping to maximize biogas production. Typically perforated PVC pipes are used within the tanks. As the waste flows through the pipes, it is exposed to the methanobacteria. Fixed-film tanks require low solid contents of around 3% to allow them to flow easily through the plastic media. The increased surface area within the digester allows higher numbers of bacteria to grow. This causes fixed-film digesters to have short retention times of 3–6 days.

## Batch

The last type of digester is known as a **batch digester**. This digester is a small-scale lined pit filled with organic waste (Figure 25-9). The pit is covered and anaerobic digestion begins. A pipe in the cover collects the biogas. Batch digesters are designed to produce biogas in batches, not in a continuous stream like large-scale digesters. The production time for the batch system depends on the type of waste being used and the size of the pit.

**fixed-film digester**—a biogas generator that uses a large tank filled with a plastic media with a large surface area. The plastic surface allows a film of bacteria to grow, maximizing biogas production

**batch digester**—a small-scale biogas generator that consists of a lined pit filled with organic waste. The pit is covered and gas collects at the top

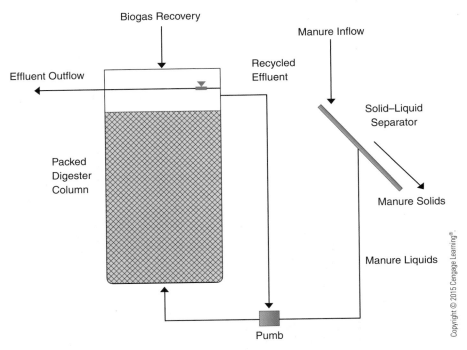

FIGURE 25-8 Fixed-film digester.

FIGURE 25-9 Batch digester.

## Gas Collection and Treatment

After digestion, the next phase in biogas production is gas collection. Gas collection uses a network of pipes attached to the digester that collects the biogas. Typically a pump is installed in the gas line to create a slight vacuum to draw the biogas out of the digester. There is also a gas regulator that adjusts the pressure of the gas. Often included with the collection system is a treatment system. Biogas treatment involves the removal of water vapor and hydrogen sulfide gas. Water vapor is removed using a condensation trap that cools the gas, causing the water vapor to condense. The water is collected and drained off. Some biogas collection systems also use a process to remove the toxic hydrogen sulfide from the gas that is corrosive to pipes and machinery. Once the gas has been collected and scrubbed, it can be used as a source of energy.

## Biogas Use

Biogas can power an internal combustion engine or a mini gas turbine that drives a generator to produce electricity. Gas boilers can also be fueled by biogas to produce heat. Gas chillers that power refrigerators and cooling tanks used on farms can also use biogas. Essentially, biogas can be used that the same as natural gas and is considered to be a renewable fuel. Biogas can also be used for transportation by modifying internal combustion engines.

## Effluent Manager

The last part of the production of biogas involves the storage and use of the **effluent**. Effluent is the solid and liquid remains left after digestion. When the effluent is removed from the digester, it can be used as a fertilizer and soil additive. Effluent is commonly stored in a large tank then applied to land. Some effluent can be dewatered and dried to be burned as a solid biomass fuel.

**effluent**—the solid and liquid waste produced as a result of an industrial process

## BENEFITS OF BIOGAS USAGE

There are many benefits to using biogas. Biogas is a renewable fuel produced by waste products. The production of biogas reduces the amount of methane released into the atmosphere. Methane is a strong greenhouse gas that contributes to global warming. The number one source of atmospheric methane is from ruminant animals and their manure. Capturing the methane and using it as a fuel, prevents it from going into the atmosphere. Although the combustion of methane produces carbon dioxide, methane is more than 20 times more effective at absorbing infrared radiation than carbon dioxide. It is the absorption of infrared radiation that heats the planet's atmosphere. Another benefit to producing biogas is that the process of anaerobic digestion prevents the emission of toxic ammonia. Normally, animal manures emit ammonia into the air when exposed to the atmosphere. The use of a digester converts the ammonia into better plant available forms of nitrogen. Biogas production reduces the negative odors associated with manure storage. Because the manure is being treated within a closed system, odors and volatile compounds are contained. Another benefit is the reduction of runoff and seepage often associated with the storage of manure. The lined lagoons and impermeable tanks used for anaerobic digestion prevent manure from infiltrating the ground and contaminating groundwater. There is also a reduction in the runoff of manure into surface waters. The digester process can also reduce the number of harmful microorganisms present in organic waste such as manure or sewage. The high temperatures reached during anaerobic digestion often kill harmful *coliform* bacteria. Last, the use of biogas effluent as a fertilizer and soil amendment enhances soil fertility.

## LANDFILL BIOGAS

**landfill**—a large deposit of buried garbage

The production of biogas from sewage or animal manures involves the use of a digester; however, there is another major source of biogas in the United States: **landfills**. A landfill is a large deposit of buried garbage. Landfills have been used for more than one hundred years as a way to dispose of our refuse. Typically a landfill begins as a large hole in the ground lined with an impermeable barrier. This is used to prevent toxic leachate from getting into groundwater. Garbage is then dumped into the hole. The garbage is covered with soil daily to control odor and prevent the spread of disease. Eventually, layer after layer of garbage builds and the hole slowly transforms into a hill. The landfill is closed when it reaches a specified height. Closing the landfill involves covering it with an impermeable barrier then covered in grass. Because the garbage is so compacted within the landfill, it is considered to be anoxic. Anoxic means there is no oxygen present. This condition causes the waste within the landfill to undergo anaerobic digestion that produces biomethane. Because there is little water within a landfill, the biogas contains a lower percentage of methane compared with biogas produced by a digester. Typically landfill biogas is about 50% methane and 50% carbon dioxide. When the landfill is closed, the production of methane begins after a few years. Normally this methane is vented into the atmosphere by a network of pipes installed within the landfill. Because landfills are the second largest source of methane, the gas captured in the pipes is burned using flares.

In higher concentrations, methane is diverted for use as a combustible gas (Figures 25-10 and 25-11). The methane vent pipes can be attached to a gas collection system that collects and scrubs the biogas. The gas can then be used like natural gas. Currently, in the United States there are 520 landfill gas energy projects (Figure 25-12). About 70% of these use the biogas to power turbines that drive electrical generators. Together, these landfill gas facilities produce more than 1.3 megawatts of electricity, equivalent to a large nuclear power plant. Thirty percent of landfill gas projects use the biogas as a source of gas heat and power. The EPA estimates that there are more than 500 additional landfills that can be tapped as sources of biogas. Once a landfill is equipped to collect biogas, it can produce biomethane for about 20 years. After this, the gas production begins to decline. Researchers are currently looking into the possibility of extending the production time of landfills to possibly 50 years or more. This may be achieved by the controlled injection of water into the landfill.

**Typical Landfill Gas-to-Electric Facility**

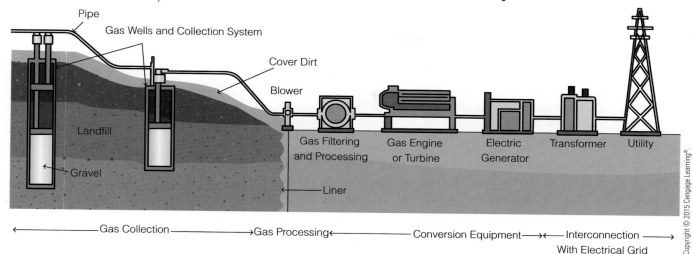

FIGURE 25-10 Gas capture from a landfill.

FIGURE 25-11 Landfill gas wells.

# BIOGAS AND AGRICULTURE

No other sector can benefit more from the production of biogas than agriculture. Livestock production is in the position to use animal manures as a source of organic material to produce biogas. Most farms already employ manure collection systems to deal with the daily accumulation of animal waste. These systems can be easily adapted to supply a biogas digester. Often this waste is mixed with high percentages of water making it an ideal feedstock for anaerobic digestion. Manure management is one of the most difficult aspects of livestock farming because of its potential to produce undesirable odors. Concerns of the effect animal manure have on air and water quality often makes it a problematic byproduct of the meat and dairy industry. Creating a clean energy resource, while reducing

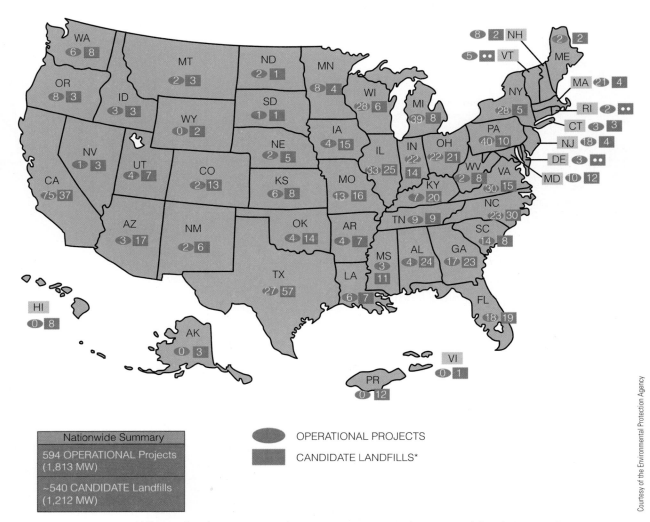

**FIGURE 25-12** Landfill gas facilities currently in operation, and potential facilities in the U. S. 2012.

biowaste's impact on the environment, makes it an attractive option for farmers. Once the biogas is produced, the effluent removed from the digester can be applied to farm fields as a soil additive and source of organic matter.

Today, there are more than 180 biogas digesters operating in the United States. Fifty percent of these are plug flow digesters, 23% complete mix systems, 20% covered lagoons, and 7% use other biogas digesters (Figure 25-13). Currently, 82% of these farms are dairy operations, followed by swine (12%), poultry (2%), and beef (1%). Together these agricultural biogas digesters produce an equivalent of more than 400 megawatts of power. Almost half of these farms use the biogas for **combined heat and power**. Combined heat and power is a system that uses a fuel to power an engine that drives a generator to produce electricity, while also capturing and using the waste heat produced by the engine. This makes the process more efficient. The waste heat can be used to heat the biogas digester or heat buildings on the farm. The other half of the biogas produced by farms in the United States was used to generate electricity or as a source of heat.

**combined heat and power—** a system that uses fuel to power an engine that drives a generator to produce electricity, while also capturing and using the waste heat produced by the engine

Courtesy of the U.S. Department of Agriculture.

**FIGURE 25-13** Covered lagoon digester.

The EPA and Department of Agriculture estimate there are more than 5,000 farms that could potentially produce biogas. If these manure resources were used, they could generate more than 1,600 megawatts of electricity that is more than the energy produced by a large nuclear power plant. In 2008, to promote this clean, renewable source of energy, the government enacted the *Food, Conservation, and Energy Act* to provide loans and grants enabling farmers to construct biogas digester systems. The EPA and the Agriculture Department joined forces to promote the development of agricultural biogas to reduce the emission of methane. In New York State, there are 25 biogas digesters used by farm operations. Figure 25-14 shows the locations of biogas digesters in other areas of the country. On average, these biogas systems are used to produce between 50 and 2,200 kilowatts of electricity. Many of these operations use combined heat and power and have been successful in producing the electricity needed to operate the farms, while also using the waste heat.

Problems associated with the production of biogas by farms include the noise associated with the engine and generators, corrosion of machinery from hydrogen sulfide gas, freezing of the intake and output pipes during extreme winter weather, low pay back prices from the local utility company for excess electricity produced, inaccurate construction estimates for building a biogas digester system, and heavy snow load that can collapse the flexible cover on a lagoon or plug system. Positive aspects of the use of farm biogas digesters includes odor reduction of manure wastes, lower energy costs, reduction of atmospheric pollutants, increased revenue by accepting other food wastes, and the ability to sell the dried effluent as a soil additive.

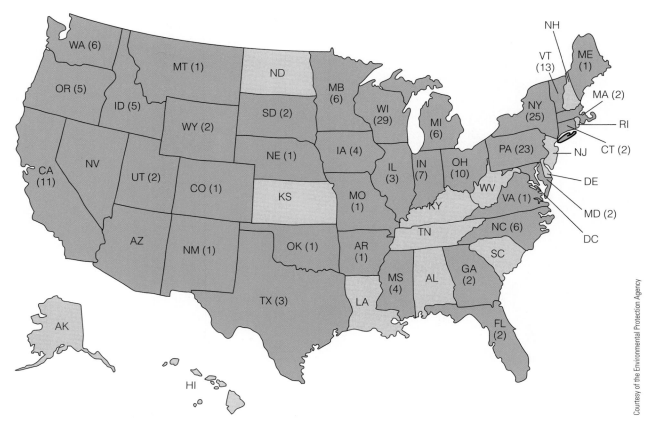

**FIGURE 25-14** Operating biogas farm digesters in the U.S. 2012.

## REVIEW OF KEY CONCEPTS

1. Anaerobic digestion involves the decomposition of organic matter by methanobacteria in the absence of oxygen to produce biogas.

2. Biogas is an odorless, colorless combustible gas produced by anaerobic digestion. It is composed of approximately 60%–70% methane and 30%–40% carbon dioxide.

3. The process of producing biogas involves waste collection, anaerobic digestion, effluent management, gas collection/treatment, and gas use. Biogas generation begins when organic waste is fed into the digester tank where the microorganisms convert it to biogas. A gas collection system then draws the gas out of the digester where it is treated to remove water and hydrogen sulfide. The gas is fed to a combustion engine or turbine that powers a generator.

4. There are five large-scale types of anaerobic digesters used on farms today: covered lagoon, plug flow, complete mix, fixed film, and batch digesters.

5. Biogas is a renewable form of fuel generated from waste. Its production reduces the amount of methane entering the atmosphere that contributes to global warming. Biogas production also reduces odors associated with animal manures, along with reduced potential for polluting surface and groundwater.

6. Biogas can act as a fuel to provide heating or cooling on the farm. The effluent produced by biogas production can be used as a soil additive and fertilizer to enhance soil fertility.

7. Some of the problems associated with the production of biogas by farms include the noise associated with the engine and generators, corrosion of machinery from hydrogen sulfide gas, freezing of the intake and output pipes during extreme winter weather, low pay back prices from the local utility company for excess electricity produced, inaccurate construction estimates for building a biogas digester system, and heavy snow load that can collapse the flexible cover on a lagoon or plug system

8. When a landfill is closed, it is covered with an impermeable barrier that is then covered in grass. Because of the lack of oxygen after the landfill is closed, the waste within the landfill undergoes anaerobic digestion that produces biomethane. This can be released into the atmosphere or captured and used as a source of energy.

## CHAPTER REVIEW

### Short Answer

1. What is anaerobic digestion?
2. Define the term biogas.
3. What are the five steps used to produce biogas?
4. Identify the five types of biogas digester systems.
5. What are some of the benefits of producing biogas from animal manures?
6. How is biogas commonly used?
7. Identify three negative aspects of producing and using biogas.
8. How do landfills produce and use biogas?
9. Briefly describe how biogas is produced by a dairy farm.

### Energy Math

1. If 8,000 swine and dairy farms produce 1,670 megawatts of electricity, how many megawatts could each farm produce?

## Multiple Choice

1. Biogas is composed of approximately:
   a. 100% methane
   b. 60% methane
   c. 25% methane
   d. 6% methane

2. Which of the following makes up proteins?
   a. Glucose
   b. Fatty acids
   c. Starch
   d. Amino acids

3. Mesophilic temperatures used for most anaerobic digesters fall between:
   a. 125°F–135°F
   b. 95°F–105°F
   c. 48°F–68°F
   d. Less than 48°F

4. Which of the following is a harmful byproduct of anaerobic digestion of animal manures?
   a. Hydrogen sulfide
   b. Carbon dioxide
   c. Water vapor
   d. Nitrogen oxide

5. What digester system has the shortest retention time?
   a. Covered lagoon
   b. Complete mix
   c. Plug flow
   d. Fixed film

6. What is the typical composition of methane in landfill biogas?
   a. 100% methane
   b. 50% methane
   c. 25% methane
   d. 6% methane

7. Which farming operation produces the most biogas in the United States?
   a. Swine production
   b. Poultry production
   c. Dairy production
   d. Beef production

8. How long can a landfill currently produce biogas?
   a. 5 years
   b. 10 years
   c. 20 years
   d. 50 years

9. Which anaerobic digester system is most widely used in the United States?
   a. Covered lagoon
   b. Complete mix
   c. Plug flow
   d. Fixed film

10. How do most farms use biogas?
    a. For heat
    b. To generate electricity
    c. To sell it as fuel
    d. For combined heat and power

## Matching

*Match the terms with the correct definitions*

a. biogas
b. methanobacteria
c. anaerobic
d. amino acids

e. covered lagoon digester
f. complete mix digester
g. plug flow digester
h. fixed-film digester

i. batch digester
j. effluent
k. landfills
l. combined heat and power

1. _____ A simple molecule that makes up proteins.

2. _____ A type of anaerobic digester that uses a long rectangular tank where waste is fed into one end and exits out the other.

3. _____ The solid and liquid remains left over from the digestion process.

4. _____ The use of a fuel to provide both electrical and thermal energy.

5. _____ A mixture of 60%–70% methane and 30%–40% carbon dioxide.

6. _____ A type of anaerobic digester that uses as tank and a mechanical mixer.

7. _____ A type of anaerobic digester that uses a covered pond.

8. _____ A large amount of buried garbage.

9. _____ A type of anaerobic digester that uses a tank filled with a plastic medium to promote the growth of bacteria.

10. _____ Lacking oxygen.

11. _____ A type of bacteria that produces methane.

12. _____ A type of anaerobic digester that uses a small lined pit that is covered.

# GLOSSARY

## A

**absolute zero**—the theoretical temperature when all atomic motion stops. This occurs at −459.67°F (−273.15°C).

**absorb**—to take in.

**acid precipitation**—a form of rain, snow or fog that has a pH of lower than 5.6.

**active solar energy**—the use of a collection device that absorbs the sun's heat energy that is then circulated by a fluid or air driven by fans or pumps.

**agriculture**—living off the land by raising crops and livestock.

**agroforestry**—the agricultural practice of growing timber.

**alkane**—a type of carbon compound made of single-bonded carbons, also known as saturated hydrocarbons.

**alternating current (AC)**—an electric current where the flow of electrons reverses between positive and negative poles many times a second.

**amino acid**—a simple compound made up of hydrogen, oxygen, nitrogen, and carbon.

**anaerobic**—without oxygen.

**anhydrous ammonia**—a form of ammonia that does not contain water. This ammonia can be used as a source of nitrogen for fertilizer, or to produce other forms of nitrogen fertilizers like urea, ammonium nitrate, and ammonium sulfate.

**anion**—an ion with a negative electric charge.

**anode**—an electrode that produces electrons.

**anoxic**—lacking oxygen.

**anthracite coal**—black, shiny, and dense; highest amount of carbon, cleanest burning with lowest amount of impurities.

**anthropology**—the study of the evolutionary development of human beings and their culture.

**anticline**—a layer of folded rock in the shape of an arch.

**autotrophs**—"self-feeders" are organisms that derive energy from sunlight or from chemical reactions.

## B

**batch digester**—a small-scale biogas generator that consists of a lined pit filled with organic waste. The pit is covered and gas collects at the top.

**biobutanol**—a type of butyl alcohol that is made from the fermentation of biomass.

**biodiesel**—a combustible, vegetable oil-based fuel composed of fatty acids.

**biogas**—a colorless, odorless, combustible gas composed of methane and carbon dioxide.

**bioheat**—home heating oil mixed with biodiesel.

**biomass**—a short term for biological mass, that is the total dry weight of an organism.

**biomass gasification**—the use of heat, steam, and pressure in a low oxygen environment to produce syngas from biomass.

**biomethanol**—methyl alcohol is produced from biomass.

**bio-oil**—also known as bio-crude, a thick, black liquid hydrocarbon that is similar to crude oil.

**biopower**—the use of the combustion of biomass to produce electricity.

**bio-refinery**—a modern manufacturing plant used for the production of biofuels.

**biosolids**—the dried, solid remains collected from wastewater treatment plants and animal manures.

**bipedalism**—the ability to walk upright.

**bitumen**—a thick, black, hydrocarbon compound much like tar.

**bituminous coal**—dark black, brittle, less dense; lower amount of carbon.

**black shale**—a sedimentary rock composed of a mixture of clay size sediments and kerogen.

**board-foot**—the unit of measure used for lumber equal to the size of a piece of wood being 1 foot long, by 1 foot wide, and 1 inch thick.

## C

**carbon-neutral**—the amount of carbon dioxide produced by combustion is removed by plants, negating the amount of carbon dioxide entering the atmosphere.

**carbon sequestration**—the process by which carbon dioxide gas produced from combustion is captured and stored to prevent it from going into the atmosphere.

**carnivores**—"meat eaters" or organisms that consume the herbivores.

**catagenesis**—the process of heat and pressure breaking the long chains of hydrocarbon molecules in kerogen into smaller ones.

**cathode**—an electrode that gains electrons.

**cation**—an ion with a positive electric charge.

**cellulose**—a polymer of glucose and the most abundant form of plant material on the Earth.

**char**—the solid remains of partially decomposed hydrocarbons.

**charcoal**—substance created from heating wood in a low oxygen environment; used as a heat fuel source.

**chemical energy**—the potential energy stored within the bonds between atoms that make up compounds.

**chemosynthesis**—the process used by organisms to convert the chemical energy stored in sulfur compounds into carbohydrates.

**civilization**—the stage of human development that is considered most advanced.

**cloud point**—the temperature that oil starts to contain small amounts of solids that make the oil appear cloudy and more viscous.

**coal**—a black or brown, rock-like material made up of hydrocarbons that came from the partially decomposed remains of plants that lived millions of years ago.

**coal gas**—a combustible gas made of composed of carbon monoxide and hydrogen.

**coal reserve**—the amount of coal that is left in the ground to still be mined.

**coal seam**—a narrow band of coal sandwiched between surrounding layers of rock.

**coal slurry**—a mixture of soil, rocks, and other impurities removed from the coal with the consistency of soft-serve ice cream.

**co-firing**—the process of burning biomass that is mixed with coal or natural gas.

**coke**—a solid hydrocarbon made from coal similar to charcoal that is produced by pyrolysis.

**cold reservoir**—the destination or sink, for the energy used in a heat engine.

**combined cycle biopower**—also known as biomass gasification, the process of converting solid biomass fuel into a combustible gas that can be burned to produce electricity.

**combined heat and power**—a system that uses fuel to power an engine that drives a generator to produce electricity, while also capturing and using the waste heat produced by the engine.

**combustion**—the process of producing heat and light by the rapid combining of oxygen with a substance.

**complete mix digester**—a biogas digester that consists of a large tank containing a mechanical mixing unit that circulates organic waste.

**compound machine**—a machine made up of two or more simple machines.

**compression**—the reduction in the volume of a substance as a result of an increase in pressure.

**concentrated solar power**—the use of mirrors or lenses to focus the energy of the sun in order to generate high temperatures.

**condensable particulate matter**—fine particles produced from combustion that are captured in a liquid during a stack test.

**condensation**—the phase change of a gas turning into a liquid.

**condenser**—a device that cools a working fluid in order to return it to a liquid state.

**conduction**—a method of energy transfer that involves the movement of energy by direct molecular contact.

**contour**—the shape of a surface feature.

**contour mining**—the overburden is removed following the contours of the coal seam within the mountain.

**convection**—the transfer of energy as a result of differences in temperature and density.

**coppicing**—the practice of cutting the stem of the sapling to spur the growth of multiple stems.

**corn stover**—the left over biomass that results from the production, harvest, and processing of corn.

**covered lagoon digester**—a type of biogas generation system composed of a large outdoor lagoon covered by a flexible impermeable barrier that traps the gas.

**cultural eutrophication**—the increase of nutrients within water by human activity that leads to the rapid growth of aquatic plants and algae.

## D

**decomposer**—organisms that break down dead organisms by consuming them.

**deep mining**—a method of extracting coal that is usually found 300 feet or more below the surface.

**dehulling**—the process of removing the hull or outer covering of seeds.

**destructive distillation**—the heating of a hydrocarbon in the absence of oxygen.

**detritivore**—organisms that eat dead things, also known as decomposers.

**dewatered**—the process of removing suspended solids from within a water solution.

**direct current (DC)**—an electric current that flows in one direction, from the negative to the positive.

**direct use geothermal**—a method of geothermal energy which utilizes hydrothermal fluids directly from the ground as a source of heat.

**distillation**—the process of separating liquids in a solution as a result of their different boiling points.

**distiller's dried grains and solubles (DDGS)**—the remaining biomass left over after the fermentation process, also known as sillage.

**domesticated**—to adapt to live in with and for the benefit of humans.

## E

**ecosystem**—the interaction between the living and non-living components in a specific area.

**effluent**—the solid and liquid waste produced as a result of an industrial process.

**effort force**—the force being applied to a machine.

**electrical energy**—a form of energy that is the result of the movement of electrons through a conductor.

**electric current**—the movement of electrons through an electric circuit.

**electricity**—the movement of electrons through a conductor.

**electrochemistry**—the study of the chemical reactions that produce electricity.

**electrolysis**—the use of an electric current to separate water into hydrogen and oxygen gas.

**electrolyte**—a solution that contains ions.

**electromagnetic energy**—energy emitted from the oscillations of the electrons surrounding the nucleus of an atom.

**endosperm**—the stored food within a plant's seed.

**endothermic reaction**—a chemical reaction that takes in thermal energy from its surroundings.

**energy**—the ability to perform work or cause change.

**energy crop**—an agricultural product grown solely for use as an external energy source.

**energy plantation**—growing a perennial dedicated biofuel feedstock.

**enhanced oil recovery**—uses steam or gases like carbon dioxide, methane, or nitrogen to drive oil out of the ground.

**entropy**—the increase of the disorder of a system.

**ethanol**—a colorless combustible hydrocarbon fuel ($C_2H_5OH$) commonly known as grain alcohol.

**evaporator**—a device that causes a fluid to rapidly change phase from a liquid to a gas.

**exothermic reaction**—a chemical reaction that produces heat.

## F

**fast pyrolysis**—exposing biomass to high heat for a short time in a low oxygen environment.

**fatty acid**—a long hydrocarbon molecule that makes up fats and oils in animals and plants.

**feedstock**—a raw material used to supply energy or an industrial process.

**fermentation**—the process of breaking down a substance by microorganisms, in the absence of oxygen.

**Fertile Crescent**—region in the Southwest part of Asia; believed to be the location where agriculture began.

**first law of thermodynamics**—the physical law that states energy cannot be created nor destroyed; it just changes from one form to another.

**Fischer–Tropsch process**—a method of converting syngas into liquid fuels by exposing biomass to high heat and a catalyst.

**fixed-film digester**—a biogas generator that uses a large tank filled with a plastic media with a large surface area. The plastic surface allows a film of bacteria to grow, maximizing biogas production.

**fluidized bed combustion**—the process of injecting coal dust into the boiler at high pressures, where the dust burns in a flowing red hot, fluid-like mixture.

**food chain**—a series of eating processes by which energy and nutrients flow from one organism to another.

**force**—an influence that changes the motion of an object, or produces motion of a stationary object.

**fractional distillation**—the process by which heat is used to separate liquids based on their different boiling point temperatures.

**frictional force**—an influence that reduces the velocity of something.

**fuel cell**—a device that uses a special membrane that combines hydrogen and oxygen gas to produce an electric current and water vapor as a byproduct.

## G

**gasifier**—a device that uses high heat in a low oxygen environment to convert substances into a gas.

**gel point**—the temperature that something begins to gel.

**geothermal gradient**—the rate the temperature changes with depth in the Earth.

**gravitational energy**—the attractive force all objects in the universe have toward one another.

## H

**hearth**—foundations of a fireplace.

**heat engine**—a device that converts thermal energy into mechanical energy by a cyclic process.

**heat exchanger**—a device that transfers external heat energy to a working fluid.

**heat pump**—a device that is used to move heat, typically from a cooler region to a warmer region using mechanical energy.

**heavy water reactor**—a nuclear reactor run with the use of water that contains 2 deuterium isotopes of hydrogen bonded to one oxygen molecule ($D_2O$).

**herbivores**—"plant eaters" or organisms that eat plants or algae as a source of energy.

**heterotrophs**—"other feeders" or organisms that derive energy from consuming other organisms.

**hominid**—classification of primates sharing physical features; early ancestors of modern humans.

***Homo erectus***—a hominid species; an early ancestor to modern day humans.

**hot reservoir**—the source of energy for a heat engine.

**hydraulic fracturing (fracking)**—the process of injecting high pressure water mixed with sand and chemicals into the ground to fracture rock, which is often associated with gas and oil recovery.

**hydrocarbon**—a molecule made up of hydrogen and carbon atoms that is plentiful in plants.

**hydrocracking**—the process of breaking long chain hydrocarbon molecules into shorter chains.

**hydroelectric power**—the use of flowing water to drive an electric turbine.

**hydrokinetic power**—the use of tidal energy, currents, wave energy, and differences in the temperature of water to supply power.

**hydrologist**—someone who studies all aspects of water.

**hydrolysis**—the separation of a substance by water.

**hydronic heat**—the use of circulating hot water heated to temperatures between 130°F–180°F (54°C–82°C) to heat a building or home.

**hydrothermal power**—the use of superheated water from the ground to produce electricity.

**hydrotreating**—the use of hydrogen gas to lower the oxygen content of bio-oil.

**hypoxia**—low levels of oxygen.

## I

**ignition point**—the temperature at which a gas will spontaneously combust.

**in-stream hydropower**—a form of hydrokinetic power that uses underwater turbines that have rotor blades that spin as a result of the flow of water.

**ion**—an atom or molecule with an electric charge.

## J

**joule**—the unit of energy required to apply the force of one newton on an object to move it a distance of one meter.

## K

**kerogen**—a compound made up of many different types of hydrocarbons that formed from organic remains.

**kerosene**—a liquid hydrocarbon compound produced from the distillation of crude oil that has a condensation point of around 400°F (204°C).

**kinetic energy**—energy that is being used to perform work.

## L

**landfill**—a large deposit of buried garbage.

**legume**—a family of plants that have seedpods and house tiny nodules in their root system that is home to nitrogen producing bacteria.

**light water reactor**—a nuclear reactor run with the use of regular water ($H_2O$), made up of 2 hydrogen atoms with an atomic mass of approximately 2 and an oxygen with a mass of 16.

**lignite coal**—brown, lowest density, least amount of carbon.

**lime**—a common term for calcium oxide (CaO).

## M

**machine**—a device used to increase or change the direction of a force in order to perform a specific task.

**mechanical advantage**—the use of a machine to do work.

**mechanical energy**—the energy produced by physical movement.

**methane hydrate**—a unique compound that consists of a methane molecule encased in a crystal lattice of ice.

**methanobacteria**—a class of anaerobic bacteria that consume organic matter and produce methane as a byproduct.

**methanol fuel cell**—a device that uses a membrane catalyst to mix oxygen with a solution of methanol to produce carbon dioxide and water vapor, while also generating an electric current.

**monoculture**—the growth of one species of crop.

## N

**nacelle**—the aerodynamically shaped enclosure of a wind turbine housing the gearbox, generator, braking system, and power conditioning systems.

**natural gas**—a flammable hydrocarbon consisting of mostly methane mixed with smaller amounts of propane and butane.

**newton**—a measurement of force roughly equal to the force of gravity pulling on an object with the mass of a baseball.

**nuclear energy**—energy that is associated with the forces that bind the nucleus of atoms together.

**nuclear fission**—the process of using nucleons to split apart an atom.

**nuclear fusion**—the process of fusing two atoms together to produce a new element and energy.

**nucleon**—the sub-atomic particles that make up an atom's nucleus, which include protons and neutrons.

## O

**Ocean Thermal Energy Conversion (OTEC)**—the use of the differences in the temperature of water in the ocean to generate electricity.

**oil seep**—an area where crude oil flows naturally out of the ground.

**oil shale**—a black, sedimentary rock composed of clay-sized particles mixed with kerogen.

**oil traps**—reservoirs of petroleum found within porous sedimentary rocks also known as reservoir rocks.

**omnivores**—"all eaters" or organisms that eat both plants and animals.

**ore**—a mineral used to make valuable metals.

**organic**—something containing carbon.

**overburden**—the rocks and soil that cover a resource.

**oxidation**—when an atom or molecule losses an electron.

**ozone layer**—a layer in the stratosphere of an unstable form of oxygen ($O_3$) that filters out ultraviolet radiation.

## P

**passive solar energy**—a means of harnessing the natural light and heat energy produced by the sun, with no other input of energy.

**peat**—partially decomposed remains of plant material that can be burned as an energy source when dried.

**petroleum**—a dark brown, flammable liquid hydrocarbon formed millions of years ago from the remains of marine organisms; also known as crude oil.

**petroleum seep**—an area where crude oil flows naturally out of the ground.

**photochemical smog**—a type of air pollution that forms respiratory irritants in the presence of sunlight.

**photosynthesis**—the chemical reaction that takes radiant energy from sunlight and combines it with carbon dioxide and water to form glucose and oxygen as a byproduct.

**photovoltaics**—the use of a semiconductor like silicon to convert the sun's energy directly into electricity.

**phytoplankton**—tiny organisms like algae that float freely in water and gain their energy from photosynthesis.

**plug flow digester**—a type of biogas generator that consists of rectangular tank and a mixing chamber. Organic waste is fed into the tank where it undergoes anaerobic digestion. As new waste enters, it pushes old waste through the tank.

**polymer**—long chains of the same molecule.

**potential energy**—energy that has the ability to perform work, but is not being used.

**power**—the amount of work done over a specific amount of time.

**primary energy source**—a naturally occurring form of energy used directly to perform work.

**primary oil recovery**—the process by which oil flows out under its own pressure to the surface to be collected.

**primary production**—the amount of chemical energy an autotroph converts from solar energy by the process of photosynthesis.

**pumped storage**—the use of excess electricity to pump water from a lower elevation reservoir into a higher elevation reservoir. When electricity is needed, the water can flow back to the lower reservoir and generate electricity.

**pyrolysis**—the heating of a hydrocarbon fuel in a low oxygen environment.

**R**

**Rankine cycle**—a type of heat engine that uses heat to produce mechanical power.

**reduction**—when an atom or molecule gains an electron.

**resistance force**—the force produced by a machine.

**respiration**—the chemical reaction that takes the energy stored in sugars and produces carbon dioxide and water.

**runoff**—the carrying away of soil and sediments by the action of water moving across a surface.

**R-value**—the ability for a substance to reduce heat loss, also known as thermal resistance.

**S**

**secondary energy source**—a form of energy used to transport the energy generated by a primary source.

**secondary oil recovery**—the process by which oil is extracted by injecting water down into the well to wash more of the oil out.

**second law of thermodynamics**—the physical law explains that with every transfer of energy, small amounts of heat are lost to the environment.

**sediment pollution**—the addition of soil and rock particles to water.

**seismic wave**—a powerful shock wave that travels through the Earth's crust, much like ripples traveling across a pond when you throw a rock in the water.

**silviculture**—the art and science of growing trees.

**smelting**—the process of melting a substance to promote a chemical change to create another more useful or valuable substance.

**sound energy**—energy that travels in the form of a wave in a solid, liquid, or gas.

**spoil pile**—the stored remains of rocks and soil leftover from the mining process.

**starch**—a common name for carbohydrate molecules composed of long chains of sugars.

**sub-bituminous coal**—dull black, low density, low carbon content, low heating value.

**sugar**—a simple carbohydrate molecule, like glucose ($C_6H_{12}O_6$).

**surface mining**—the removal of material covering a resource that is located close to the ground surface; also known as opencast mining.

**sustainable**—the ability to be maintained at a certain rate or level.

**syngas**—a combustible gas composed of carbon monoxide and hydrogen.

**T**

**tallow**—rendered animal fat from processing meat.

**tar sands**—rocks composed of sand, clay, water, and bitumen.

**taxonomic**—a system of classifying living organisms.

**thermal conductivity**—the ability for a substance to conduct heat.

**thermal energy**—the kinetic energy of motion of vibrating atoms, also known as heat energy.

**thermal pollution**—the lowering or increase of the temperature of a body of water, causing adverse effects.

**thermochemical conversion**—the use of high heat and a catalyst to transform chemical compounds.

**thermochemistry**—the study of the energy absorbed or given off during a chemical reaction.

**thermocline**—the rapid change in the temperature of water with depth.

**thermodynamics**—the study of the movement of heat and energy.

**tidal power**—the use of the rise and fall of tides to generate electricity.

**timber**—wood used for producing a product.

**transesterfication**—process of mixing the triglycerides that make up vegetable oil with a methanol-catalyst solution to break them into smaller fatty acids.

**treadwheel** or **treadmill**—device used to power grinding stones or to move water.

**turbine**—a circular wheel or rotor, that is forced to spin at high speed when water, air, steam or any moving fluid transfers its energy to a series of angled blades or vanes attached to a central pivot point.

## U

**unconventional natural gas**—natural gas that is not associated with oil deposits, or that is not easily extracted from depths close to the surface.

## V

**vaporization**—the phase change of a liquid turning into a gas.

**Venturi effect**—the increase in the velocity of water as it flows through a constricted area.

**viscous**—a liquid that resists flow.

## W

**watershed**—the total land area that drains into a specific river, river system or body of water.

**waterwheel**—a device use to produce power through the use of a wheel, axle, and flowing water.

**watt**—a unit of energy defined as the use of one joule over the time period of one second.

**wave power**—the use of the rise and fall of ocean waves to generate electricity.

**white grease**—unprocessed fat from meat processing.

**wind**—the horizontal movement of air from areas of high atmospheric pressure to low.

**wind farm**—a group of wind turbines in a concentrated area that are linked together to produce large amounts of electricity.

**windmill**—a device used to create power by the use of wind.

**wind turbine**—a device that converts the force of the wind into rotational motion that is used to turn an electric generator.

**woody residue**—the left over biomass that results from the production, harvest, and processing of timber.

**work**—a change in position caused by a force.

**work animal**—an animal used to accomplish tasks such as plowing; also known as beasts of burden.

## Y

**yellow grease**—used vegetable oil from the food service industry.

# INDEX

Note: Page numbers referencing figures are followed by an *"f"*. Page numbers referencing tables are followed by a *"t"*.